FATE OF PESTICIDES IN THE ATMOSPHERE: IMPLICATIONS FOR ENVIRONMENTAL RISK ASSESSMENT

Fate of Pesticides in the Atmosphere: Implications for Environmental Risk Assessment

Proceedings of a workshop organised by The Health Council of the Netherlands, held in Driebergen, The Netherlands, April 22–24, 1998

Edited by

HARRIE F.G. VAN DIJK
Health Council of the Netherlands,
The Hague, The Netherlands

W. ADDO J. VAN PUL
National Institute of Public Health and the Environment (RIVM),
Bilthoven, The Netherlands

PIM DE VOOGT
University of Amsterdam
Amsterdam, The Netherlands

GEZONDHEIDSRAAD

Reprinted from *Water, Air, and Soil Pollution* 115: 1–4, 1999

KLUWER ACADEMIC PUBLISHERS
Dordrecht/Boston/London

A C.I.P. Catalogue record for this book is available from the Library of Congress.

ISBN 0-7923-5994-1

Published by Kluwer Academic Publishers
P.O. Box 17, 3300 AA Dordrecht, The Netherlands

Sold and distributed in North, Central and Latin America
by Kluwer Academic Publishers
101 Philip Drive, Norwell, MA 02061, U.S.A.

In all other countries, sold and distributed
by Kluwer Academic Publishers
P.O. Box 322, 3300 AH Dordrecht, The Netherlands

Printed on acid-free paper

Printed in the Netherlands

TABLE OF CONTENTS

ATMOSPHERIC TRANSPORT OF PESTICIDES: ASSESSING ENVIRONMENTAL RISKS

Current global food requirements have made great demands on agricultural production, including the need for efficient weed and pest control. In the second half of the 20th century, this had led to an ever increasing use of pesticides. Pesticides are a special case inasmuch as they are applied directly in the environment for the purpose of eliminating pests. Due to their obvious inherent toxicity, strict regulations exist throughout the world regarding their registration.

However, that non-target areas, even in remote places, can be exposed to chemicals that have become airborne elsewhere and that this exposure can cause unwanted effects, has only come to light in the last two decades. The global atmospheric transport of 'common' gaseous pollutants like sulphur dioxide and freons was already known at that time. Yet it was the observation of the presence of organochlorine pesticides (such as lindane and toxaphene) and industrial chemicals (like the polychlorinated biphenyls) in for example, animals living in the Arctic, which spurred serious scientific and political interest in the long-range transport of substances. Surprisingly, the monitoring efforts made in remote areas as a result of this awareness, have shown that compounds (perhaps) initially believed to be degraded quite rapidly after application, may still be found far away from the application area.

International bodies and governmental agencies, such as the UNEP and the UN-ECE, have launched activities to tackle the issue of the long-range transport of chemicals. Most of these initiatives focus on the so called persistent organic pollutants (POPs) or persistent, bio-accumulating and toxic (PBT) substances. This raises the obvious question of definitions. The issue of how to incorporate a chemical's potential for being transported over long distances, into the regulations for registration, of both industrial chemicals and modern pesticides, is an issue which has received relatively little attention to date. For pesticides, this is all the more surprising when one considers their typical use, where a relatively high potential for becoming airborne is obvious.

The Health Council of The Netherlands took the initiative to organise a workshop on the issue of 'Fate of pesticides in the atmosphere; implications for risk assessment'. About forty experts were invited to discuss this subject with the aims of reviewing the current scientific knowledge and possible risk assessment approaches and exploring possible statutory, environmental criteria that could be incorporated into pesticide regulation.

This special issue of Water, Air and Soil Pollution publishes the proceedings of the workshop held in Driebergen, The Netherlands on April 22-24, 1998. First, an extended summary of the deliberations of the workshop is presented followed by keynote papers that were presented during it. Finally discussion papers summarise the several topics discussed. Contributions were

Water, Air, and Soil Pollution **115:** 3–4, 1999.

received from invited keynote speakers, or drafted from the minutes taken and audio recordings made during the discussions. All contributions were subjected to the usual peer review procedure of WASP.

The editors wish to express their gratitude, to all the participants of the workshop, who have worked hard to deliver what has become this special issue. It is our sincere hope that this issue will become a major starting point for further efforts to estimate environmental risks associated with the long-range transport of pesticides.

H.F.G. van Dijk, W.A.J. van Pul, P. de Voogt
The Hague, The Netherlands

ENVIRONMENTAL RISK ASSESSMENT FOR PESTICIDES IN THE ATMOSPHERE; THE RESULTS OF AN INTERNATIONAL WORKSHOP

ROBERT GUICHERIT[1], DICK J. BAKKER[1], PIM DE VOOGT[2], FREDERIK VAN DEN BERG[3], HARRIE F. G. VAN DIJK[4*] and W. ADDO J. VAN PUL[5]

[1] TNO Institute of Environmental Sciences, Energy Research and Process Innovation, P. O. Box 342, NL-7300 AH Apeldoorn, the Netherlands, [2] Amsterdam Research Institute for Substances in Ecosystems (ARISE), University of Amsterdam (UvA), Nieuwe Achtergracht 166, NL-1018 WV Amsterdam, the Netherlands, [3] Winand Staring Centre for Integrated Land, Soil and Water Research (SC-DLO), P. O. Box 125, NL-6700 AC Wageningen, the Netherlands, [4] Health Council of the Netherlands, P. O. Box 16052, NL-2500 BB The Hague, the Netherlands, [5] National Institute of Public Health and the Environment (RIVM), P. O. Box 1, NL-3720 BA Bilthoven, the Netherlands

(author for correspondence, e-mail: hfg.van.dijk@gr.nl)*

Abstract. The Health Council of the Netherlands organised an international workshop on the fate of pesticides in the atmosphere and possible approaches for their regulatory environmental risk assessment. Approximately forty experts discussed what is currently known about the atmospheric fate of pesticides and major gaps in our understanding were identified. They favoured a tiered approach for assessing the environmental risks of atmospheric dispersion of these chemicals. In the first tier a pesticide's potential for emission during application, as well as its volatilisation potential should be assessed. Estimates of the former should be based on the application method and the formulation, estimates of the latter on a compound's solubility in water, saturated vapour pressure and octanol/water partition coefficient. Where a pesticide's potential for becoming airborne exceeds critical values, it should be subjected to a more rigorous second tier evaluation which considers its toxicity to organisms in non-target areas. This evaluation can be achieved by calculating and comparing a predicted environmental concentration (PEC) and a predicted no-effect concentration (PNEC). By applying an extra uncertainty factor the PNEC can be provisionally derived from standard toxicity data that is already required for the registration of pesticides. Depending on the distance between the source and the reception area, the PEC can be estimated for remote areas using simple dispersion, trajectory type models and for nearby areas using common dispersion models and standard scenarios of pesticide use. A pesticide's atmospheric transport potential is based on factors such as its reaction rate with OH radicals. It should be used to discriminate between those compounds for which only the risks to nearby ecosystems have to be assessed, and those for which the risks to remote ecosystems also have to be determined. The participants were of the opinion that this approach is, in principle, scientifically feasible, although the remaining uncertainties are substantial. Further field and laboratory research is necessary to gain more reliable estimates of the physico-chemical properties of pesticides, to validate and improve environmental fate models and to validate the applicability of standard toxicity data. This will increase both the accuracy of and our confidence in the outcome of the risk assessment.

Keywords: atmospheric fate, atmospheric transport, deposition, emission, long-range transport, pesticides, registration, remote area, risk assessment, transformation

Water, Air, and Soil Pollution **115:** 5–19, 1999.

1. Introduction

Around the world, pesticides are used on a large scale in agriculture. However, the last decades have seen an increase in public and political concern about the possible side effects of pesticide use and this has resulted in many countries implementing legislation to control the registration of pesticides. This legislation requires risk assessments on human health and environmental impact to be conducted prior to registration. To date risk evaluation procedures have given scant consideration to the atmospheric fate of pesticides despite many having been detected in the air and in rainwater, not only in agricultural areas, but also in nearby and remote nature reserves. Atmospheric transport is generally considered to be an important contamination pathway of such non-target areas. This applies not only to the older organochlorine insecticides, the use of which has been abandoned or severely restricted in many industrialised countries, but also to many modern ones.

The Health Council of the Netherlands organised a workshop on 'Fate of pesticides in the atmosphere; implications for risk assessment', which was held at Driebergen, the Netherlands, from the 22nd to the 24th of April, 1998. Participation was based solely on personal expertise in the area, and not on affiliation. Almost 40 experts (see the list of participants elsewhere in this issue) from industry, governmental organisations and academia were invited to discuss what is presently known about emissions of pesticides to the atmosphere, their transformations and transport in the atmosphere, and their exchange between the atmosphere and surfaces on the earth. The discussions focused on these physico-chemical processes rather than on the possible human and environmental impact in non-target areas. It is recognised that these two aspects are different fields of expertise, which is why the organising committee focused on the physico-chemical aspects during the workshop. However, impact issues were taken into account when possible approaches to risk assessment were considered.

The objectives of the workshop were (i) to review the current scientific understanding of the atmospheric fate of pesticides, (ii) to consider possible risk assessment approaches, that can constitute a basis for the development of statutory environmental criteria for use in registration procedures which limit the dispersion of pesticides via the atmosphere, and (iii) to determine if the current level of knowledge is sufficient for these criteria to be feasible. This paper presents the main findings and conclusions of the workshop.

2. Background information

Participants were given an informative introduction to the topic of atmospheric transport of pesticides through a series of keynote lectures. In these, invited speakers summarised the available evidence from monitoring studies of non-target areas contaminated by airborne pesticides and also deliberated on how the organisms and ecosystems may be affected. Hites presented results on the atmospheric fate of organochlorine pesticides. By analysing tree bark samples from around the world, he could confirm the hypothesis that more volatile compounds like hexachlorobenzene and hexachlorocyclohexanes are subject to global fractionation and cold condensation, resulting in their accumulation at higher latitudes. He also presented long-term measurements from the Great Lakes area, which show that organochlorine concentrations in the air are steadily decreasing. From this he extrapolated that by the middle of next century they will be virtually eliminated. This data demonstrates the success of regulatory measures, which restricted or banned many of these older pesticides.

From a regulatory viewpoint, modern pesticides are probably of more interest than those already banned, although knowledge of the atmospheric fate of organochlorine pesticides will help to increase our understanding of the fate of the modern compounds. Van Dijk and Guicherit (this issue) summarised the evidence from monitoring studies which show that modern pesticides are also subject to atmospheric dispersion. It is clear that many of them, such as triazine, acetanilide and chlorophenoxy herbicides, and organophosphorus insecticides, commonly occur in air, rainwater and fog in Europe and North America and probably throughout the world. Concentrations in the air range from several pg/m^3 to many ng/m^3 and in rainwater from a few ng/L to several µg/L. Concentrations in fog are even higher. Deposition levels are in the order of several mg/ha/yr to a few g/ha/yr. From their presence in rather remote areas it can be inferred that several of these compounds are transported over hundreds, and in some cases probably thousands, of kilometres, since input other than via atmospheric routes can be ruled out.

Van Straalen stressed the scantiness of information relating to the effects of modern pesticides on organisms and ecosystems in non-target areas (Van Straalen and Van Gestel, this issue). He suggested that the risks in remote areas could be assessed in a similar way to those at and near application sites, *i.e.* by calculating a PEC (predicted environmental concentration) over PNEC (predicted no-effect concentration) ratio. The PNEC could be derived from the toxicity data (a part of the standard information required to support the registration of a product). The PEC in a non-target area could be estimated from the recommended application rate, emission factors, atmospheric residence time and persistence in soil and water. Van Straalen showed that the concentrations in soil or water in the non-target area can only be higher than at the application site if the residence time in the non-target area is one to two orders of magnitude longer. To compensate for the large uncertainty he proposed that the maximum acceptable ratio between PEC and PNEC be increased by an extra safety factor.

3. Setting the stage

The workshop focused on our current understanding of physical and chemical processes involved in the emission, atmospheric dispersion and deposition of pesticides, and our ability to describe them quantitatively by means of models for the purpose of risk assessment. Four experts were invited to give a keynote lecture on 1. emission, 2. transformation, transport and exchange processes, 3. modelling, and 4. risk assessment. This set the stage for small group discussions on these topics later in the workshop.

Majewski presented an overview of the micrometeorological techniques currently used to estimate post-application volatilisation fluxes of pesticides (Majewski, this issue). He explained both the advantages and disadvantages of these methods and stressed that we must be aware of their assumptions and limitations. Most importantly, he pointed out that, due to the lack of a reliable validation technique, it is not certain whether they, or any other method in current use, accurately describe what is occurring in the field. Very labour intensive and expensive mass balance experiments may be the only means of supplying an answer to this accuracy question.

Partitioning of pesticides between gas and particle phases governs the physical and chemical removal processes. Nevertheless, Bidleman (Bidleman, this issue) has pointed out that these studies have only been done occasionally for organochlorine pesticides and almost none have been done for modern pesticides. This information is required for both urban and rural environments and for carbon-rich and mineral aerosols. As gas-phase reactions of pesticides with hydroxyl radicals are probably the single most important chemical removal process, experimental methods must be developed to measure the reaction rate constants as a function of temperature. Reactions of particle-bound pesticides also deserve more attention. Physico-chemical properties of pesticides, such as vapour pressure, the octanol-air partition coefficient and Henry's law constant, have been shown to be useful for describing their exchange between environmental compartments. Precise and accurate values are therefore needed for more pesticides, in particular the polar ones. This should also include information on their temperature dependency. In order to estimate air-water and air-soil gas exchange fluxes for organochlorine pesticides more data on their levels in soils and surface waters is needed, especially for the open ocean and regional seas. Arctic air monitoring should be continued to check the efficacy of international controls on persistent organic pollutants (POPs) and to provide observations on the transport of modern pesticides.

Van Jaarsveld provided an overview of the models currently available to predict the atmospheric fate of pesticides (Van Jaarsveld and Van Pul, this issue). He made a distinction between fate models and regional scale models. Fate models are meant to show where a substance will end up, *i. e.* its partitioning between the environmental compartments. Examples are the Mackay type fugacity models. Regional scale models usually just consider the atmospheric compartment and describe the transport and deposition on a certain spatial scale within a certain period of time. They can be divided into Lagrangian or trajectory models and Eulerian or 3D grid models. Van Jaarsveld presented the main results from modelling studies on pesticides, which have been conducted to date in Europe and North America. Far more data is needed to validate calculations from these models, but there is some evidence of a good correlation between measurements and calculations. However, the absolute levels of predicted and measured concentrations still show considerable differences. From a critical appraisal of all the uncertainties involved, it was concluded that the largest uncertainties are in the emissions.

A concise overview of authorisation procedures for pesticides within the European Union was presented by Gilbert (Gilbert, this issue). One of the requirements is the prediction of residue levels in air, to which humans and the environment may be exposed. The EPPO/CoE decision-making schemes for the environmental risk assessment of plant protection products also require potential air exposure data, which can be taken from the sub-scheme for air. The objective is to enable a comparison between the predicted environmental concentration (PEC) and the predicted no-effect concentration (PNEC). However, according to Gilbert, the current level of scientific knowledge hampers the operation of predictive environmental exposure models. He advocated a tiered approach, in which simple criteria are used to screen out low risk substances and the remaining products are subjected to a more complex, predictive exposure assessment.

4. In-depth discussions

On the second day of the workshop the participants were split into four working groups, according to their main field of expertise. Each group considered current knowledge about one of the following topics: (i) emission into the air, (ii) transformation processes in the atmospheric compartment, (iii) transport and exchange processes, and (iv) possible approaches for assessing the risks of atmospheric transport.

4.1 Emission

The participants in the working session on emission concluded that the extent to which pesticides are lost directly to the atmosphere during application is largely unknown (Van den Berg *et al.*, this issue). It depends on the application technique, the weather conditions and the formulation of the pesticide product. They pointed out that more data and model development are needed to reliably predict this loss term. Losses of pesticides after application depend primarily on the properties of the substance, the soil, the crop and the environmental conditions. They estimated that up to 50%, or even more, of the amount applied may end up in the atmosphere. The volatilisation of soil fumigants and soil incorporated pesticides following application can be reasonably estimated using the models currently available. However, improvements in the reliability of these models and the measurement methods require further development and testing of the models. For many soil surface-applied pesticides, the rate and extent of volatilisation remains uncertain and the participants agreed upon the necessity of further model development and testing. The volatilisation from plant surfaces is not well understood. More field experiments have to be conducted to study the fate processes on plants and to provide information for the development of predictive models.

4.2 Transformation

The participants in the working session on atmospheric transformation inferred from the available field data that extensive atmospheric transformation occurs with many pesticides, but that direct evidence of this is usually lacking (Atkinson *et al.*, this issue). Indeed, it is unclear whether the transformation products observed in the air are formed there or on a surface where they are then subjected to volatilisation or wind erosion. Furthermore, it is not apparent whether airborne transformation products are formed in the vapour phase or on particles. Most of the current understanding concerns vapour phase reactions; our understanding of particle phase reactions in the atmosphere is poor and experimental data is only available for a limited number of pesticides. The most important atmospheric transformation reaction of pesticides is the reaction with OH radicals, except for the reactions of isolated and activated unsaturated C-C bonds, for which the reaction with NO_3 radicals and ozone may be important. For those pesticides for which reaction rates (of homologues) are available, the major uncertainty in estimating their atmospheric lifetime lies in the ambient OH radical concentration. The working group members expressed the opinion that this uncertainty is at least a factor of 2. If the reactions of the parent compounds in the atmosphere are rapid, relevant reaction products should be subjected to the same studies and/or evaluations as the parent compound. There is currently no data available on the reaction of first generation products from atmospheric pesticide transformations.

4.3 Transport and deposition

The participants in the session on transport and deposition concluded that the mechanisms for the transport and dispersion of pesticides can be described in the same way as those of other air pollution components. These mechanisms are rather well known (Van Pul *et al.*, this issue). The uncertainty in the transport model calculations is about 30-50% and is caused by modelling concepts and meteorological variability. Exchange fluxes across the interfaces between atmosphere/soil, atmosphere/water and atmosphere/vegetation depend on the concentrations in these compartments, the properties of the receiving surfaces (soil organic carbon and plant lipids content) and the partitioning of a pesticide between the gaseous, particle and aqueous phases. This phase partitioning is mainly governed by the physico-chemical properties of the pesticide, such as the saturated vapour pressure and the water solubility. There is still considerable uncertainty in the estimates for these parameters. More accurate data is needed, especially on the temperature dependency. Hardly any data on water, plant and soil concentrations on a regional scale is available for use in flux calculations. The overall uncertainty in the modelled deposition values is tentatively estimated at a factor of 3 to 4. However, this does not include the uncertainties in emissions and transformations.

4.4 Environmental Risk Assessment

On discussing possible approaches to incorporating atmospheric transport in the regulatory environmental risk assessment of pesticides, the members of the fourth working group concluded that a tiered approach would be best (Bakker *et al.*, this issue). They recommended that the risks to non-target areas lying beyond the immediate vicinity of the application site should only be assessed for those pesticides that successfully passed the existing local risk assessment. The first tier should be directed towards assessing the total emission potential of a pesticide, *i.e.* the potential for becoming airborne during and after the application. It should be based on a pesticide's intrinsic compound properties, such as the saturated vapour pressure, Henry's Law constant or the octanol/air partition coefficient. A trigger value for this potential should be set, beyond which a more complex second tier assessment would be necessary. This should at least involve the calculation of a PEC/PNEC ratio for non-target areas at short (10-50 km) distances from the application site. If the pesticide's atmospheric transport potential is expected to exceed a certain trigger value, then the calculation of a hazard quotient should also be performed for remote (>1000 km) areas. The transport potential can be estimated from intrinsic compound properties and the rates of removal processes. The working group members recommended that investigations should be conducted to assess which (combinations of) compound properties would be most reliable in estimating the emission and atmospheric transport potentials. Furthermore, they recommended to elaborate the PEC over PNEC calculations for nearby and remote areas.

5. Risk assessment and scientific feasibility

The final day of the workshop was spent trying to reach a consensus on the best procedure for environmental risk assessment, to elaborate on some of its elements and to investigate its feasibility in relation to the current level of knowledge.

The workshop participants agreed that a tiered approach is the best way of assessing the risks of the atmospheric transport of pesticides (see Figure 1). For

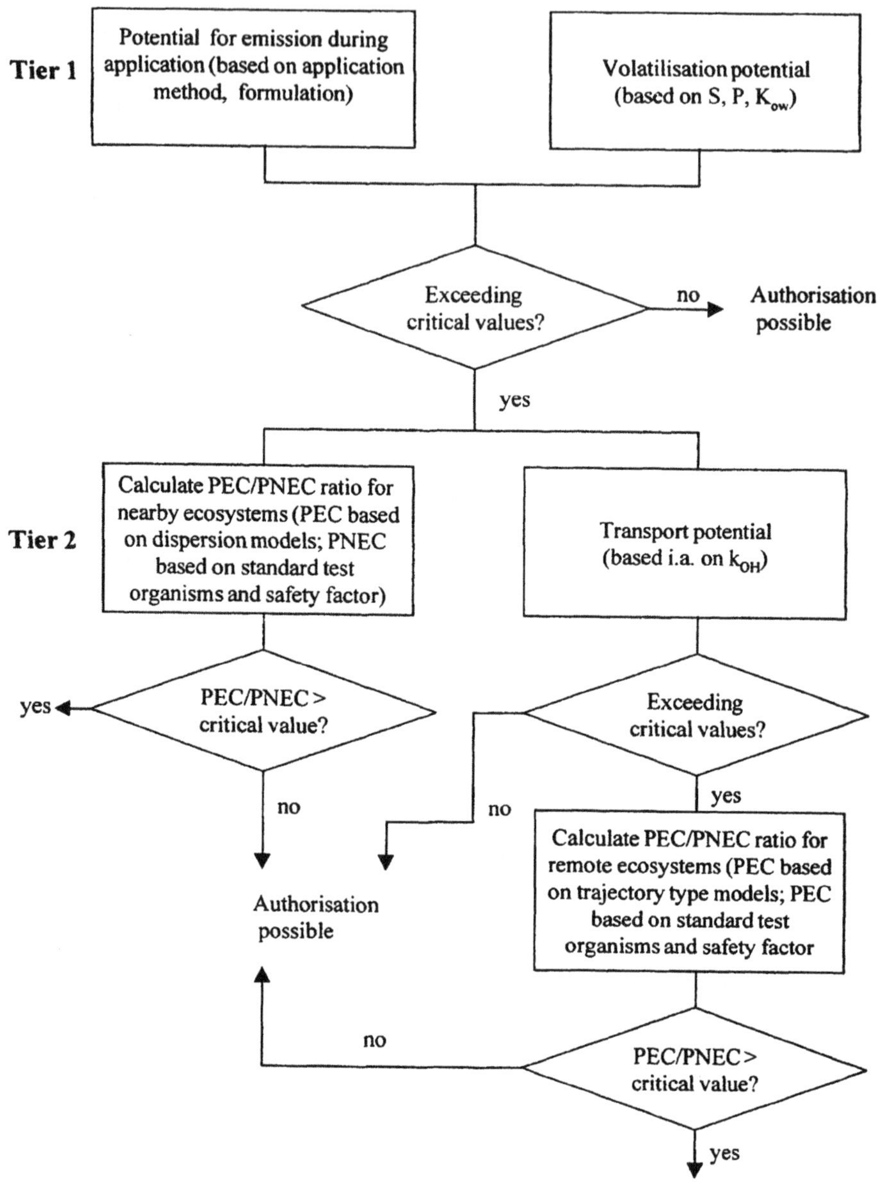

Figure 1. The proposed scheme for assessing the environmental risks of atmospheric transport of pesticides (for explanation of the abbreviations, see the main text)

the initial screening of problem compounds, a pesticide's potential for becoming airborne during application, as well as its volatilisation potential after reaching the target surfaces should be determined. The potential for becoming airborne during application depends primarily on the application method and the formulation of the pesticide. Existing drift models, developed to estimate the fraction of the applied dose that settles down within the immediate vicinity of the application site (*e.g.* within 1 km), may be adjusted to estimate the fraction of the dose that remains airborne and is therefore available for atmospheric transport over larger distances (see Van den Berg *et al.*, this issue).

A compound's potential for volatilisation from water, soils and plants can be predicted from its physico-chemical properties, *i.e.* its solubility in water (S), saturated vapour pressure (P) and n-octanol/water partition coefficient (K_{ow}) (see *e. g.* McLachlan *et al.*, 1995; Simonich and Hites, 1995; Smit *et al.*, 1997; Woodrow *et al.*, 1997; Hippelein and McLachlan, 1998; Smit *et al.*, 1998). If the degradation in soil, water or on plants is ignored, then the estimated volatilisation potential represents a worst case scenario.

The persistence of a pesticide and most of its physico-chemical properties are usually determined at one temperature, typically between 20 and 25 °C. There is a need for data at other temperatures, but the participants considered that it was only feasible to require manufacturers to submit this data if a potential for long-range transport has been demonstrated. However, for screening purposes the temperature dependence of the vapour pressure and the water solubility can already be estimated with sufficient accuracy.

It was concluded that numerical values should be set for the emission (during and after application) potentials. These will be used as triggers to discriminate between those compounds that are considered safe and those that require a more thorough evaluation of the risks in a higher tier. The participants believe that this is scientifically feasible. In fact, it is already being done in the decision making schemes of the BBA in Germany (Gottschild *et al.*, 1990), the AEA (1993), the EPPO/CoE (1998) and the UN-ECE (1998). The participants considered such schemes to be an appropriate framework, although thought would need to be given to the values used.

The participants agreed that those pesticides that were recognised as problem compounds, based on an evaluation of their potentials for emission during and after application, should be subjected to a more rigorous evaluation, that takes into account a compound's toxicity. They proposed to calculate a predicted environmental concentration (PEC) and to compare it with a predicted no-effect concentration (PNEC). However, the difficulties which are encountered when calculating a PEC in a non-target area, depend on the spatial scale of the transport. Different approaches may be necessary for estimating exposure levels in nearby and remote areas. As most pesticides have the potential to travel at least a few kilometres through the air, this PEC/PNEC calculation for nearby (within 50 km of the application sites) ecosystems should probably be performed whenever a pesticide has an emission potential that exceeds the trigger value.

A pesticide's transport potential may be used to discriminate between those compounds for which only the risks to nearby ecosystems have to be assessed and those for which the risks to remote ecosystems also need assessing. This atmospheric transport potential can be estimated from the calculated atmospheric residence time. It depends on the rates of removal processes, *i.e.* atmospheric degradation by OH radicals, and dry and wet deposition, including the degradation in water and soil compartments. It can be estimated by following the procedures proposed by Mackay (1991) or Van Pul *et al.* (1998). However, for this the rate constants for OH-radical reactions in the gas phase (k_{OH}) are needed and these are not currently part of the data requirements for registration. Moreover, as these constants are difficult to measure, they may have to be determined by using models or data on homologous compounds. Therefore, the participants advocated to require data on k_{OH} values only when the potential for emission during application or the volatilisation potential exceed critical values.

For the estimation of a PEC in a nearby non-target area, a scenario approach was proposed (see Bakker *et al.*, this issue). This scenario should include predetermined environmental conditions regarding climate and landscape (soil type, type and amount of vegetation) and a predetermined distribution pattern of application sites within a scenario area (area surrounding the non-target area with a radius of 50 km), based on land use or crop type. By using the models presented earlier during the workshop (Van Jaarsveld and Van Pul, this issue) it is possible to quantify the total loading of a non-target area from the atmosphere, the contribution of each application site within the scenario area and the background contribution from sources outside this area, assuming that the application volume of a pesticide is known. Apart from these models, further models are needed to derive a PEC in soil, water or vegetation from this input estimate. For these models and for the calculation of input-output balances, information is required on the environmental conditions in the receptor area (*e.g.* temperature), because these greatly affect degradation rates. Some of the participants in the workshop tentatively estimated the overall uncertainty in the PEC calculated for a nearby receptor area to be in the order of a factor of 10.

Simple dispersion, trajectory type (Lagrangian) models could be used to assess the atmospheric transport to remote areas (at a distance of more than 1000 km from the sites of application). When calculating a PEC, particular attention should be given to special environmental conditions in remote areas, such as low temperatures in the Arctic, which may exert a strong influence on the fate of a pesticide and hence on the extent and duration of exposure. An alternative approach to estimate the PEC in remote areas was suggested by Van Straalen and Van Gestel (this issue).

There was a consensus that a predicted no-effect level (PNEC) for nearby and remote areas should provisionally be based on the same standard test organisms as are currently used in the local risk assessment. The application of an additional uncertainty factor bears consideration. Meanwhile, research should be carried out to ascertain that this approach offers adequate protection to species in nearby, and especially in remote, pristine areas with vulnerable ecosystems.

To maximise the efficiency of the risk assessment it would be wise to assess the risks to nearby and remote non-target areas only for those compounds, that have already passed the existing local risk assessment and have been shown to pose no threat to the environment in the immediate vicinity of the application sites (Bakker *et al.*, this issue). Similarly, a pesticide's atmospheric transport potential and risks to remote areas may be determined only for those chemicals that have been established to be (probably) safe for nearby areas.

There was a general agreement that transformation products may affect non-target areas. Those considered relevant should be subject to the same risk evaluation procedures as their parent compounds. In current legislation, only transformation products which arise in amounts of at least 10% of the parent chemical are considered. The participants favoured a more flexible approach in setting this percentage and advocated a lower percentage whenever metabolites are very toxic or persistent.

The participants felt that the approach outlined is, in principle, scientifically feasible, although substantial uncertainties remain. Further field and laboratory research is necessary to validate and improve our environmental fate models, to gain more reliable estimates of the physico-chemical properties of pesticides and to validate the applicability of standard toxicity data. This will increase both the accuracy of and our confidence in the outcome of the risk assessment. Finally, monitoring in non-target areas was regarded as an essential ultimate safety net in the assessment of risks posed by pesticides to organisms living in non-target areas.

6. Conclusions and recommendations

1. Atmospheric transport of pesticides to non-target areas is an established fact. For remote areas it is the most important route of contamination.

2. Total emissions of surface-applied pesticides into the atmosphere during and after application are highly variable and uncertain due to the fact that:
   a. little quantitative information is available on the fraction of the dosage which does not reach the target surface; this may typically be up to 20-30% and sometimes more than 50%. It is dependent upon the application technique and environmental conditions.
   b. the rate and extent of volatilisation of the applied pesticide is uncertain; depending upon a pesticide's physico-chemical properties, soil and crop characteristics and environmental conditions up to 50-60% of the dose, and sometimes even more, may enter the atmosphere by volatilisation. To improve the accuracy of the emission estimates, further development and testing of the models along with accompanying field tests are necessary.

3. The distance travelled by a pesticide through the atmosphere is determined by the rate of its removal from the atmosphere. Removal processes may be either physical (deposition) or chemical (transformation). OH radical reactions are the most important transformations of pesticides in the gaseous phase, except for pesticides containing unsaturated C-C bonds, for which the reaction with NO_3 radicals and O_3 may be important.

4. A limited data set on pesticide transformation rates and transformation products in air is available. Protocols need to be developed for measuring transformation rates and products in air under laboratory conditions. Such protocols already exist for other environmental compartments.

5. If the reactions are rapid, the reaction products, which may be more persistent in the environment and even more toxic than the parent compound, should be subject to further evaluations similar to those of the parent pesticide.

6. Only for those pesticides for which reaction rates of homologues are available, extrapolations or estimations may be within an uncertainty margin that is acceptable for risk assessment. The variability of the OH radical concentration in the atmosphere has an additional uncertainty, which is at least a factor of 2.

7. Sinks of airborne pesticides are poorly characterised. Deposition and re-emission are very important input and output processes in the total budget of pesticides; this is particularly the case for water bodies such as large lakes and oceans.

8. In principle, gaseous exchange of pesticides is reversible. Hence, for those compounds that tend to accumulate, concentrations in surface waters and the soil must be known on a regional scale for the deposition- and re-emission fluxes to be estimated.

9. (Re-)emission factors of historically used organochlorine pesticides are largely unknown due to uncertainties in the concentrations of soil residues.

10. The processes of transport and dispersion are similar to those for other air pollutants (such as ozone and SO_2), that have been studied more intensively over the last decades. From these studies it can be concluded that the uncertainty in the transport mechanism in models is about 30-50%. This is caused by the modelling concepts and meteorological variability.

11. The uncertainty in the deposition of pesticides is tentatively estimated at a factor of 3. This uncertainty is added to the above uncertainty in modelling of the transport and dispersion of pesticides. The overall uncertainty in the modelled deposition values is thus estimated to be a factor of 3-4. This does not include the uncertainties in the emissions and transformations.

12. For regulatory purposes a tiered approach is the most efficient manner of assessing the environmental risks of pesticides in the atmosphere. In the first tier simple criteria should be used to select problem compounds, which will then be evaluated more rigorously in a second tier.

13. The first tier should include the potential for
1. emission into the air during application and
2. emission after application (volatilisation potential).
Relevant factors are application method, formulation, vapour pressure, solubility in water, octanol/water partition coefficient (K_{ow}), and half-lives in soil and water (DT_{50}s). Measured values should preferably be used for these properties.

14. The second tier should be a PEC/PNEC approach, using
1. commonly available dispersion models for the short range (<50 km) and simple trajectory-type models for the long range (>1000 km) to estimate concentrations in air and precipitation in non-target areas,
2. environmental fate models to derive a PEC in soil and water, taking into account the environmental conditions in non-target areas,
3. toxicity data of standard test species to estimate a PNEC, using an additional uncertainty factor and
4. a pesticide's transport potential, based on factors such as the OH radical reaction rate (k_{OH}), to discriminate between compounds for which only the risks to nearby ecosystems should be assessed and those for which the risks to remote ecosystems also have to be evaluated.

15. The outlined approach is, in principle, scientifically feasible. However, the uncertainties are still substantial. More research, both in the field and the laboratory, is necessary:
a. to validate and improve our emission, transport, deposition and other environmental fate models,
b. to gain more reliable estimates of the physico-chemical properties of pesticides and
c. to validate the applicability of standard toxicity data.
This will increase the accuracy of and our confidence in the outcome of the risk assessment.

16. Monitoring the presence of pesticides, including their transformation products (in air and other environmental compartments) in non-target

areas should provide the ultimate safety net in the post-registration process.

References

AEA, 1993: *Priority setting for long-range transboundary air pollution by persistent organic chemicals.* AEA Technology report, Environment and Energy, Harwell Laboratory, Oxfordshire, UK.

Atkinson, R., Guicherit, R., Hites, R., Palm, W.-U., Seiber, J. N. and De Voogt, P.: 1999, *Water Air Soil Pollut.* this issue.

Bakker, D. J., Gilbert, A. J., Gottschild, D., Kuchnicki, T., Laane, R. W. P. M., Linders, J. B. H. J., Van de Meent, D., Pino, J., Montforts, M. H. M. M., van Straalen, N. M. and Pol, W.: 1999, *Water Air Soil Pollut.* this issue.

Bidleman, T. F.: 1999, *Water Air Soil Pollut.* this issue.

European and Mediterranean Plant Protection Organisation / Council of Europe (EPPO/CoE): 1998, *Decision making schemes for the environmental risk assessment of plant protection products. Chapter 12. Sub-scheme for air.* EPPO, Paris.

Gilbert, A. J.: 1999, *Water Air Soil Pollut.* this issue.

Gottschild, D., Siebers, J. and Nolting, H.-G.: 1990, *Richtlinien für die Prüfung von Pflanzenschutzmitteln im Zulassungsverfahren, Teil IV, 6-1, Prüfung des Verflüchtigungsverhaltens und des Verbleibs von Pflanzenschutzmitteln in der Luft.* Biologische Bundesanstalt für Land- und Forstwirtschaft, Bundesrepublik Deutschland, Saphir Verlag, Ribbesbüttel.

Hippelein, M. and McLachlan, M. S.: 1998, *Environ. Sci. Technol.* **32**, 310-316.

Mackay, D.: 1991, *Multimedia environmental models: the Fugacity approach.* Lewis Pub. Co. Michigan, U. S. A..

Majewski, M. S.: 1999, *Water Air Soil Pollut.* this issue.

McLachlan, M. S., Welschpausch, K. and Tolls, J.: 1995, *Environ. Sci. Technol.* **29**, 1998-2004.

Simonich, S. L. and Hites, R. A.: 1995, *Environ. Sci. Technol.* **29**, 2905-2914.

Smit, A. A. M. F. R., Van den Berg, F. and Leistra, M.: 1997, *Estimation method for the volatilization of pesticides from fallow soils*, Environmental Planning Bureau series 2, DLO Winand Staring Centre, Wageningen, The Netherlands.

Smit, A. A. M. F. R., Leistra, M. and Van den Berg, F.: 1998, *Estimation method for the volatilization of pesticides from plants*, Environmental Planning Bureau series 4, DLO Winand Staring Centre, Wageningen, The Netherlands.

UN-ECE: 1998, *UN-ECE/EB.air/WG.5/52. Annex II: Draft executive body decision 1998/2 on information to be submitted and the procedure for adding substances to Annexes I, II, or III to the POPs protocol.*

Van den Berg, F., Kubiak, R., Benjey, W. G., Majewski, M. S., Yates, S. R., Reeves, G. L., Smelt, J. H. and Van der Linden A. M. A.: 1999, *Water Air Soil Pollut.* this issue.

Van Dijk, H. F. G. and Guicherit, R.: 1999, *Water Air Soil Pollut.* this issue.

Van Jaarsveld, J. A. and Van Pul, W. A. J.: 1999, *Water Air Soil Pollut.* this issue.

Van Pul, W. A. J., Bidleman, T. F., Brorström-Lundén, E., Builtjes, P. J. H., Dutchak, S., Duyzer, J. H., Gryning, S.-E., Jones, K. C., Van Dijk, H. F. G. and Van Jaarsveld, J. A.: 1999, *Water Air Soil Pollut.* this issue.

Van Pul, W. A. J., De Leeuw, F. A. A. M., Van Jaarsveld, J. A., Van der Gaag, M. A. and Sliggers, C. J.: 1998, *Chemosphere* **37**, 113-141.

Van Straalen, N. M. and Van Gestel, C. A. M.: 1999, *Water Air Soil Pollut.* this issue.

Woodrow, J. E. Seiber, J. N. and Baker, L. W.: 1997, *Environ. Sci. Technol.* **31**, 523-529.

ATMOSPHERIC DISPERSION OF CURRENT-USE PESTICIDES: A REVIEW OF THE EVIDENCE FROM MONITORING STUDIES

HARRIE F. G. VAN DIJK[1*] and ROBERT GUICHERIT[2]

*[1]Health Council of the Netherlands, P. O. Box 16052, NL-2500 BB The Hague, The Netherlands, [2]TNO Institute of Environmental Sciences, Energy Research and Process Innovation, P. O. Box 342, NL-7300 AH Apeldoorn, the Netherlands (*corresponding author, e-mail: hfg.van.dijk@gr.nl)*

Abstract. Recently, evidence has accumulated that the extensive use of modern pesticides results in their presence in the atmosphere at many places throughout the world. In Europe over 80 current-use pesticides have been detected in rain and 30 in air. Similar observations have been made in North America. The compounds most often looked for and detected are the organochlorine insecticide lindane and triazine herbicides, especially atrazine. However, acetanilide and phenoxyacid herbicides, as well as organophosphorus insecticides have also frequently been found in rain and air. Concentrations in air normally range from a few pg/m^3 to many ng/m^3. Concentrations in rain generally range from a few ng/L to several µg/L. In fog even higher concentrations are observed. Deposition varies between a few mg/ha/y and more than 1 g/ha/y per compound. However, these estimates are usually based on the collection and analysis of (bulk) precipitation and do not include dry particle deposition and gas exchange. Nevertheless, model calculations, analysis of plant tissue, and first attempts to measure dry deposition in a more representative way, all indicate that total atmospheric deposition probably does not normally exceed a few g/ha/y. So far, little attention has been paid to the presence of transformation products of modern pesticides in the atmosphere, with the exception of those of triazine herbicides, which have been looked for and found frequently.

Generally, current-use pesticides are only detected at elevated concentrations in air and rain during the application season. The less volatile and more persistent ones, such as lindane, but to some extent also triazines, are present in the atmosphere in low concentrations throughout the year. In agricultural areas, the presence of modern pesticides in the atmosphere can be explained by the crops grown and pesticides used on them. They are also found in the air and rain in areas where they are not used, sometimes even in remote places, just like their organochlorine predecessors. Concentrations and levels are generally much lower there. These data suggest that current-use pesticides can be transported through the atmosphere over distances of tens to hundreds, and sometimes even more than a thousand kilometres. The relative importance of these atmospheric inputs varies greatly. For mountainous areas and remote lakes and seas, the atmosphere may constitute the sole route of contamination by pesticides. In coastal waters, on the other hand, riverine inputs may prevail. To date, little is known about the ecological significance of these aerial inputs.

Keywords: atmospheric transport, current-use pesticides, deposition, field measurements, long-range transport, monitoring, organochlorine pesticides, pesticides, rainwater

Water, Air, and Soil Pollution **115:** 21–70, 1999.

1. Introduction

Global pesticide use is currently estimated at approximately 2.5 billion kg per year (Pimentel *et al.*, 1998). To be effective, pesticides need to persist for a certain period of time. However, the longer their persistence, the greater the potential for transport of a fraction of the amount applied away from the target area. Pesticides are dispersed in the environment by water currents, wind, or biota. Pesticides can directly contaminate ground and surface waters by leaching, surface run-off and drift. Pesticides can also enter the atmosphere during application by evaporation and drift of small spray droplets, that remain airborne. Following application, pesticides may volatilise from the crop or the soil. Finally, wind erosion can cause soil particles and dust loaded with pesticides to enter the atmosphere. The extent to which pesticides enter the air compartment is dependent upon many factors: the properties of the substance in question (*e.g.* vapour pressure), the amount used, the method of application, the formulation, the weather conditions (such as wind speed, temperature, humidity), the nature of the crop and soil characteristics. Measurements at application sites reveal that sometimes more than half of the amount applied is lost into the atmosphere within a few days (Spencer and Cliath, 1990; Taylor and Spencer; 1990; Van den Berg *et al.*, this issue).

Although it is clear that the atmosphere is a major receptacle and transport vehicle for pesticide residues, it is the environmental compartment about which we know least regarding the fate of pesticides (Seiber and Woodrow, 1995). Once in the atmosphere, pesticides can exist as a gas and adsorbed onto particles. By in- and below-cloud scavenging both can enter the aqueous phase. The distribution among these phases depends on the physico-chemical properties of the compound in question, such as water solubility and vapour pressure. It is also influenced by environmental factors, especially temperature, humidity and the nature and concentration of dust particles in the air. This phase distribution greatly influences the residence time of a pesticide in the atmosphere, since the removal processes (deposition and degradation) are different for the various phases (De Voogt and Jansson, 1993; Majewski and Capel, 1995; Van Pul et al., 1998). Clearly, the longer a substance persists in the atmosphere, the further it can be transported. Deposition from the atmosphere in recipient areas can occur in the form of wet deposition (dissolved or suspended in rainwater, snow, and fog) and in the form of dry deposition, either sorbed on particles or by gas exchange. Once they have been deposited on the soil, vegetation or water surface, persistent compounds can reenter the atmosphere, leapfrogging from place to place until they finally break down or reach their permanent sink (Majewski and Capel, 1995; Wania and Mackay, 1996).

The presence of pesticides in the atmosphere was first demonstrated in the 1960s with organochlorine insecticides, such as DDT, drins (aldrin, dieldrin), chlordane, toxaphene, hexachlorocyclohexanes (HCHs) and hexachlorobenzene (HCB). For example, Tarrant and Tatton (1968) showed that these compounds could be detected in rainwater throughout the year over the whole of the United Kingdom. Soon it became evident that their occurrence in the atmosphere was not limited to areas of extensive use. They turned up in areas where they never had been used and even in remote and hitherto pristine places like the world's oceans and the polar regions. That the atmosphere was the main route of transport was first shown by Risebrough et al. (1968) and Prospero and Seba (1972). Since then, numerous investigations confirmed the ubiquitous presence of these compounds in all environmental compartments and demonstrated their accumulation in biota through foodwebs. Although most organochlorine pesticides have been banned or restricted in the industrialised countries since the 1970s, many of them are still being used in developing countries in both agriculture and vector control. Some, notably endosulfan and HCH, are still registered in several European and North American countries, albeit that in most western countries technical grade HCH, consisting predominantly of the less effective α-isomer, has been replaced by lindane, which consists almost entirely of the γ-isomer, after the 1970s. For more information regarding atmospheric transport of organochlorines, we refer to a comprehensive review by Bidleman, elsewhere in this issue. In this paper, we will summarise evidence of atmospheric transport of current-use pesticides, which has resulted from monitoring studies. To prevent disruption of the text, the rather extensive tables included in this paper, have been placed at the end.

2. Current-use pesticides in the atmosphere

Probably the first data on levels of modern pesticides in rainwater, notably the herbicides atrazine and 2,4-D, were reported by Cohen and Pinkerton as early as 1966. More research into the atmospheric behaviour of modern pesticides was conducted in the United States and Canada in the 1970s and the early-1980s (*e.g.* Grover, 1974; Arthur *et al.*, 1976; Kutz, 1976; Robinson and Fox, 1978; Wu, 1981). Most studies were aimed at 2,4-D, of which in addition to less volatile forms, highly volatile butyl esters were used. These butyl esters rapidly enter the atmosphere after application and can cause damage to crops at considerable distance from the application sites. For this reason, highly volatile forms were restricted (Sandmann *et al.*, 1991). However, it was not until the late 1980s that the occurrence of modern pesticides in the atmosphere received more widespread attention. A comprehensive overview of all the local, state-wide, multistate and national monitoring studies within the United States and Canada was compiled by Majewski and Capel (1995). Since then, additional data from these countries were presented by Waite *et al.* (1995), Baker *et al.* (1996), Hatfield *et al.* (1996), Hawthorne *et al.* (1996), Goolsby *et al.* (1997), McConnell *et al.* (1997), Majewski *et al.* (1998), McConnell *et al.* (1998), Rawn *et al.* (1998), Zhu *et al.* (1998), Rawn *et al.* (1999a) and Waite *et al.* (1999c). An overview of the results from European monitoring studies is presented in tables I and II. An earlier list of pesticides detected in precipitation was drawn up by Siebers *et al.* (1991). It contains five additional pesticides (bromoxynil, chlortoluron, metobromuron, monuron and phenmedipham) that were detected in rain in Germany and that are not included in table I, as no reports documenting these findings were available. Very recently, 28 European monitoring studies were reviewed by Dubus and Hollis (1999). Only a few studies were conducted outside Europe and North America, *e.g.* in Australia (Beard *et al.*, 1995), Japan (Haraguchi *et al.*, 1994 and 1995; Suzuki, 1996) and South Africa (Sandmann *et al.*, 1991; De Beer *et al.*, 1992).

In general, monitoring studies have been conducted on an *ad hoc* basis and are characterised by a small number of sampling sites, covering a small geographical area (often only a small part of a country) and a limited period of time. In the USA, however, some large, nation-wide studies have been conducted. The lack of consistency in sampling methodologies, sampling site selection, collection time and duration, selected analytes, analytical methods and detection limits, which was observed by Majewski and Capel (1995) for the Canadian and U.S. studies, holds also for the European studies. In contrast to the research in the USA and Canada most European monitoring studies have been focused on rain rather than air. So far, at least over 80 pesticides have been detected in precipitation in Europe (Table I) and 30 in air (Table II). This is probably more than in America (see Majewski and Capel, 1995). The actual number of pesticides found in the atmosphere may be even higher, as undoubtedly more studies have been conducted than those reported here. Often, the results are not published in international journals and consequently not easily accessible.

Apart from lindane, herbicides are the current-use pesticides most frequently looked for and detected in air and precipitation. This holds for both Europe and North America. Most attention has been focused on the triazines, such as simazine, cyanazine, terbuthylazine and in particular atrazine (see reviews by Dörfler and Scheunert, 1997; Grover *et al.*, 1999), but also acetanilides (alachlor, metolachlor) and phenoxyherbicides (2,4-D, MCPA, dichlorprop, mecoprop) have been targeted frequently. Among the insecticides, organophosphorus compounds (parathion, malathion, diazinon, chlorpyrifos) have been looked for most often. The occurrence of other groups of pesticides in air and rain has generally been poorly investigated (Majewski and Capel, 1995).

Concentrations of modern pesticides in air often range from a few pg/m^3 to many ng/m^3. In rain, concentrations have been measured from a few ng/L to several µg/L. However, concentrations in precipitation depend not only upon the amount of pesticides present in the atmosphere, but also on the amounts, intensity, and timing of rainfall. The highest concentrations are observed during light rainfall events, especially after extended dry periods (Tsai *et al.*, 1991; Gath *et al.*, 1993; Dankwardt *et al.*, 1994; Goolsby *et al.*, 1997). Accordingly, concentrations are high in the first millimetres of a precipitation event and decrease over the course of the event (Oberwalder *et al.*, 1991; Trautner *et al.*,

TABLE I
Pesticides detected in precipitation in Europe

pesticide	chemical structure	type	country	reference
alachlor	AA	H	D, CH, I, NL, GR	10, 13, 27, 33, 36, 38
aldicarb	C	I, N	F	19, 30
ametryn	T	H	F	7
atratone	T	H	D	26
atrazine	T	H	S, D, F, NL, N, CH, GR, A, I, UK, FIN, B	1, 3, 4, 5, 6, 7, 8, 9, 10, 11, 12, 13, 14, 15, 16, 17, 18, 19, 20, 21, 22, 23, 24, 25, 26, 27, 28, 29, 30, 31, 32, 33, 35, 38, 39, 40, 42, 43, 46
azinphos-ethyl	OP	I	NL	34, 48
azinphos-methyl	OP	I	NL	31, 34, 48
bentazone	o	H	S, NL, FIN, DK	1, 31, 32, 40, 52
bitertanol	TO	F	D	37
bromacil	o	H	NL	31
captan	o	F	NL	31
carbaryl	C	I	I	27
carbofuran	C	I	GR	14, 36
chlorfenvinphos	OP	I	NL	34, 48
chloridazon	o	H	D, B	37, 42
chlorothalonil	o	F	NL, GR, D, UK	31, 33, 36, 37, 43
chlorpropham	C	H	NL	31, 33
chlorpyrifos	OP	I	UK	46
cyanazine	T	H	S, F	1, 5, 7
2,4-D	PA	H	S, NL, D, CH, FIN, DK	1, 6, 11, 13, 23, 31, 40, 52

TABLE I
(*continued*)

pesticide	chemical structure	type	country	reference
deltamethrin	PYR	I	UK	46
demeton-S-methyl	OP	I	NL	34, 48
diazinon	OP	I	D, GR, I, NL	3, 4, 14, 27, 34, 48
dicamba	o	H	S	1
dichlobenil	o	H	I, NL	27, 31, 33, 34, 48
dichloropropene	o	N	NL	31
dichlorprop	PA	H	S, N, D, CH, I, Fi, DK	1, 8, 10, 11, 13, 15, 17, 22, 23, 24, 27, 40, 41, 52
dichlorvos	OP	I	NL	6, 31, 34, 48
dimethenamid	o	H	CH	13
dimethoate	OP	I	S, NL, UK	1, 31, 43
dinoseb	NP	H	B, CH	42, 45
dinoterb	NP	H	B	42
disulfoton	OP	I	NL	34
diuron	U	H	NL, B	31, 42
DNOC	NP	H, I	D, B, CH, DK	11, 42, 45, 52
endosulfan I, II	OC	I	NL	6, 34, 48
ethofumesate	o	H	GR	36
etrimfos	OP	I	D	17
fenpropathrin	PYR	I	F	19, 30
fenpropimorph	o	F	D, UK	3, 4, 46
fenthion	OP	i	NL	34, 48
flusilazole	TO	F	D	4
heptenophos	OP	I	NL	34, 48

TABLE I
(*continued*)

pesticide	chemical structure	type	country	reference
iprodione	o	F	NL	34
isoproturon	U	H	D, F, UK, B, DK	12, 19, 20, 21, 22, 30, 39, 42, 43, 44, 46, 52
lindane (γHCH)	OC	I	S, DK, D, F, NL, N, UK, FIN	1, 2, 3, 4, 5, 6, 8, 11, 12, 15, 17, 19, 20, 21, 22, 23, 28, 30, 31, 34, 38, 39, 40, 43, 46, 47, 48, 49, 50, 51
linuron	U	H	D	26
malathion	OP	I	NL	34, 48
MCPA	PA	H	S, N, D, CH, I, NL, FIN, DK	1, 8, 10, 11, 13, 23, 27, 31, 40, 41, 52
mecoprop (MCPP)	PA	H	S, NL, D, CH, F, FIN, DK, UK	1, 6, 10, 11, 13, 15, 17, 19, 23, 24, 30, 31, 40, 41, 43, 52
metalaxyl	o	F	S, CH, D	1, 13, 37
metamitron	o	H	D, NL, DK	10, 31, 37, 52
metazachlor	AA	H	D, CH, NL	3, 4, 13, 31
methabenz-thiazuron	U	H	DK	52
metolachlor	AA	H	D, CH, NL, GR, B	3, 4, 13, 33, 36, 37, 38, 42
metoxuron	U	H	NL	31
metribuzin	o	H	D	3, 4
mevinphos	OP	I	NL	6, 34, 48
molinate	o	H	GR	36
organotin	o	F	NL	31
parathion(-ethyl)	OP	I	NL, D, GR, I	6, 10, 12, 14, 21, 22, 26, 27, 31, 34, 48

TABLE I
(*continued*)

pesticide	chemical structure	type	country	reference
parathion-methyl	OP	I	GR, F, NL	14, 19, 30, 34, 36, 48
pendimethalin	o	H	D	3, 4, 9, 10
pentachloro-phenol	OC	F	NL	6
phorate	OP	I	I	27
pirimicarb	C	I	S, D, NL	1, 12, 20, 21, 22, 31, 37
pirimiphos-methyl	OP	I	NL	34
procymidone	o	F	NL	34
prometryn	T	H	F, D	7, 26
propachlor	AA	H	CH, NL	13, 31, 33
propanil	o	H	GR	36
propazine	T	H	D, CH	3, 13, 26, 28
propiconazole	TO	F	D	3, 4, 12, 37
propoxur	C	I	D	12
pyrazophos	OP	F, I	NL	34, 48
sebuthylazine	T	H	D	26
simazine	T	H	S, D, F, NL, CH, GR, UK, B	1, 3, 4, 5, 6, 7, 10, 13, 14, 15, 16, 17, 23, 24, 26, 28, 29, 31, 32, 33, 37, 38, 39, 42, 46
tebuconazole	TO	F	D	37
tebutam	o	H	D	37
terbuthylazine	T	H	S, D, F, CH, DK	1, 3, 4, 5, 9, 10, 11, 12, 13, 16, 17, 26, 29, 37, 38, 52
terbutryn	T	H	F, D, CH	7, 26, 29
tetrachlorvinphos	OP	I	NL	34, 48

TABLE I
(*continued*)

pesticide	chemical structure	type	country	reference
tolclofos-methyl	OP	F	NL	33, 34, 48
triadimenol	TO	F	D	4, 20, 21
tri-allate	o	H	S, D, NL, UK	1, 3, 4, 10, 33, 43
triazophos	OP	I	NL	34, 48
trifluralin	o	H	NL, I, UK	6, 27, 46
vinclozolin	o	F	D, NL	12, 31, 33, 34, 48

Chemical structure: AA = acetanilide, C = carbamate, NP = nitrophenol, o = others, OC = organochlorine, OP = organophosphorus, PA = phenoxy acid, PYR = pyrethroid, T= triazine, TO = triazole, U = urea compound

Type: F = fungicide, H = herbicide, I = insecticide, N = nematicide

Countries: A = Austria, B = Belgium, CH = Switzerland, CRO = Croatia, D = Germany, DK = Denmark, F = France, FIN = Finland, GR = Greece, I = Italy, N = Norway, NL = The Netherlands, S = Sweden, UK = United Kingdom

References: 1 = Kreuger, 1995; 2 = Cleemann *et al.*, 1995; 3 = Scharf *et al.*, 1992; 4 = Scharf and Bächmann, 1993; 5 = Chevreuil *et al.*, 1996; 6 = Baas and Duyzer, 1997; 7 = Chevreuil and Garmouma, 1993; 8 = Lode *et al.*, 1995a,b, 9 = Oberwalder and Hurle, 1993; 10 = Oberwalder *et al.*, 1992; 11 = Geissler and Schöler, 1993; 12 = Siebers *et al.*, 1994; 13 = Bucheli *et al.*, 1998; 14 = Albanis *et al.*, 1998; 15 = Hurle *et al.*, 1987; 16 = Wendlandt *et al.*, 1989; 17 = Oberwalder *et al.*, 1991; 18 = Neururer and Womastek, 1991; 19 = Millet *et al.*, 1997; 20 = Gath *et al.*, 1992; 21 = Gath *et al.*, 1993; 22 = Siebers *et al.*, 1991; 23 = Stähler, 1993; 24 = Herterich, 1991; 25 =Trautner *et al.*, 1992; 26 = Bester *et al.*, 1995; 27 = Trevisan *et al.*, 1993; 28 = Schössner and Koch, 1991; 29 = Buser, 1990; 30 = Millet *et al.*, 1996; 31 = Van Boom, 1993; 32 = Van Zoonen *et al.*, 1991; 33 = Bakker, 1996; 34 = Provincie Zuid-Holland, 1994; 35 = Dankwardt *et al.*, 1994; 36 = Charizopoulos and Papadopoulou-Mourkidou; 37 = Hüskes and Levsen, 1997; 38 = Jager *et al.*, 1998; 39 = Harris *et al.*, 1992; 40 = Hirvi, 1994 and Hirvi and Rekolainen, 1995; 41 = Kirknel and Felding, 1995; 42 = Quaghebeur, 1995; 43 = Gomme *et al.*, 1991; 44 = Johnson *et al.*, 1996; 45 = Leuenberger *et al.*, 1988; 46 = Turnbull, 1995; 47 = Hühnerfuss *et al.*, 1997; 48 = Provincie Zuid-Holland, 1991; 49 = CCRX, 1993; 50 = Fingler *et al.*, 1994; 51 = Villeneuve and Cattini, 1986; 52 = Felding *et al.*, 1999

TABLE II
Pesticides detected in air in Europe

pesticide	chemical structure	type	country	reference
alachlor	AA	H	I	8
aldicarb	C	I, N	F	6, 9
ametryn	T	H	F	3
atrazine	T	H	D, F, NL, UK	1, 3, 5, 6, 7, 9, 10, 11, 12
chlorothalonil	o	F	NL	10
chlorpropham	C	H	NL	10
chlorpyrifos	OP	I	UK	12
cyanazine	T	H	F	3
2,4-D	PA	H	NL	4
deltamethrin	PYR	I	UK	12
dichlobenil	o	H	I	8
dichlorvos	OP	I	NL	4
endosulfan I, II	OC	I	NL	4
fenpropathrin	PYR	I	F	6, 9
fenpropimorph	o	F	UK	12
isoproturon	U	H	F, UK	6, 9, 11, 12
lindane (γHCH)	OC	I	N, F, NL, UK, S	2, 3, 4, 6, 9, 12, 13, 14
mecoprop (MCPP)	PA	H	NL, F	4, 6, 9
metolachlor	AA	H	NL	10
mevinphos	OP	I	NL	4
parathion(-ethyl)	OP	I	NL, I	4, 8
parathion-methyl	OP	I	F	6, 9
pentachlorophenol	OC	F	NL	4
phorate	OP	I	I	8
simazine	T	H	F, NL, UK	3, 4, 11, 12

TABLE II
(*continued*)

pesticide	chemical structure	type	country	reference
terbumeton	T	H	F	3
terbuthylazine	T	H	F	3
tri-allate	o	H	NL	10
trifluralin	o	H	NL, I, UK	4, 8, 12
vinclozolin	o	F	NL	10

For abbreviations: see table 1. References: 1 = Elling *et al.*, 1987; 2 = Haugen *et al.*, 1998; 3 = Chevreuil *et al.*, 1996; 4 = Baas and Duyzer, 1996; 5 = Chevreuil *et al.*, 1993; 6 = Millet *et al.*, 1997; 7 = Herterich, 1991; 8 = Trevisan *et al.*, 1993; 9 = Millet *et al.*, 1996; 10 = Bakker, 1996; 11 = Harris *et al.*, 1992; 12 = Turnbull, 1995; 13 = Brorström-Lundén *et al.*, 1994; 14 = Brorström-Lundén, 1996.

1992; Wu 1981, Nations and Hallberg, 1992, Goolsby *et al.*, 1997; Bucheli *et al.*, 1998). The reason is that the initial part of a precipitation event tends to scavenge most of the pesticides from the atmosphere, especially those associated with particulate material and gaseous pesticides with high water solubility. Rainfall occurring later during the event dilutes the concentration of the pesticides that were deposited during the early part of the rainfall event (Tsai *et al.*, 1991; Goolsby *et al.*, 1997). Oberwalder *et al.* (1991) regard this as an indication that the presence of pesticides in rainwater is predominantly the result of below-cloud scavenging. This view is not shared by Gath *et al.* (1993): they believe that in-cloud scavenging is more important, because the annual average concentrations of pesticides in rainwater at their sampling locations were independent of the amounts of precipitation. According to Tsai *et al.* (1990) the scavenging mechanism depends *i. a.* upon the particle size: in-cloud scavenging is the dominant removal process for submicrometer particles, whereas for coarse particles below-cloud scavenging is most important.

Several authors have reported the presence of pesticides in fog. Glotfelty *et al.* (1987), Glotfelty *et al.* (1990a), Schomburg *et al.* (1991) and Seiber *et al.* (1993) found high concentrations (many ng/L up to as much as 100 µg/L) of organophosphorus insecticides, both in wintertime fog formed under stagnant inversion conditions in the Central Valley, California, and in fog along California's Pacific coast in late summer, resulting from the advection of oceanic air over land surfaces. By measuring the air-water distribution coefficients of the pesticides between the liquid fog and the interstitial gas phase they showed that the pesticides were enriched manifold in the suspended liquid fog droplets compared to equilibrium distributions expected from Henry's Law coefficients for pure aqueous solutions. This makes fog a good indicator for the contamination of the atmosphere with pesticides (Seiber 1995). Schomburg *et al.* (1991) hypothesised that non-filterable dissolved organic carbon present in fogwater exists in a fine particle or colloidal form and has properties similar to activated carbon in being highly sorptive, thus causing the apparent pesticide enrichment. Alternatively, the enrichment may be caused by the presence of an organic film on the surface of the droplets (Majewski and Capel, 1995), or the high surface area-to-volume ratio of fogwater, allowing the surface to act as a significant third phase (Hoff *et al.*, 1993). In spite of the enrichment in the liquid phase, the bulk of the organophosphorus pesticides was present in the interstitial air phase, either as vapour or adsorbed to aerosol particles (Glotfelty *et al.*, 1990a). High concentrations of pesticides in fog were also observed during autumn and winter in the northeastern part of France by Millet *et al.* (1997).

Chernyak *et al.* (1996), Rice (1996) and Rice and Chernyak (1997) detected chlorpyrifos, trifluralin, metolachlor, chlorothalonil, terbufos and endosulfan in arctic marine fog in the Bering and Chukchi Seas in concentrations (on the order of a few ng/L up to 147 ng/L) several times higher than in adjacent water or ice. Concentrations in this remote area were generally much lower than in agricultural areas in California.

Rice and Chernyak (1997) and Seiber and Woodrow (1995) pointed out the great similarities between fog and clouds: fog is simply a cloud with its base at ground level or very close to it. Therefore, chemical and physical processes are probably very similar in both clouds and fog, but much more easy to study in the latter. Cloud chemistry in relation to long- and medium-range atmospheric transport is considered one of the weakest areas of knowledge concerning fate processes in the atmosphere (Seiber and Woodrow, 1995).

3. Deposition

Estimated deposition of pesticides, based on the analysis of precipitation collected with bulk or wet-only samplers, is generally on the order of a few mg/ha/y up to more than 1 g/ha/y for individual pesticides (Wendlandt *et al.*, 1989; Buser, 1990; Nations and Hallberg, 1992; Oberwalder *et al.*, 1991; Sandmann *et al.*, 1991; Gath *et al.*, 1992; Oberwalder *et al.*, 1992; Gath *et al.*, 1993; Geissler and Schöler, 1993; Chevreuil and Garmouma, 1993; Oberwalder and Hurle, 1993; Scharf and Bächmann, 1993; Stähler, 1993; Zabik and Seiber, 1993; Dankwardt *et al.*, 1994; Provincie Zuid-Holland, 1994; Siebers *et al.*, 1994; Bester *et al.*, 1995; Waite *et al.*, 1995; Bakker, 1996; Chevreuil *et al.*, 1996; Hatfield *et al.*, 1996; Goolsby *et al.*, 1997; Bucheli *et al.*, 1998). In table III this is illustrated for lindane. This organochlorine pesticide has higher wet deposition rates in central Europe than other current-use pesticides due to its high concentrations in rainwater and its presence throughout the year (Scharf and Bächmann, 1993; Siebers *et al.*, 1994). Tables IV to VI present some data on wet deposition of atrazine, mecoprop and parathion.

Considerable between-year variation, sometimes amounting to more than a factor of 10, in the deposition of a certain pesticide at a specific sampling site may exist (Wu, 1981; Glotfelty *et al.*, 1990b; Kreuger, 1995; Lode *et al.*, 1995a,b; Waite *et al.*, 1995; Hatfield *et al.*, 1996; McConnell *et al.*, 1998, Rawn *et al.,* 1999c). This may be explained by differences in application amounts and atmospheric conditions, including the timing of rainfall events relative to application (Hatfield *et al.*, 1996).

By comparing the results of bulk and wet-only samplers, some authors (Glotfelty *et al.*, 1990b; Oberwalder *et al.*, 1991; Nations and Hallberg, 1992; Gath *et al.*, 1993; Siebers *et al.*, 1994) conclude that the contribution of dry deposition to total atmospheric pesticide fluxes is of minor importance. In agreement with this is, that deposition collected with bulk collectors during dry periods is small compared to deposition during periods with precipitation (Nations and Hallberg, 1992; Oberwalder *et al.*, 1992). However, dry deposition on a funnel is probably not representative of that on the more complex surfaces of water, soil and vegetation. Moreover, exposure to sunlight and the atmosphere may result in photodegradation, volatilisation and removal by wind (Waite *et al.*, 1999). Finally, gas exchange is not included.

TABLE III

Wet deposition of lindane as determined with bulk or wet-only precipitation samplers

location[1]	period[2]	maximum concentration ng/l	mean or median conc.[3] ng/l	deposition[4] g/ha/y	ref.[5]
Evo, Mekrijärvi, FIN	5-11/91,5-10/92	20	n.r.	n.r.	40
Abisko, northern S	6/92-7/92	n.r.	n.r.	0.003 (2 m)	1
Lurbo, central S	5-9/90-92	29	7	0.009 (5 m)	1
Ekeröd, southern S	5-9/90-92	73	13	0.016 (5 m)	1
As, southern N	6-9/92,5-9/93	43	n.d.	0.025 (4-5m)	8
Lista, southern N	6-9/92,5-9/93	84	n.d.	0.018 (4-5m)	8
drilling platform Gullfaks, central North Sea	2-12/91	n.r.	n.r.	0.009	47
Husby, DK	1/90-12/91	85	17	0.114	2
Ulborg, DK	1/90-2/93	120	14	0.132	2
Bagenkop, DK	3/92-12/92	95	16	0.098	2
Anholt, DK	3/92-2/93	70	15	0.091	2
Hailfingen, D	5/85-11/86	550	50	0.550	17
Braunschweig, D	3/90-3/92	400	117	0.43	12
Helmstedt, D	3/90-3/92	310	130	n.r.	12
Goslar, D	3/90-3/92	710	171	0.70	12
Gelsenkirchen-Buer, D	3-9/90	505	131	n.r.	28
Gelsenkirchen Stadtmitte, D	5-9/90	70	40	n.r.	28
Klein Feldberg, D	3/90-3/91	270	n.r.	0.6	20
Klein Feldberg, D	10/90-10/91	277	52	0.372	21
Neustadt, D	10/90-10/91	160	36	0.163	21
Hortenkopf, D	10/90-10/91	297	48	0.352	21
Schauinsland, D	6/90-8/91	833	208	n.r.	3

TABLE III
(*continued*)

location[1]	period[2]	maximum concentration ng/l	mean or median conc.[3] ng/l	deposition[4] g/ha/y	ref.[5]
Deuselbach, D	6/90-8/91	760	151	n.r.	3
Bensheim, D	5/90-4/91	183	116	n.r.	3
Schauinsland, D	1-11/91	n.r.	n.r.	0.292 (11m)	4
Deuselbach, D	1-11/91	n.r.	n.r.	0.421 (11m)	4
Meierhof, D	4-11/91	n.r.	n.r.	0.309 (8m)	4
Hohenfinow, D	4-12/91	260	n.r.	0.120 (9m)	23
Kleinmachnow, D	4-12/91	250	n.r.	0.240 (9m)	23
Krielow, D	4-12/91	250	n.r.	0.100 (9m)	23
NeuglobsoW, D	4-12/91	160	n.r.	0.190 (9m)	23
Niedergörsforf, D	4-12/91	360	n.r.	0.130 (9m)	23
Ruhlsdorf, D	4-12/91	220	n.r.	0.090 (9m)	23
Berlin-Dahlem, D	4-12/91	120	n.r.	0.030 (9m)	23
Dikopshof, D	4-12/91	n.r.	n.r.	0.309 (9m)	11
Venusberg, D	4-12/91	n.r.	n.r.	0.195 (9m)	11
Hau, D	4-12/91	n.r.	n.r.	0.176 (9m)	11
Hannover, n. D	3-11/96	130	41	n.r.	38
Flevoland, NL	9-12/90	30	n.r.	n.r.	31
Westland, NL	2/90-1/91	110	27	0.165	48
De Bilt, NL	91	n.r.	n.r.	0.30	49
Naaldwijk, NL	11/91-10/92	200	29	0.23	34
Nieuwerkerk, NL	11/91-10/92	190	47	0.25	34
Hillegom, NL	11/91-10/92	240	56	0.30	34
Korendijk, NL	11/91-10/92	160	30	0.12	34
Naaldwijk, NL	8-12/96	35	25	n.r.	6

TABLE III
(*continued*)

location[1]	period[2]	maximum concentration ng/l	mean or median conc.[3] ng/l	deposition[4] g/ha/y	ref.[5]
De Zilk, NL	9-12/96	45	26	n.r.	6
Noordwijk, NL	11/96-1/97	28	n.r.	n.r.	6
Rosemaund, UK	10/92-7/93	250	50	0.35	46
Brimstone, UK	1/91-8/93	400	55	0.35	46
Menton, F	3-5/?	76	31	0.009	51
Colmar, F	12/91-12/92	800	160	n.r.	19
Paris, F	1/92-9/93	130	8	0.299	5
La Ferté sous Jouarre, F	1/92-9/93	350	63	0.905	5
Zagreb, CRO	12/90-6/92	38	14	n.r.	50
Burnt Island, ON, CAN	90-92	n.r.	2	n.r.	60
Point Petre, ON, CAN	90-92	n.r.	3	n.r.	60
Eagle Harbor, MI, USA	90-92	n.r.	1	n.r.	60
Sleeping Bear Dunes, MI, USA	90-92	n.r.	2	n.r.	60
Sturgeon Point, NY, USA	90-92	n.r.	1	n.r.	60
Sequioa National Park, CAL, USA	12/95-4/96	2	n.r.	n.r.	61

[1] For abbreviations of country names see Table 1, CAN = Canada, USA = United States of America; [2] format used: month/year-month/year or month-month/year-year; [3] A direct comparison between these data should be considered with caution as methods of dealing with values below limits of detection will vary; [4] (xm) indicates that the amount of lindane was deposited in x months; [5] see Table 1 for the references, 60 = Hoff *et al.*, 1996, 61 = McConnell *et al.*, 1998; n.d. = not detected, n.r. = not reported.

TABLE IV

Wet deposition of atrazine as determined with bulk or wet-only precipitation samplers

location[1]	period[2]	maximum concentration ng/l	mean or median conc.[3] ng/l	deposition g/ha/y	ref.[4]
Evo, Mekrijärvi, FIN	5-11/91,5-10/92	10	n.r.	n.r.	40
Lurbo, Uppsala, S	5-9/90-92	60	10	0.004	1
Ekeröd, Scania, S	5-9/90-92	160	16	0.021	1
As, N	5-10/93	84	n.d.	0.041	8
Lista, N	5-10/93	86	n.d.	0.068	8
Hailfingen, D	4-8/85	510	140	n.r.	15
Hailfingen, D	5/85-11/86	650	55	0.450	17
Stuttgart, D	3-8/88	1110	80	0.225	17
Bühl, D	3/89-10/90, 3-7/91	n.r.	n.r.	0.462	10
Enzklösterle, D	3-10/90, 3-7/91	n.r.	n.r.	0.201	10
Calw, D	3/89-10/90, 3-7/91	n.r.	n.r.	0.149	10
Hohenheim, D	3/89-10/90, 3-7/91	n.r.	n.r.	0.191	10
Stuttgart, D	5-10/90, 3-7/91	n.r.	n.r.	0.147	10
Bruchsal, D	3/89-10/90, 3-7/91	n.r.	n.r.	0.212	10
Heilbronn, D	3/89-10/90, 3-7/91	n.r.	n.r.	0.158	10
Schäbisch Hall, D	3/89-10/90, 3-7/91	n.r.	n.r.	0.144	10
Überlingen-SPW, D	1-12/89	400	27	0.195	16
Überlingen-SB, D	5-12/89	480	44	n.r.	16
Lindau, D	?-?/89	120	13	n.r.	16

TABLE IV
(*continued*)

location[1]	period[2]	maximum concentration ng/l	mean or median conc.[3] ng/l	deposition g/ha/y	ref.[4]
Wank (Alps), D	4-7/89-90	420	n.r.	0.120	24
Fichtelgebirge, D	4-7/89-90	1600	n.r.	1.300	24
Gelsenkirchen-Bür, D	3-9/90	437	n.r.	n.r.	28
Gelsenkirchen-Stadtmitte, D	5-9/90	339	n.r.	n.r.	28
Dürnast, D	6-11/90	3290	160	1.80	35
Braunschweig, D	4/90-1/91	82	n.d.	0.100	22
Rotenkamp, D	4/90-1/91	240	n.d.	0.240	22
Neuenkirchen, D	4/90-1/91	430	n.d.	0.130	22
Bensheim, D	5/90-4/91	39	25	n.r.	3
Deuselbach, D	6/90-11/91	134	36	0.059	3,4
Schauinsland, D	6/90-11/91	135	76	0.024	3,4
Darmstadt, D	4-11/91	150	68	0.118	4
Braunschweig, D	3/90-3/92	113	44	0.020	12
Rotenkamp, D	3/90-3/92	240	80	n.r.	12
Neuenkirchen, D	3/90-3/92	430	105	n.r.	12
Klein Feltberg, D	3/90-3/91	430	<20	0.250	20
Klein Feltberg, D	10/90-10/91	80	n.d.	0.093	21
Hortenkopf, D	10/90-10/91	140	n.d.	0.098	21
Neustadt, D	10/90-10/91	270	n.d.	0.103	21
Wesseling, D	4-12/91	n.r.	n.r.	0.071	11
Bonn, D	4-12/91	n.r.	n.r.	0.056	11
Siegkreis, D	6-12/91	n.r.	n.r.	n.d.	11
Hau, D	4-12/91	n.r.	n.r.	0.047	11

TABLE IV
(*continued*)

location[1]	period[2]	maximum concentration ng/l	mean or median conc.[3] ng/l	deposition g/ha/y	ref.[4]
Hohenfinow, D	4-12/91	n.d.	n.d.	n.d.	23
Kleinmachnow, D	4-12/91	100	n.d.	n.r.	23
Krielow, D	4-12/91	n.d.	n.d.	n.d.	23
Neuglobsov, D	4-12/91	n.d.	n.d.	n.d.	23
Niedergörsdorf, D	4-12/91	n.d.	n.d.	n.d.	23
Ruhlsdorf, D	4-12/91	n.d.	n.d.	n.d.	23
Berlin-Dahlem, D	4-12/91	n.d.	n.d.	n.d.	23
Dürnast, D	4-11/91-92	1490	65	0.250	35
Waldhof, D	4-11/91-92	490	80	0.125	35
Bayer. Wald, D	6-10/91-92	310	55	0.125	35
Berchtesgaden, D	6-10/91-92	50	35	0.040	35
Lichtenau, D	4-7/92	n.r.	n.r.	0.350	9
Bühl, D	4-7/92	n.r.	n.r.	0.328	9
Enzklösterle, D	4-7/92	n.r.	n.r.	0.223	9
Hohenheim, D	4-7/92	n.r.	n.r.	0.103	9
Heligoland, D	4-7/93	125	n.r.	n.r.	26
Norderstedt, D	5-7/93	120	n.r.	n.r.	26
Hannover, D	3-11/96	90	21	n.r.	38
Flevoland, NL	9-12/90	200	n.r.	n.r.	31
Flevoland, NL	4-12/91	900	n.r.	n.r.	31
Eibergen, NL	5-10/88	740	<10	n.r.	32
Huijbergen, NL	5-10/88	160	20	n.r.	32
Kloosterburen, NL	5-10/88	160	<20	n.r.	32
Leiduin, NL	5-10/88	140	<10	n.r.	32

TABLE IV
(*continued*)

location[1]	period[2]	maximum concentration ng/l	mean or median conc.[3] ng/l	depositio g/ha/y	ref.[4]
Vredepeel, NL	5-10/88	690	40	n.r.	32
Wieringerwerf, NL	5-10/88	190	<20	n.r.	32
De Groote Peel, NL	7/95-7/96	900	21	0.240	33
Naaldwijk, NL	9-12/96	<5	<5	n.r.	6
De Zilk, NL	9-11/96	6	5	n.r.	6
Noordwijk, NL	11/96	17	17	n.r.	6
5 sites Flanders, B	5/92-10/94	1620	n.r	n.r.	42
Colmar, F	12/91-12/92	5000	n.d.	n.r.	19, 3
Vosges, Col du Hantz, F	5-7/91	137	22	n.r.	25
Vosges, Kreuzberg, F	5-7/91	67	18	n.r.	25
Paris, F	1/92-9/93	400	50	0.770	5
La Ferté-sous-Jouarre, F	1/92-9/93	380	50	0.680	5
Paris, F	3/91-12/91	140	17	n.r.	7
La Ferté-sous-Jouarre, F	3/91-2/92	350	45	0.280	7
Brimstone, UK	1/91-8/93	69	15	n.r.	46
Rosemaund, UK	2/92-8/93	15	1.5	n.r.	46
Wädenswil, CH	2/88-5/89	193	n.r.	0.200	29
Zürich, CH	7-8/88	3	n.r.	n.r.	29
Bachtel, CH	7-8/88	2	n.r.	n.r.	29
Lägern, CH	9/88-5/89	600	n.r.	n.r.	29
Grüze (Zürich), CH	2-10/96	903	33	0.139	13

TABLE IV

(continued)

location[1]	period[2]	maximum concentration ng/l	mean or median conc.[3] ng/l	deposition g/ha/y	ref.[4]
Vallombrosa, I	5-10/88	1130	n.r.	n.r.	27
Imathia, GR	5/96-4/97	7	n.r.	n.r.	14
Maryland, USA	3-7/81-82, 5-9/84	3300	n.r.	0.54-1.03	72
>80 sites, Midwestern and Northeastern USA	3/90-9/91	>1000	n.r.	<0.1 - >1.0	71
Walnut Creek, IA, USA	91-94	154000	700	0.845	70

[1] For abbreviations of country names see Table 1, USA = United States of America; [2] format used: month/year-month/year or month-month/year-year; [3] A direct comparison between these data should be considered with caution as methods of dealing with values below limits of detection will vary; [4] see Table 1 for the references, 70 = Hatfield *et al.*, 1996, 71 = Goolsby *et al.*, 1997, 72 = Glotfelty *et al.*, 1990b; n.d. = not detected, n.r. = not reported.

TABLE V

Wet deposition of mecoprop as determined with bulk or wet-only precipitation samplers

location[1]	period[2]	maximum concentration ng/l	mean or median conc.[3] ng/l	deposition g/ha/y	ref.[4]
Evo, Mekrijärvi, FIN	5-11/91,5-10/92	32	n.r.	n.r	40
Lurbo, Uppsala, S	5-9/90-92	32	9.7	0.006	1
Ekeröd, Scania, S	5-9/90-92	46	1.0	0.003	1
Ulborg, DK	4/93-4/94	119	n.d.	n.r.	41
Gadevang, DK	4/93-4/94	n.d.	n.d.	n.d.	41
Sjaelland, DK	5-11/96,4-11/97	145	n.d.	0.30	52
Sjaelland, DK	3-12/98	n.d.	n.d.	n.d.	52
Hailfingen, D	4-8/85	1390	60	n.r.	15
Hailfingen, D	5/85-11/86	1390	n.d.	n.r.	17
Hohenheim, D	3-7/88	150	n.d.	0.017	17
Bühl, D	3/89-10/90, 3-7/91	n.r.	n.r.	0.073	10
Enzklösterle, D	3-10/90, 3-7/91	n.r.	n.r.	0.020	10
Calw, D	3/89-10/90, 3-7/91	n.r.	n.r.	0.020	10
Hohenheim, D	3/89-10/90, 3-7/91	n.r.	n.r.	0.033	10
Stuttgart, D	5-10/90, 3-7/91	n.r.	n.r.	n.r.	10
Bruchsal, D	3/89-10/90, 3-7/91	n.r.	n.r.	0.067	10
Heilbronn, D	3/89-10/90, 3-7/91	n.r.	n.r.	0.111	10
Schäbisch Hall, D	3/89-10/90, 3-7/91	n.r.	n.r.	0.080	10
Hohenfinow, D	4-12/91	n.d.	n.d.	n.d.	23

TABLE V

(*continued*)

location[1]	period[2]	maximum concentration ng/l	mean or median conc.[3] ng/l	deposition g/ha/y	ref.[4]
Kleinmachnow, D	4-12/91	n.d.	n.d.	n.d.	23
Krielow, D	4-12/91	n.d.	n.d.	n.d.	23
Neuglobsov, D	4-12/91	n.d.	n.d.	n.d.	23
Niedergörsdorf, D	4-12/91	140	n.d.	0.020	23
Ruhlsdorf, D	4-12/91	n.d.	n.d.	n.d.	23
Berlin-Dahlem, D	4-12/91	n.d.	n.d.	n.d.	23
Wesseling, D	4-12/91	n.r.	n.r.	0.097	11
Bonn, D	4-12/91	n.d.	n.d.	n.d.	11
Siegkreis, D	6-12/91	n.d.	n.d.	n.d.	11
Hau, D	4-12/91	n.r.	n.r.	0.104	11
Naaldwijk, NL	9-12/96	19	n.d.	n.r.	6
De Zilk, NL	9-11/96	12	1	n.r.	6
Noordwijk, NL	11/96	n.d.	n.d.	n.d.	6
Flevoland, NL	9-12/90	n.d.	n.d.	n.r.	31
Flevoland NL	4-12/1991	1000	n.r.	n.r.	31
Grüze, CH	2-10/96	50 (19)[5]	10 (10)[5]	0.039 (0.030)[5]	13
Colmar, F	12/91-12/92	60000	15000	n.r.	19, 30

[1] For abbreviations of country names see Table 1; [2] format used: month/year-month/year or month-month/year-year; [3] A direct comparison between these data should be considered with caution as methods of dealing with values below limits of detection will vary; [4] see Table 1 for the references; [5] R-form (S-form); n.d. = not detected, n.r. = not reported.

TABLE VI

Wet deposition of parathion-ethyl as determined with bulk or wet-only precipitation samplers

location[1]	period[2]	maximum concentration ng/l	mean or median conc.[3] ng/l	deposition g/ha/y	ref.[4]
Bühl, D	3/89-10/90, 3-7/91	n.r.	n.r.	0.059	10
Enzklösterle, D	3-10/90, 3-7/91	n.r.	n.r.	n.d.	10
Calw, D	3/89-10/90, 3-7/91	n.r.	n.r.	0.007	10
Hohenheim, D	3/89-10/90, 3-7/91	n.r.	n.r.	0.023	10
Stuttgart, D	5-10/90, 3-7/91	n.r.	n.r.	0.063	10
Bruchsal, D	3/89-10/90, 3-7/91	n.r.	n.r.	0.044	10
Heilbronn, D	3/89-10/90, 3-7/91	n.r.	n.r.	0.185	10
Schäbisch Hall, D	3/89-10/90, 3-7/91	n.r.	n.r.	0.012	10
Braunschweig, D	3/90-3/91	320	n.d.	0.34	22
Helmstedt, D	3/90-3/91	190	n.d.	0.29	22
Goslar, D	3/90-3/91	(520)[5]	n.d.	(0.97)[5]	22
Braunschweig, D	3/90-3/92	320	122	n.r.	12
Helmstedt, D	3/90-3/92	190	117	n.r.	12
Goslar, D	3/90-3/92	(569)[5]	(254)[5]	n.r.	12
Klein Feldberg, D	10/90-10/91	n.d	n.d.	n.d.	21
Neustadt, D	10/90-10/91	160	n.d.	n.d.	21
Hortenkopf, D	10/90-10/91	n.d.	n.d.	n.d.	21
Westland, NL	2/90-1/91	100	18	0.100	48
Flevoland, NL	9-12/90	100	n.r.	n.r.	31

TABLE VI
(*continued*)

location[1]	period[2]	maximum concentration ng/l	mean or median conc.[3] ng/l	deposition g/ha/y	ref.[4]
Flevoland, NL	4-12/1991	300	n.r.	n.r.	31
Naaldwijk, NL	11/91-10/92	50	6.3	0.052	34
Nieuwerkerk, NL	11/91-10/92	60	12	0.076	34
Hillegom, NL	11/91-10/92	40	10	0.053	34
Korendijk, NL	11/91-10/92	190	18	0.120	34
Naaldwijk, NL	9-12/96	15	7	n.r.	6
De Zilk, NL	9-11/96	26	5	n.r.	6
Noordwijk, NL	11/96	n.d.	n.d.	n.d.	6
Renon, I	5-10/88	170	n.d.	n.r.	27
Imathia, GR	5/96-4/97	4	n.r.	n.r.	14

[1] For abbreviations of country names see Table 1; [2] format used: month/year-month/year or month-month/year-year; [3] A direct comparison between these data should be considered with caution as methods of dealing with values below limits of detection will vary; [4] see Table 1 for the references; [5] application of parathion near the sampling site; n.d. = not detected, n.r. = not reported.

Waite *et al.* (1999) tried to measure dry deposition rates of some pesticides more representatively with a new sampler, called the 'dust collector', that can collect dry (particle deposition plus gas absorption) and wet deposition separately. It consists of a moving sheet of water, which is recirculated through an absorbing resin. It simulates an open water surface, which is the natural surface that is most easily reproduced. They determined dry deposition of lindane, dicamba and 2,4-D to amount to 3.27, 0.69 and 2.76 mg/ha/day, respectively, in Regina, Saskatchewan, during five weeks in May and June. The wet deposition during the same period was 1.89 mg/ha/d for 2,4-D. Concentrations of lindane and dicamba in precipitation remained below the limits of quantification. From these limits it can, however, be inferred that the wet deposition was at most 2.15 mg/ha/d for lindane and 1.07 mg/ha/d for dicamba. This indicates that dry deposition, including gas exchange, of pesticides is quantitatively important relative to wet deposition and should not be neglected when total atmospheric loads are estimated. However, the sampler has been designed to resemble an open water surface. Consequently, dry deposition flux rates found by Waite *et al.* (1999) are representative only for deposition to water surfaces, not land surfaces. Moreover, the sampler may overestimate actual net dry deposition, especially gas absorption. All pesticide molecules trapped in the sheet of water of the sampler are removed immediately. As a result, the concentration gradient between water and air is kept at a maximum and revolatilisation is prevented. In reality, concentrations of pesticides in surface waters are often not zero. High surface water concentrations in combination with low air concentrations may result in a net gas exchange flux that is directed towards the atmosphere. Consequently, total net atmospheric deposition is not necessarily higher than wet deposition alone. Data from the Great Lakes, *e.g.*, indicate that the concentrations of many organochlorine pesticides, including HCHs, in air and lakewater are more or less in equilibrium. This implies that the input into the lakes from wet and particle dry deposition is compensated by gas transfer out of the lake (Hillery *et al.*, 1998).

McConnell *et al.* (1997) determined the exchange of chlorpyrifos between surface water and air in Chesapeake Bay. Based on simultaneous measurements of concentrations of chlorpyrifos in air and surface water and using a Mackay fugacity type model, they showed that net gas exchange strongly dominated wet deposition and dry particle fluxes. In March and April, a net flux of 0.04-0.4 mg/ha/d was directed from the water to the air, driven by high surface water concentrations and low air concentrations. In June and September, high air concentrations resulted in a net flux of up to 0.3 mg/ha/d towards the water. Within this study, sample collection did not focus on the particle phases; particle-phase concentrations were estimated with the model. The investigators pointed out that the quality of their predictions would improve when particle-phase concentrations could be determined accurately.

Precipitation and gas exchange across the air-water boundary were the dominant pathways for chlorpyrifos and dacthal in the South Tobacco Creek Watershed in Manitoba, Canada (Rawn et al., 1999c). Just like in Chesapeake

Bay, the direction of the net gas exchange was shown to vary with the season. Estimates of net gas exchange for several herbicides across the air-water interface of the remote Lake 375 in Ontario were in the same order of magnitude as those in Chesapeake Bay. However, for atrazine, wet deposition inputs were higher than inputs resulting from net gas exchange (Muir and Grift, 1995; Rawn *et al.*, 1998).

By analysing throughfall water from forests and rainwater from open fields, Dankwardt *et al.* (1994) observed that wet deposition of atrazine was higher in forests than in open fields. They attributed this to the filtering effect of the tree canopies: during dry periods airborne pesticides adsorb to the large surface of needles and leaves. During the next precipitation event, part of the substance is washed off and can be found in the rainwater. According to Herterich (1991) the true deposition rate in a forest stand is usually higher than recorded, because part of the pesticides remains adsorbed to the foliage.

From the analysis of pine needles, Aston and Seiber (1997) estimated air-to-foliage dry deposition (including gas exchange) of chlorpyrifos and its oxon in Sequoia National Park, California, to be 0.1 g/ha during the period from May through October. This represents approximately one half to two thirds of the total annual flux (McConnell *et al.*, 1998). Concentrations of chlorpyrifos in the air much closer to the application sites were expected to result in a (approximately) 65 times higher foliar uptake (Aston and Seiber, 1997). These results indicate that foliar uptake may be substantial (see also Aston and Seiber, 1996).

It can be concluded that the first estimates of dry deposition, including gas exchange, of currently used pesticides are available now. The uncertainty in these figures, however, is still substantial. Nevertheless, it seems reasonable to suppose that total net atmospheric deposition in non-target areas generally won't exceed a level of a few grammes per hectare per year per pesticide.

4. Vapour phase - particle phase partitioning

Generally, the distribution of pesticides between the different phases in ambient air may be predicted from their physico-chemical properties, notably vapour pressure and water solubility (see *e.g.* Majewski and Capel, 1995). However, measurement data, necessary to validate the predictions, are scarce, especially for modern pesticides (Dörfler and Scheunert, 1997; Bidleman, this issue). In order to completely understand the processes of wet and dry deposition, it is necessary to develop a sampling strategy that not only samples wet and dry deposition separately, but also discriminates between gaseous and particulate phases in the atmosphere, and between dissolved and particulate phases in rainwater (Glotfelty *et al.*, 1990b). Distinguishing between dissolved and particulate phases in rainwater requires the water to be filtered through a highly efficient filter as it is collected. To separate gaseous and particulate compounds in air, air samplers are used in which the particles are retained by a quartz fibre filter and the gaseous phase is trapped on polyurethane foam (PUF) or a resin (XAD). Glotfelty *et al.* (1990b) found that atrazine, alachlor and metolachlor were present in the atmosphere of the Wye River area (Maryland, USA) almost completely in the vapour phase during May and June, and that the less volatile simazine was for 75% in the vapour phase. In the colder winter months the filter-retained fractions for both atrazine and simazine were higher, approximately 30 and 40%, respectively. Similarly, Haraguchi *et al.* (1994) concluded that in spring and summer, many pesticides, including atrazine and simazine, exist in a gaseous rather than a solid state in urban air in Japan. Only a few non-volatile pesticides and chlorothalonil were detected at higher amounts on the filter than in the resin trap. Also Millet *et al.*, 1997 reported the dominance of the vapour-phase fraction.

Turnbull (1995) found lindane almost exclusively in the vapour phase in air in rural England. This is consistent with observations from the Great Lakes area (Lane et al., 1992). Also the volatile pesticides trifluralin and chlorpyrifos were detected at greater than 95% in the vapour phase. On the other hand deltamethrin and fenpropimorph, which have low vapour pressures, were almost exclusively found adsorbed to particulates in the atmosphere. The partitioning of atrazine, simazine and isoproturon to particles was intermediate: 62%, 26% and 53%, respectively. In comparing them with organochlorines, Turnbull (1995) pointed out that modern pesticides showed a tendency towards particle phase partitioning, revealing their less volatile nature. Very similar results were obtained by Hawthorne *et al.* (1996), who sampled air in rural North Dakota. They also found that more than 97% of trapped trifluralin was associated with the PUF. The distribution of atrazine seemed to vary with temperature: in August 75% of atrazine was collected on the PUF sorbent, whereas all of the atrazine was collected on the filter in September and November. Carbofuran, chlorothalonil, dicamba, MCPA en 2,4-D were only detected on the filter. Scanning electron microscope analysis of the filter samples revealed that the vast majority of the particulates were mineral dust (quartz, clay, etc.) of <20 μm size and that a considerable fraction consisted of pollen and spore grains, which were generally smaller than 15 μm. Such small particles have long atmospheric lifetimes, and long-range atmospheric transport of the pesticides associated with them seems likely (Hawthorne *et al.*, 1996). Rawn *et al.* (1999a) reported that 2,4-D, dichlorprop, MCPA and bromoxynil were present in the atmosphere in both gas and particle phases in an agricultural area in Manitoba, although these herbicides are considered non-volatile. Bromoxynil was found in both the gas and particle phase throughout the season, whereas the phenoxyacids were restricted mainly to the vapour phase. The partitioning varied, however, considerably from year to year. At the same sampling site, the fraction of atrazine associated with the particle phase in 85 air samples varied from 0 to 100%, with a mean value of 40% (Rawn *et al.*, 1998). According to Herterich (1991) atrazine was almost completely bound to particulate matter in air above mountain forests in southern Germany. However, all these monitoring data should be considered with caution: the phase distribution in the samples does not necessarily reflect ambient vapour-/particulate-phase distributions, since particulate-bound pesticides can be stripped during sampling and vapour-phase pesticides can be sorbed onto dust-containing filters (Cotham and Bidleman, 1992).

Indirect evidence on the phase distribution was gained by Waite *et al.* (1999) in a rural area in Saskatchewan. By simultaneously measuring total (particulate plus gas) atmospheric concentrations of pesticides and dry deposition, they determined the deposition velocity of lindane, dicamba and 2,4-D to be 0.29, 1.2 and 1.5 cm/s, respectively. These values represent maximum estimates, as the concentrations in the water sheet of the sampler were kept at zero. The lower value for lindane is typical for gaseous compounds. Lindane has already been shown to exist in the atmosphere almost exclusively in the gas phase (Lane *et al.*, 1992; Turnbull, 1995) and gas exchange is commonly considered to dominate the fluxes between air and surface waters (Hillery *et al.*, 1998). The relatively high values for the herbicides suggest that these pesticides are, at least in part, adsorbed to particles. This is consistent with their molecular chemistry, which would favour adsorption to particles. Finally, Scharf and Bächmann (1993) concluded from the presence of low volatile pesticides in the air that they must be bound to particles. They saw further evidence in the atmospheric behaviour of fenpropimorph, which reached peak levels in precipitation during three distinct periods: April/May, July and October. The pesticide's presence during the first period is explained by its application. The second peak is caused by harvesting and the third by sowing winter cereals. These agricultural activities cause soil and plant particles with old residues of the pesticide to enter the atmosphere. The second and the third peak cannot be explained by volatilisation of unbound residues, according to Scharf and Bächmann (1993).

It can be concluded that rather conflicting results are reported. This may reflect the variation in environmental conditions, such as temperature and concentration and nature of dust particles. Differences in sampling and analytical techniques may, however, also play a role (Rawn *et al.*, 1999a).

5. Seasonal and spatial variation

In general, the occurrence of many pesticides in air and rainwater shows a distinct seasonal variation, with peak concentrations occurring during application periods (Arthur *et al.*, 1976; Elling *et al.*, 1987; Hurle *et al.*, 1987; Richards *et al.*, 1987; Buser, 1990; Glotfelty *et al.*, 1990b; Gath *et al.*, 1992; Nations and Hallberg, 1992; Oberwalder *et al.*, 1992; Scharf *et al.*, 1992; Chevreuil and Garmouma, 1993; Gath *et al.*, 1993; Geissler and Schöler, 1993; Scharf and Bächmann, 1993; Stähler, 1993; Dankwardt *et al.*, 1994; Provincie Zuid-Holland, 1994; Siebers *et al.*, 1994; Bester *et al.*, 1995; Lode *et al.*, 1995a; Muir and Grift, 1995; Waite *et al.*, 1995; Bakker, 1996, Chevreuil *et al.*, 1996; Hatfield *et al.*, 1996, Suzuki, 1996; Goolsby *et al.*, 1997; Hüskes and

Levsen, 1997; Bucheli *et al.*, 1998; Jager *et al.*, 1998; Rawn *et al.*, 1999a). The occurrence pattern also depends upon the pesticides' persistence in soil, water and air, as well as their volatility. Relatively volatile compounds with low persistence, such as alachlor and metolachlor, often exhibit high peak concentrations, which are strictly limited to the application period, whereas more persistent, less volatile pesticides, such as lindane, atrazine, simazine and 2,4-D, show lower peak concentrations, but remain in the air for a much longer time after they have been applied (Glotfelty *et al,* 1990b, Oberwalder *et al.*, 1991; Goolsby *et al.*, 1997; Bucheli *et al.*, 1998).

In Europe, lindane is detectable in rainwater throughout the year (Lode *et al.*, 1995a; Oberwalder *et al.*, 1991; Scharf and Bächmann, 1993; Stähler, 1993; Provincie Zuid-Holland, 1994; Chevreuil *et al.*, 1996; Jager *et al.*, 1998), but often peak concentrations are observed in spring and early summer, when the pesticide is applied (Siebers *et al.*, 1991; CCRX, 1993; Gath *et al.*, 1993; Stähler, 1993; Provincie Zuid-Holland, 1994; Cleemann *et al.*, 1995; Turnbull 1995; Chevreuil *et al.*, 1996). Concentrations in air follow the same pattern (Chevreuil and Garmouma, 1993; Turnbull 1995; Chevreuil *et al.*, 1996). Triazine herbicides exhibit a pattern very similar to that of lindane, but less pronounced. Glotfelty *et al.* (1990b) found atrazine and simazine to be present in the atmosphere at all times of the year, albeit the winter values were much lower than the summer values. Wu (1981) even found atrazine to be present in rainwater in Maryland in high concentrations throughout the year. This is consistent with the findings of Millet *et al.* (1997), who observed roughly constant contaminant levels in both air and precipitation throughout the year in the upper Rhine Valley in France. Several other investigations revealed the presence of pesticides in air and rainwater outside the application period (Chevreuil and Garmouma, 1993; Scharf and Bächmann, 1993; Baas and Duyzer, 1997). This may be the result of either medium- and long-range transport (see section 8) or the entrance into the atmosphere of old residues, either as vapours or adsorbed to soil dust or crop material, *e.g.* due to agricultural activities (Scharf and Bächmann, 1993).

The time profile of lindane concentrations in air at Listå, Norway, was characterised by a low, fairly steady baseline level of 40 pg/m^3 with short episodes of clearly elevated concentrations of up to several hundred pg/m^3 resulting from atmospheric transport events (Haugen *et al.*, 1998). Most of these episodes occurred in spring and early summer, which is the main application period for lindane in central Europe. During these brief transport episodes, concentrations at Listå reached levels similar to those in the areas of application. Hoff *et al.* (1992) and Burgoyne and Hites (1993) showed that an increase in air temperature during the summer months is accompanied by an increase in a variety of organochlorine insecticides (HCHs, DDTs, endosulfan, chlordane, dieldrin, methoxychlor, and toxaphene) even though many of these chemicals are no longer used. The measured concentrations in air are thought to be caused by the volatilisation of old residues remaining in the soil. In Listå, however, this temperature-dependent revolatilisation only marginally influenced air concentrations (Haugen *et al.*, 1998).

Often, the presence of pesticides in air and rainwater in a certain area can be explained by the crops grown there and the pesticides used on them (Richards *et al.*, 1987; Nations and Hallberg, 1992; Oberwalder *et al.*, 1992, Provincie Zuid-Holland, 1994; Majewski *et al.*, 1998). Many pesticides are predominantly found in air and precipitation in agricultural areas, but some investigations reveal higher concentrations of some organophosphorus insecticides, such as chlorpyrifos, diazinon, malathion and methyl-parathion, in air and precipiation in urban areas (Nations and Hallberg, 1992; Majewski *et al.* 1998). These pesticides are extensively used in and around urban homes and gardens.

Modern pesticides do not only occur in the atmosphere in agricultural and urban areas where they are applied. They are also found in areas where no pesticides are or ever have been used, such as mountains (Buser, 1990; Herterich, 1991; Neururer and Womastek, 1991; Zabik and Seiber, 1993; Dankwardt *et al.*, 1994; Datta *et al.*, 1998), isles (Bester *et al.*, 1995), boreal forests (Muir and Grift, 1995; Rawn *et al.*, 1998) and even arctic seas (Chernyak *et al.*, 1996; Rice *et al.*, 1997). In general, atmospheric deposition of pesticides is lower there, but there is evidence that the more environmentally stable pesticides, such as lindane, are accumulating in the cooler regions of the globe (Wania and Mackay, 1996). Bester *et al.* (1995) found concentrations of the triazine herbicide terbuthylazine in rainwater during the main application period, to be five times lower at the isle of Heligoland in the German Bight than in an agricultural area in Northern Germany. This is in good agreement with the results of Oberwalder *et al.* (1992), Dankwardt *et al.* (1994) and Siebers *et al.* (1994) in Germany, Waite *et al.* (cited by Muir and Grift, 1995) in Canada and Goolsby *et al.* (1997) in the USA, who compared atmospheric input of pesticides in agricultural and non-agricultural areas. Zabik and Seiber (1993) reported that pesticide concentrations in air and rainwater decreased with distance and elevation from the major pesticide use area in the Californian Central Valley to the Sierra Nevada mountains.

Concentrations of lindane in air are much higher in countries where the pesticide is still used, *e.g.* France and the Netherlands, than in areas were it was banned some years ago, for instance the Scandinavian countries. Chevreuil and Garmouma (1993) and Chevreuil *et al.* (1996) reported average ambient air concentrations for lindane of 1.5 ng/m^3 for the Paris area. Concentrations were highest (6 ng/m^3) in April and May, when lindane was used. During the rest of the year, the concentration was approximately 0.5 ng/m^3. This is consistent with the median air concentration of 0.4-0.6 ng/m^3 reported by Baas and Duyzer (1997) for the western part of the Netherlands between August and December. At Listå, Norway, where lindane was taken off the market in 1991 (Lode *et al.*, 1995a), average ambient air concentrations between 1991 and 1995 were much lower: 0.048 ng/m^3 (Haugen *et al.*, 1998). This is similar to concentrations measured at the west coast of Sweden (Brorström-Lundén *et al.*, 1994; Brorström-Lundén, 1996), where lindane is also not registered for use. In both Scandinavian countries wet deposition of lindane was an order of magnitude lower than in central Europe (see table III). Within Sweden, lindane concentrations in rainwater decrease with increasing latitude (Kreuger, 1995).

6. Transformation products

For several pesticides it has been demonstrated that the parent compound is not the only form that is present in air, precipitation and fog. Transformation products of these pesticides have been detected in the atmosphere (see Table VII and Atkinson *et al.*, this issue). Transformation products of triazine herbicides are the most frequently reported transformation products in rainwater, particularly desethylatrazine. Deisopropylatrazine has been found less often and usually in lower concentrations (Oberwalder *et al.*, 1992; Majewski and Capel, 1995; Goolsby *et al.*, 1997; Bucheli *et al.*, 1998; Jager *et al.*, 1998). Sometimes, it was looked for, but not detected (Chevreuil *et al.*, 1996). Desethyldesisopropylatrazine was found in one rainwater sample by Jager *et al.* (1998). Reported values for the ratio of desethylatrazine to atrazine - often referred to as DAR - in rainwater are highly variable. Hurle *et al.* (1987), Wendlandt *et al.* (1989) and Oberwalder *et al.* (1991, 1992) found atrazine concentrations in Germany to be much higher than those of desethylatrazine, both in agricultural and non-agricultural areas. In agreement with this, Bester *et al.* (1995) found that the parent compound was dominant in precipitation at sampling locations in northern Germany. However, the transformation product prevailed on the isle of Heligoland in the German Bight (Bester *et al.*, 1995). In rainwater samples collected near Hannover (Germany), the DAR varied between 0.4 and 2.0 (Jager *et al.*, 1998). Highly variable ratios, with the transformation product prevailing frequently over the parent compound, were also observed by Scharf *et al.* (1992), Scharf and Bächmann (1993) in Germany and by Chevreuil *et al.* (1996) in France. Precipitation samples from Greece also had, on average, DAR-values above unity (Albanis *et al.*, 1998). In Sweden, however, the DAR was less than 1 (Kreuger, 1995). Goolsby *et al.* (1997) reported the ratio of desethylatrazine to its parent compound in rainwater in the midwestern and northeastern USA to be on average 0.5.

Peak concentrations of desethylatrazine and desisopropylatrazine in rainwater are often observed with a delay of few weeks with respect to the parent compound (Wendlandt *et al.*, 1989; Oberwalder *et al.*, 1991, Bucheli *et al.*, 1998; Jager *et al.*, 1998). Consequently, the DAR changes markedly over the season. Bucheli *et al.* (1998) reported an average value of 0.36 in the samples taken before June, 30, and 0.70 in the samples collected after that date. In spite of the high variability, it is evident that DAR values in precipitation are normally much higher than 0.1, a value commonly observed in surface run-off (Goolsby *et al.*, 1997), and it is speculated that atrazine is desethylated in the atmosphere (Scharf *et al.*, 1992; Goolsby *et al.*, 1997; Albanis *et al.*, 1998; Jager *et al.*, 1998).

Desethylatrazine has also been detected in air: it was found in 60% of the air samples taken along the Mississippi River in concentrations generally about 10 times less than atrazine (Majewski *et al.*, 1998). However, in air samples collected near Paris, the transformation product frequently prevailed over the parent compound (Chevreuil *et al.*, 1996).

Desethylterbuthylazine appears to be commonly present in rain in Northern Germany (Bester *et al.*, 1995; Jager *et al.*, 1998). According to Bester *et al.* (1995) this transformation product and its parent compound displayed the same pattern, albeit less pronounced, than atrazine and desethylatrazine: ratio's above 1 at Heligoland and below 1 over the German mainland. However, Jager *et al.* (1998) observed ratios between 0.1 and 4 in precipitation collected near Hannover. Acetyl-terbuthylazine was detected by Hüskes and Levsen (1994) in air and rain samples collected in Lower Saxony. Palm *et al.* (1997) found this to be the major transformation product of the OH radical reaction of terbuthylazine in their smog chamber experiments.

Abiotic degradation of triazine herbicides in the atmosphere is mainly attributed to OH radical attack (Palm *et al.*, 1997; see also Atkinson *et al.*, this issue). The concentration of these radicals in cloud and rain droplets is low. Therefore, it seems likely that the transformation products of the triazines are formed in the gas phase rather than the liquid phase of the atmosphere (Jager *et al.*, 1998). Because most oxidative reaction products are more polar than their parent compounds (Majewski and Capel, 1995; Jager *et al.*, 1998), they are more water soluble and more readily removed by wet deposition processes or air-water exchange. Indeed, the solubility of desethylatrazine is approximately two orders of magnitude higher than that of atrazine (Mills and Thurman, 1994). Dealkylation products of triazine herbicides deserve special attention as they are phytotoxic (Kaufman and Kearney, 1970; Belfroid *et al.*, 1998).

Diethylaniline, a transformation product of alachlor, occurred in 20% of the air samples along the Mississippi River (Majewski *et al.*, 1998). However, the collection efficiency of this metabolite on the PUF was poor and its real presence may have been underestimated.

S-enantiomers of the optically active herbicides mecoprop and dichlorprop were observed quite frequently in rainwater by Bucheli *et al.* (1998) in Switzerland. They are known to be formed in the environment from the active R-forms by enantiomerisation (Müller and Buser, 1997). These compounds appear a few weeks later in precipitation than the parent compounds, as was observed for the triazine transformation products. As a result the R/S ratio of mecoprop was 2.05 before June 30 and 0.35 thereafter (Bucheli *et al.*, 1998).

Oxones, which result from the oxidation of sulphur containing organophosphorus insecticides, are also very frequently observed in air, rainwater and fog (see Table VII). Schomburg *et al.* (1991) noticed that the ratio of the oxones to their respective parent compounds, called thions, in fog water collected near the Californian coast, were higher in non-agricultural areas than in agricultural areas. In the former, concentrations of the oxones

TABLE VII

Transformation products of current-use pesticides detected in air, rain or fog

transformation product	parent compound	matrix	reference
desethylatrazine	*e.g.* atrazine	rain, air	1, 2, 3, 4, 5, 6, 7, 8, 9, 10, 11, 12, 13, 14, 15, 16, 17, 18, 19, 30
desisopropylatrazine (= desethylsimazine)	*e.g.* atrazine, simazine, terbuthylazine	rain	5, 6, 8, 12, 15, 17, 18
desethyldesisopropyl-atrazine	*e.g.* atrazine, simazine, terbuthylazine	rain	18
desethylterbuthylazine	terbuthylazine	rain	10, 18, 20
acetyl-terbuthylazine	terbuthylazine	air, rain	20, 21
diethylaniline	alachlor	air	19
s-dichlorprop	r-dichlorprop	rain	17
s-mecoprop	r-mecoprop	rain	17
endosulfan sulphate	endosulfan	rain	22
chlorpyrifos oxon	chlorpyrifos	air, fog	23, 24, 25, 26, 27
diazinon oxon	diazinon	air, fog, rain	23, 24, 25, 26, 28
malaoxon	malathion	air, fog	25
methidathion oxon	methidathion	air, fog	23, 24, 25, 26, 27
paraoxon-methyl	parathion-methyl	fog, rain	23, 25, 29
paraoxon(-ethyl)	parathion(-ethyl)	air, fog, rain	20, 23, 24, 25, 26, 28

References: 1 = Hurle *et al.*, 1987; 2 = Wendlandt *et al.*, 1989; 3 = Oberwalder *et al.*, 1991; 4 = Schössner and Koch, 1991; 5 = Oberwalder *et al.*, 1992; 6 = Scharf *et al.*, 1992; 7 = Oberwalder and Hurle, 1993; 8 = Scharf and Bächmann, 1993; 9 = Dankwardt *et al.*, 1994; 10 = Bester *et al.*, 1995; 11 = Kreuger, 1995; 12 = Majewski and Capel, 1995; 13 = Bakker, 1996; 14 = Chevreuil *et al.*, 1996; 15 = Goolsby *et al.*, 1997; 16 = Albanis *et al.*, 1998; 17= Bucheli *et al.*, 1998; 18 = Jager *et al.*, 1998; 19 = Majewski *et al.*, 1998; 20 = Hüskes and Levsen, 1994; 21 = Palm et al., 1997; 22 = Provincie Zuid-Holland, 1994; 23 = Glotfelty *et al.*, 1987; 24 = Glotfelty *et al.*, 1990a; 25 = Schomburg *et al.*, 1991; 26 = Seiber *et al.*, 1993; 27 = Aston and Seiber, 1997; 28 = Zabik and Seiber, 1993; 29 = Charizopoulos and Papadopoulou-Mourkidou, 1998; 30 = Rawn *et al.* , 1998.

frequently exceeded those of the thions. Very similar results were obtained in California's Central Valley, were the oxon to thion ratios were highest in fog and air samples collected at greater distance from the sites of application (Glotfelty *et al.*, 1990a; Aston and Seiber, 1997). Oxon air concentrations tended to be higher relative to the parent thion concentrations in day-light versus night samples, suggesting that they are formed in the atmosphere by photooxidation (Glotfelty *et al.*, 1990a; Seiber *et al.* 1993). Oxon to thion ratios in tree drip fogwater were significantly higher than those in fogwater collected by a Teflon strand collector (Seiber *et al.* 1993). This finding suggests that tree surfaces (needles, leaves, branches) catalise the conversion of thion to oxon. In general, the importance of the transformation products relative to their parent compounds seems to increase with the remoteness of the receptor area. This is of particular interest, since oxones are more toxic than their parent thions (Aston and Seiber, 1997).

Felding *et al.* (1999) found sporadically 6-hydroxybentazone (a transformation product of bentazone), isopropylphenylurea and isopropylphenylmethylurea (two transformation products of isoproturon) and 2-hydroxyterbuthylazine (a transformation product of terbuthylazine) in rainwater samples from Sjaelland, Denmark. However, analysis of standard solutions placed at the sampling sites showed that the samples were not stable, although the sample bottles contained acid as a preservative. Therefore, it could not be ruled out that transformation products were formed in the sampler during the collection period (approximately two weeks), especially in the summer season or when rain fell at the beginning of the collection period. Felding *et al.* (1999) point out that more advanced sampling equipment is needed, in which the samples are cooled during the collection period.

Hüskes and Levsen (1994) discovered photochemical transformation products of a wide variety of pesticides in air and rain in Lower Saxony. However, their preliminary results have not yet been confirmed.

7. Pesticides in surface waters due to atmospheric transport

An atmospheric pesticide input of up to a few grammes per hectare per year constitutes normally (far) less than 1% of the recommended application rate (Glotfelty *et al.*, 1990b; Waite *et al.*, 1995; Hatfield *et al.*, 1996). Nevertheless, in non-target areas this input may be significant. The actual contribution of atmospheric deposition of pesticides to the total loading of an area varies per area and per pesticide. In areas where no pesticides are used and that are not connected to use areas by surface or ground-waters, the total load can be attributed to atmospheric deposition. Especially the presence of pesticides in remote lakes or those lakes with very limited drainage areas makes a strong case for atmospheric transport. Buser (1990) discovered triazine herbicides in Swiss Alpine lakes, situated at a higher altitude than the agricultural areas were the pesticides were applied. Lakes on Isle Royale in Lake Superior also contain atrazine, although no pesticides have been used on the isle (Aga and Thurman, 1993 in Muir and Grift, 1995; Goolsby *et al.*, 1997). Muir and Grift (1995) reported that the water of Lake 375 in Ontario, Canada, was contaminated by atrazine, alachlor, metolachlor, 2,4-D, dicamba, trifluralin and two organochlorine pesticides, although pesticides never have been used in the drainage area of the lake.

In other areas, varying percentages of the total pesticide loading to surface waters can be brought in via surface or subsurface water flows. Concentrations of the herbicides 2,4-D, MCPA, dichlorprop and bromoxynil in the water of the South Tobacco Creek in Manitoba, Canada, were not related to runoff losses. Instead, they corresponded to elevated levels of these pesticides in precipitation and air measured within the watershed (Rawn *et al.*, 1999a, 1999b). Deposition directly to the water surface through precipitation contributed significantly (generally between 5% and 50%) to the herbicide discharge of the creek. Additionally, gas absorption and particle dry deposition may have contributed to herbicide loadings in the creek water.

In Maryland, Glotfelty *et al.* (1990b) determined the load of contaminants to the Why River from both runoff and precipitation. They estimated that pesticide input from precipitation into the Wye River was approximately 3% and 20% of the total pesticide input for atrazine and alachlor, respectively. Wet deposition of atrazine by rainfall was on the order of 10% of the riverine input in Chesapeake Bay. Muir and Grift (1995) compared atmospheric and riverine inputs of pesticides in Lake Erie and concluded that the contribution of wet and dry deposition to the total input was 9.3% for atrazine, 52% for alachlor, 35% for metolachlor and 22% for endosulfan. Gas exchange, however, which may be substantial, was not taken into account in these calculations (Glotfelty *et al.*, 1990b).

Bester *et al.* (1995) and Hühnerfuss *et al.* (1997) made similar comparisons for triazine inputs into the German Bight and the North Sea, based on the analysis of bulk precipitation and riverwater. They found that atmospheric inputs by precipitation were negligible as compared to riverine inputs near the coast, but that they are essential in the central parts of the German Bight and the North Sea. In Maryland, USA, atmospheric loadings and riverine inputs of chlorpyrifos into Chesapeake Bay were compared by McConnell *et al.* (1997). In March and April, riverine inputs were the most important source of chlorpyrifos to the Bay, with approximately 10% loss to the atmosphere. However, during June and September, riverine inputs decreased and air concentrations increased. As a result atmospheric inputs became more important and caused the surface water budget to be increased by approximately 25% per month.

Moore *et al.* (1998) and Alegria and Shaw (1999) reported the presence of plumes of freshened water in the South Atlantic Bight, off the coast of South Carolina. Salinity in the plumes was depressed 1-2 parts per thousand compared to surrounding shelf waters. Measurement of short-lived Ra isotopes and Si revealed that the plumes were not caused by river discharge, but were due to an intense rain event a few days prior to the investigation. One of the plumes contained atrazine, metolachlor and trifluralin in concentrations that were 2 to 6 times higher than in coastal waters outside the plume, thus providing direct evidence of rain deposition of these pesticides. Alegria and Shaw (1999) estimated that the single storm event deposited atrazine, metolachlor and trifluralin in amounts of approximately 4-10%, 3-5% and 1-3%, respectively, of the estimated yearly riverine input of these pesticides into the South Atlantic Bight.

8. Source - receptor relationships

Pesticides detected in air or precipitation at a certain sampling site may originate from source areas located nearby or further away. By comparing the results from several sampling sites, inference can be made about the distance a pesticide has travelled (Baas and Duyzer, 1997). If local sources dominate, a pesticide's concentration in air or precipitation may be highly variable, both in space and time. For instance, Geissler and Schöler (1993) concluded from precipitation analysis at four sampling sites near Cologne in Germany, all located within 50 km of each other, that local sources must be dominant, because input patterns differed greatly from site to site. Another example is the presence of endosulfan in air in the Province of South Holland, in the Netherlands in 1996 (Baas and Duyzer, 1997). Air concentrations at one sampling site were much higher than at two other sites located within 40 km. Baas and Duyzer (1997) concluded that there must be a local souce, although this pesticide is not registered anymore in the Netherlands.

If pesticides originate from up to several hundreds of kilometres, deposition patterns at sampling sites within, *e.g.* 50 km of each other, are more similar. Distinct time trends may be visible. Trifluralin may serve as an example. Its use is not allowed in the Netherlands, but it was present in air samples collected in autumn at three sites in the Province of South Holland in 1996 (Baas and Duyzer, 1997). Concentrations and time trends were very similar at all sampling sites and Baas and Duyzer (1997) speculated that the source areas may be located in Belgium, where trifluralin is used on Brussels sprout and winter corn. Cleemann *et al.* (1995) concluded from the similarity in concentrations and seasonal trends of lindane in precipitation in different parts of Denmark, that the source areas must be located outside the country, probably in European countries south and west of Denmark, because the wind came mainly from these directions.

Finally, pesticides that originate from remote sources at many hundreds or even thousands of kilometres distance, will show very steady and gradual trends in concentrations in air and precipitation, both in space and in time. Alternatively, such a pattern may result from a very slow but prolonged release from residues in the soil (Baas and Duyzer, 1997). Often, these concentrations are referred to as background concentrations. Not only persistent organochlorines show such behaviour, but also current-use pesticides, such as atrazine and 2,4-D, which may be present in the atmosphere in low concentrations throughout the year outside the application season (Glotfelty *et al.*, 1990b, Baas and Duyzer, 1997).

A linear regression of local use (within 5 km) data and wet deposition amounts resulted in a good correlation for four pesticides (r-values between 0.75 and 0.84, Oberwalder and Hurle, 1993). The intersect of the regression line and the y-axis indicates the deposition when no pesticides are used locally. It was much higher for atrazine (155 mg/ha/y) than for triallate, pendimethalin and metamitron (3-23 mg/ha/y). The slope of the regression line was also highest for atrazine, which indicates that an increase in the amount locally used, has the strongest impact on the wet deposition.

An easy way to obtain information about the origin of a pesticide detected in air or precipitation is to simply look for the nearest agricultural or urban area in which it is used in substantial amounts. From the distance between the sampling site and this possible source area inference can be made as to the minimum distance the pesticide has travelled through the atmosphere. For several pesticides these minimum travelling distances through the atmosphere are reported in tabel VIII. Oberwalder *et al.* (1992) and Oberwalder and Hurle (1993) found atrazine in rainwater in Baden-Württemberg, Germany, to be reduced by 30-40% after its ban in 1991. They concluded that in the past, approximately two third of the atrazine in precipitation must have come from abroad. Rainwater samples collected in 1992 along a transect across the southern part of Germany, from the border with France to the border with Austria, revealed that atrazine concentrations were highest near the borders and lowest in the central part of Germany. Assuming that atrazine is not used illegally, Oberwalder and Hurle (1993) inferred from this that atrazine is transported through the atmosphere over at least 100-200 km. Atrazine, simazine and alachlor were all found in rainwater in Hannover, Germany, in 1996 (Jager *et al.*, 1998). They are not registered in Germany. For the triazines the nearest use areas are situated in the Netherlands, approximately 200 km to the west. Alachlor, which is not registered in the Netherlands, must have travelled even further. Atrazine in rainwater in the northern part of Germany, including the isle of Heligoland, originated probably from the Netherlands, some 100 km to the south west (Bester *et al.*, 1995). In the southern part of Norway, atrazine was found in rainwater in 1993, four years after it was banned (Lode *et al.*, 1995a,b). The nearest possible source areas (Denmark) are located at a distance of approximately 200 kilometres. Lindane, which was banned in Norway in 1991, was also present in precipitation. Haugen *et al.* (1998) demonstrated that countries in western central Europe were the main source of lindane detected in air in southern Norway between 1991 and 1995. Long-range atmospheric transport of lindane and other organochlorine pesticides is well established and extensively documented (see *e.g.* the review by Bidleman elsewhere in this issue), therefore, it has not been included in table VIII. Kreuger (1995) found atrazine, 2,4-D, and lindane very frequently in precipitation in Scania (southern Sweden) and Uppsala, 500 km further north. As these compounds are not registered in Sweden they must have come from source areas at tens to hundreds of kilometres distance. Traces of atrazine, dichlorprop and MCPA were also found in precipitation collected at Abisko, 200 km above the arctic circle. MCPA and dichlorprop were also detected in precipitation even more to the north, in Tromsø, Norway, in 1994 by Lode *et al.* (1995b). Glotfelty *et al.* (1990b) pointed out that the low concentrations of atrazine and simazine in rainwater in Maryland in January and February coincide with their application to sweet corn in Florida, suggesting regional-scale transport to occur over distances on the order of 1000 kilometres. Wu (1981) had arrived at similar conclusions. Some authors speculate that the presence of atrazine in the atmosphere before the start of the application season might be the result of residues in the soil entering the atmosphere due to increasing temperatures, moist soils and dust forming agricultural activities in spring. Alternatively, the application of atrazine on industrial areas and railroad tracks in early spring may be the reason (Buser, 1990; Oberwalder *et al.*, 1991; Oberwalder *et al.*, 1992; Dankwardt *et al.*, 1994).

Muir and Grift (1995) found eight current-use pesticides in the water of a small lake in Ontario. They were also detected in air and precipitation samples collected near the lake. To the investigators' best knowledge, no pesticides have ever been used within the lake drainage area. The lake is located at least 100 km from the nearest agricultural area and 1000 km from the 'corn belt' in the midwestern United States. Chernyak *et al.* (1996) and Rice and Chernyak (1997) reported the occurrence of atrazine, chlorpyrifos, chlorothalonil, terbufos, metolachlor and trifluralin in fog, air, ice and the microlayer of the surface water in the remote Bering and Chukchi Seas, at least hundreds, and probably several thousand kilometres from likely use areas. Trifluralin was also found by Halsall *et al.* (1998) in air samples from the remote Canadian and Russian Arctic. Baas and Duyzer (1997) detected trifluralin in air over the Province of South Holland in the Netherlands in 1996. Since the end of 1993, this herbicide is not registered anymore in the Netherlands and the nearest area where it is being used is Belgium, about 70 km to the south. In New Brunswick, Canada, fenitrothion was used during 1978 forest spraying operations against spruce budworm (Pearce *et al.*, 1979). It was detected in low concentrations in rainwater collected at a site 85 km from the nearest place of application. De Beer *et al.* (1992) detected 2,4-D in the Tala Valley, were its use has been banned to prevent damage to non-target crops. The most likely source area is located at a distance of 50 km. Although they only detected particle-bound, polar forms in Tala Valley, they speculated that these stem from volatile iso-octyl esters, which were heavily used in the source area and that were hydrolised whilst transported.

It is obvious that modern pesticides can travel over distances of tens to hundreds of kilometres. Some of them may travel even further. Quite surprisingly, this includes compounds which are hardly volatile, such as the acid forms of 2,4-D and dicamba, or photo-chemically unstable, such as trifluralin. The fact that certain pesticides are registered in some countries, whereas their use is prohibited in others, greatly enhances the possibility to study long-range transport of these compounds, although illegal use can never be completely ruled out.

For most current-use pesticides the atmospheric lifetime is not known. However, for atrazine and terbutylazine the rate constant for the reaction with OH radicals (k_{OH}) have recently been determined: k_{OH} is $14*10^{-12}$ cm^3 s^{-1} for atrazine (Klöpfer and Kohl, 1990 cited by Jager *et al.*, 1998) and $11*10^{-12}$ cm^3 s^{-1} for terbuthylazine (Palm *et al.*, 1997). Given an average OH radical concentration of $5\text{-}10*10^5$ cm^{-3}, atmospheric half-lives of about 1 day can be inferred from these values (Palm *et al.*, 1997; Jager *et al.*, 1998). With mean wind speeds of 3-5 m s^{-1} (typical for Northern Germany) these compounds may travel 250-500 km within a day (Hüskes and Levsen, 1997). Van Pul *et al.* (1998) calculated typical travelling distances (X_t) for several pesticides, based on estimates of the atmospheric residence time. This is the distance at which 50% of the emitted mass of a pesticide is still airborne. For atrazine they estimated X_t to be 27 km. This would mean that the amount of atrazine in an air parcel would decrease by approximately 3 orders of magnitude over a travelling distance of 270 km. The evidence presented in table VIII seems to indicate that Van Pul's estimate of X_t for atrazine is too low.

Another approach for obtaining rough information about possible source areas is to look at the wind direction at the moment pesticides are occurring in air or rainwater (De Beer *et al.*, 1992; Trautner *et al.*, 1992; Oberwalder and Hurle, 1993; Lode *et al.*, 1995; Cleemann *et al.*, 1995). Air mass back trajectory studies have successfully been used to prove that long-range atmospheric transport of organochlorine pesticides, including lindane, from Europe and North America to the Arctic (Oehme, 1991, Oehme *et al.*, 1996) as well as from South America to Antarctica (Kallenborn *et al.*, 1998) is occurring. With the same method Haugen *et al.* (1998) demonstrated that lindane detected in air in southern Norway originated predominantly from countries in central western Europe. These were also the source areas of lindane and atrazine in Swedish precipitation (Kreuger, 1995). Dicamba and 2,4-D, on the other hand, were shown to originate from Baltic states.

In the case of hexachlorocyclohexane (HCH) the ratio between the α and γ (lindane) isomers can also provide information on the origin of the pesticide residues (Oehme, 1991), as in some parts of the world the technical grade HCH, in which the α isomer dominates, is still being used, whereas in other areas it has been replaced by lindane, which is almost exclusively made up of the γ isomer. In fact, the co-occurrence of several pollutants or substances in air samples and rainwater can provide a clue to the source area (Oehme, 1991). By combining several of the above mentioned methods, the strongest evidence on the origin of pesticides can be obtained.

Table VIII

Minimum travelling distances of modern pesticides, based on the distance between the sampling site and the nearest possible source area

pesticides	matrix	sampling site[1]	minimum distance[1,2,3]	ref.
atrazine dichlorprop MCPA	rainwater	Abisko, S	100s km - >1000 km	1
dichlorprop MCPA	rainwater	Tromso, N	100s km - >1000 km	2
atrazine 2,4-D	rainwater	Ekeröd, Lurbo, S	10s-100s km [distance to DK and FIN]	1
atrazine	rainwater	Lista and As, N	200 km [distance to DK]	2
atrazine simazine terbuthylazine	rainwater	Heligoland, D Germany	50 km [distance to D, terbuthylazine] 100 km [distance to NL, atrazine, simazine]	3
atrazine simazine alachlor	rainwater	Hannover, D	200 km [distance to NL; atrazine, simazine] >200 km [alachlor]	4
alachlor atrazine dichlorprop MCPA mecoprop pendimethalin simazine terbuthylazine	rainwater	Enzklösterle, D	20 km	5
atrazine mecoprop parathion(-ethyl) trifluralin	air, rainwater	North Sea near Noordwijk, NL	70 km [distance to B, trifluralin] 12 km [distance to mainland, others]	6
trifluralin	air	Tagish, Alert, CAN Dunai, RUS	100s-1000s km	7

Table VIII
(*continued*)

pesticides	matrix	sampling site[1]	minimum distance[1,2,3]	ref.
fenitrothion	rainwater	Lamèque, CAN	85 km	8
alachlor atrazine 2,4-D dicamba endosulfan metolachlor trifluralin	rainwater, lake water	Lake L375, Ontario, CAN	100 km	9
chlorpyrifos diazinon diazinon oxon parathion(-ethyl) paraoxon(-ethyl)	air, rainwater	Sierra Nevada Mountains, California, USA	25 km (50 km)[4]	10 (11)
paraoxon	air, rainwater	Sierra Nevada Mountains, California, USA	39 km (150 km)[4]	10 (11)
chlorpyrifos, chlorothalonil, metolachlor, terbufos, trifluralin, atrazine	fog, air, ice, surface water microlayer	Bering Sea, Chukchi Sea	100s-1000s km	12, 13
2,4-D	air	Tala Valley, SA	50 km	14

[1] For abbreviations of country names see table I; RUS = Russia, SA = South Africa; [2] Either reported by the authors or estimated by ourselves from the distance to the nearest country where the pesticide is registered for use; [3] 10s km, 100s km, 1000s km means tens, hundreds, thousands of kilometres, respectively; [4] greater distance reported by Seiber and Woodrow, 1995.
References: 1 = Kreuger, 1995; 2 = Lode *et al.*, 1995a,b; 3 = Bester *et al.*, 1995; 4 = Jager *et al.*, 1998; 5 = Oberwalder *et al.*, 1992; 6 = Baas and Duyzer, 1997; 7 = Halsall *et al.*, 1998; 8 = Pearce *et al.*, 1979; 9 = Muir and Grift, 1995; 10 = Zabik and Seiber, 1993; 11 = Seiber and Woodrow, 1995; 12 = Chernyak *et al.*, 1996; 13 = Rice and Chernyak, 1997; 14 = De Beer *et al.*, 1992

9. Concluding remarks

Most of the pesticides currently used were specifically designed to have reduced environmental persistence as compared to the older organochlorine pesticides. Therefore, it is not surprising that monitoring studies have revealed a generally more localised and episodic contamination of the atmosphere with modern pesticides. Nevertheless, it is evident that they display, to some extent, the same phenomenon of atmospheric transport, if not at a global scale, then at least at a regional scale. Triazines, acetanilides, phenoxyacids, organophosphorus compounds and probably also many other pesticides are sufficiently stable, particularly in relatively unpolluted air, for atmospheric transport to occur over intermediate distances, typically 100 - 1000 km. The same applies to their transformation products. It is obvious that in small-scaled landscapes, like in many European countries, the distances between agricultural source areas and nature reserves are too small to prevent the latter from being contaminated by airborne pesticides. Little is know about the environmental fate of pyrethroid insecticides and sulfonylurea herbicides. These pesticides are so extremely potent, that their use is on the order of g/ha rather than kg/ha, as is the case for most other pesticides. These compound classes should exhibit behaviour similar to that of other pesticides, but our analytical capabilities are not sensitive enough to measure them in the environment (see *e.g.* Turnbull, 1995).

Monitoring in reception areas is an important way to obtain valuable information about the extent and scale of transport of pesticides that are currently used. Once these substances enter the air, we have lost control of them. It is difficult to defend against the charge that we are unable to account for a large part of the applied material. The least we could do is sort out were they wind up, especially as they are highly toxic and designed to kill organisms. Monitoring programmes also increase our understanding of the underlying processes and provide information on the variation in observed levels caused by such factors as distance from application areas to sampling sites, meteorological conditions and quantity and timing of applications. Little is known of the clouds as potential accumulators and/or reactors for airborne pesticides. Air needs to be more thoroughly sampled, not only horizontally, but also vertically. Information on the ambient phase distribution of current-use pesticides is scarce. Relatively little attention is payed to dry deposition (on vegetation) including gas exchange, particularly in Europe, where monitoring studies aiming at current-use pesticides predominantly have been focused on wet deposition. Metabolites deserve more attention, because they may be more toxic, persistent or mobile then the parent compounds. There is a lack of consistency in the methodologies for sampling and analysis, which hampers a comparison of the results. Often, the results of monitoring studies are published only in 'grey literature', such as institutional reports or national journals, thus reducing their availability to the international scientific community. There is also a need for information about the fate of pesticides once they are deposited in reception areas, often nature reserves. Especially in remote and hitherto pristine areas, environmental conditions may differ greatly from those in agricultural areas, with serious implications for the persistence and bioavailability of these compounds. The ecological significance of airborne pesticides is even less clear.

Monitoring provides the information necessary for the development and the calibration of emission, atmospheric transport and deposition models. It enables us to validate their predictions, thereby increasing our confidence in predictions regarding the atmospheric dispersal of new pesticides, prior to registration. Finally, it provides information about the impact of policy measures, such as the cancellation of certain products and about the effectiveness of mitigating measures aimed at reducing pesticide emissions from source areas. Obviously, in addition to experimental laboratory and field research and modelling efforts, monitoring studies are essential if we want to increase our understanding of the atmospheric fate of current-use pesticides.

Acknowledgments

We thank Igor Dubus, Gitte Felding, Andy Gilbert, Arne Helweg, Derek Muir, Werner Pol, Greet Schoeters and Alan Turnbull for providing valuable information, Eva VandenBussche-Parmeus for the translation of a Swedish article, Marja van Kan for preparing the camera-ready typescript and Ton van der Linden, Michael Majewski, Addo van Pul and Pim de Voogt for their valuable comments on an earlier draft of this paper.

References

Albanis, T. A., Hela, D. G., Sakellarides, T. M. and Konstantinou, I. K.: 1998, *J. Chromatogr.* **A823**, 59-71.

Alegria, H. A. and Shaw, T. J.: 1999, *Environ. Sci. Technol.* **33**, 850-856.

Arthur, R. D., Cain, J. D. and Barrentine, B. F.: 1976, *Bull. Environ. Contam. Toxicol.* **15**, 129-134.

Aston, L. S. and Seiber, J. N.: 1996, *J. Environ. Sci. Health* **B31**, 671-698.

Aston, L. S. and Seiber, J. N.: 1997, *J. Environ. Qual.* **26**, 1483-1492.

Atkinson, R., Guicherit, R., Hites, R. A., Palm, W.-U., Seiber, J. N. and De Voogt, P.: 1999, *Water Air Soil pollut.* this issue.

Baas, J. and Duyzer, J. H.: 1997, *Concentrations of pesticides in air and rainwater. A pilot study into concentrations of pesticides in air and rainwater in Naaldwijk, de Zilk and site Noordwijk (in Dutch)*, report R97/113, TNO Institute of Environmental Sciences, Energy Research and Process Innovation, Apeldoorn, 52p.

Baker, L. W., Fitzell, D. L., Seiber, J. N., Parker, T. R., Shibamoto, T., Poore, M. W., Longley, K. E., Tomlin, R. P., Propper, R. and Duncan, D. W.: 1996, *Environ. Sci. Technol.* **30,** 1365-1368.

Bakker, D. J.: 1996, *The occurrence of atrazine in National Park "De Groote Peel"*, report R96/408, TNO Institute of Environmental Sciences, Energy Research and Process Innovation, Apeldoorn, 38p.

Beard, J., Westley-Wise, V. and Sullivan, G.: 1995, *Austr. J. Public Health* **19**, 357-362.

Belfroid, A. C., van Drunen, M., Beek, M. A., Schrap, S. M., van Gestel, C. A. M. and van Hattum, B.: 1998, *Sci. Total Environ.* **222**, 167-183.

Bester, K., Hühnerfuss, H., Neudorf, B. and Thiemann, W.: 1995, *Chemosphere* **30**, 1639-1653.

Bester, K. and Hühnerfuss, H.: 1996, *Chemosphere* **32**, 1919-1928.

Bidleman, T. F.: 1999, *Water Air Soil Pollut.* this issue.

Brorström-Lundén, E.: 1996, J. Sea Res. 35, 81-90.

Brorström-Lundén, E., Lindskog, A. and Mowrer, J.: 1994, Atmos. Environ. 28, 3605-3615.

Bucheli, T. D., Müller, S. R., Heberle, S. and Schwarzenbach, R. P.: 1998. *Environ. Sci. Technol.* **32**, 3457-3464.

Burgoyne, T. W. and Hites, R. A.: *Environ. Sci. Technol.* **27**, 910-914.

Buser, H.-R.: 1990, *Environ. Sci. Technol.* **24**, 1049-1058.

CCRX, 1993. *Metingen van radioactiviteit en xenobiotische stoffen in het biologisch milieu in Nederland 1991.* (in Dutch), Coördinatie Commissie voor de metingen van Radioactiviteit en Xenobiotische stoffen.

Charizopoulos, E. and Papadopoulou-Mourkidou, E.: 1998, *9th International Congress Pesticide Chemistry - The food-environment challenge. Book of Abstracts.* The Royal Society of Chemistry & The International Union of Pure and Applied Chemistry, Abstract 6E-022.

Chernyak, S. M., Rice, C. P. and McConnell, L. L.: 1996, *Marine Pollut. Bull.* **32**, 410-419.

Chevreuil, M. and Garmouma, M.: 1993, *Chemosphere* **27**, 1605-1608.

Chevreuil, M., Garmouma, M., Teil, M. J. and Chesterikoff, A.: 1996, *Sci. Total Environ.* **182,** 25-37.

Cleemann, M., Poulsen, M. E. and Hilbert, G.: 1995, *Chemosphere* **30**, 2039-2049.

Cohen, J. M. and Pinkerton, C.: 1966, In: *Organic Pesticides in the environment. Advances in Chemistry Series 60*, R. F. Gould (ed.), American Chemical Society, Washington D. C., 163-176.

Cotham, W. E. and Bidleman, T. F.: 1992, *Environ. Sci. Technol.* **26**, 469-478.

Dankwardt, A., Wüst, S., Elling, W., Thurman, E. M. and Hock, B.: 1994, *Environ. Sci. Pollut. Res.* **1**, 196-204.

Datta, S., Hansen, L., McConnell, L., Baker, L., LeNoir, J. and Seiber, J. N.: 1998, *Bull. Environ. Contam. Toxicol.* **60**, 829-836.

De Beer, P. R., Smit, C. and Van Dyk, L. P.: 1992, *Chemosphere* **24**, 719-733.

De Voogt, P. and Jansson, B. : 1993, *Rev. Environ. Contam. Toxicol.* **132**, 1-27.

Dörfler, U. and Scheunert, I.: 1997, *Chemosphere* **35**, 77-85.

Dubus, I. G. and Hollis, J. M.: 1999, *Pesticides in rainfall in Europe - a review*, Soil Survey and Land Research Centre, Cranfield University, Silsoe, UK, 28p.

Elling, W., Huber, S. J., Bankstahl, B. and Hock, B.: 1987, *Environ. Pollut.* **48**, 77-82.

Felding, G., Spliid, N. H. and Asman, W. A. H.: 1999, *Findings and effects of pesticides in air and precipitation, Danish Institute of Agricultural Sciences*, Slagelse, Denmark.

Fingler, S., Tkalcevic, B., Fröbe, Z. and Drevenkar, V.: 1994, *Analyst* **119**, 1135-1140.

Gath, B., Jaeschke, W., Ricker, I. and Zietz, E.: 1992, *Nachrichtenbl. Deut. Pflanzenschutzd.* **44**, 57-66.

Gath, B., Jaeschke, W., Kubiak, R., Ricker, I., Schmider, F. and Zietz, E.: 1993, *Nachrichtenbl. Deut. Pflanzenschutzd.* **45**, 134-143.

Geissler, A. and Schöler, H. F.: 1993, *Vom Wasser* **80**, 357-370.

Glotfelty, D. E., Seiber, J. N. and Liljedahl, L. A.: 1987, *Nature* **325**, 602-605.

Glotfelty, D. E., Majewski, M. S. and Seiber, J. N.: 1990a, *Environ. Sci. Technol.* **24**, 353-357.

Glotfelty, D. E., Williams, G. H., Freeman, H. P. and Leech, M. M.: 1990b, In: *Long range transport of pesticides.* D. A. Kurtz (ed.), Lewis Publishers, Inc., Chelsea, MI, 199-221.

Gomme, J. W., Shurvell, S., Hennings, S. M. and Clark, L.: 1991, *J. Inst. Water Environ. Management* **5**, 546-552.

Goolsby, D. A., Thurman, E. M., Pomes, M. L., Meyer, M. T. and Battaglin, W. A.: 1997, *Environ. Sci. Technol.* **31**, 1325-1333.

Grover, R.: 1974, *Chemistry in Canada* **summer**, 36-38.

Grover, R., Cessna, A. J. and Waite, D. T.: 1999, in LeBaron, H. M., Gianessi, L. P., McFarland, J. and Burnside, O. C. (eds.), *The triazine herbicides*, Elsevier Science BV, Amsterdam, in press.

Halsall, C. J., Bailey, R., Stern, G. A., Barrie, L. A., Fellin, P., Muir, D. C. G., Rosenberg, B., Rovinsky, E. Ya., Kononov, E. Ya. and Pastukhov, B.: 1998, *Environ. Pollut.* **102**, 51-62.

Haraguchi, K., Kitamura, E., Yamashita, T. and Kido, A.: 1994, *Atmos. Environ.* **28**, 1319-1325.

Haraguchi, K., Kitamura, E., Yamashita, T. and Kido, A.: 1995, *Atmos. Environ.* **29**, 247-253.

Harris, G. L., Turnbull, A. B., Gilbert, A. J., Christian, D. G. and Mason, D. J.: 1992, *Proceedings Brighton Crop Protection Conference, Pests and Diseases.* Vol. 2, pp 477-87.

Hatfield, J. L., Wesley, C. K., Prueger, J. H. and Pfeiffer, R. L.: 1996, *J. Environ. Qual.* **25**, 259-264.

Haugen, J.-E., Wania, F., Ritter, N. and Schlabach, M.: 1998, *Environ. Sci. Technol.* **32**, 217-224.

Hawthorne, S. B., Miller, D. J., Louie, P. K. K., Butler, R. D. and Mayer, G. G.: 1996, *J. Environ. Qual.* **25**, 594-600.

Herterich, R.: 1991, *UWSF-Z. Umweltchem. Ökotox.* **3**, 196-200.

Hillery, B. R., Simcik, M. F., Basu, I., Hoff, R. M., Strachan, W. M. J., Burniston, D., Chan, C. H., Brice, K. A., Sweet, C. W. and Hites, R.: 1998, *Environ. Sci. Technol.* **32**, 2216-2221.

Hirvi, J.-P.: 1994, *COST 5th International Workshop on Environmental Behaviour of Pesticides and Regulatory Aspects*, Brussels, April 26-29, 1994, Summaries of the papers and posters, 36.

Hirvi, J.-P. and Rekolainen, S.: 1995, in Helweg, A. (ed.), *Pesticides in precipitation and surface water. TemaNord 1995:558*, Nordic Council of Ministers, Copenhagen, p12-18.

Hoff, J. T., Mackay, D., Gillham, R. and Shiu, W.Y.: 1993, *Environ. Sci. Technol.* **27**, 2174-2180.

Hoff, R. H., Muir, D. C. G. and Grift, N. P.: 1992, *Environ. Sci. Technol.* **26**, 266-275.

Hoff, R. M., Strachan, W. M. J., Sweet, C. W., Chan, C. H., Shackleton, M., Bidleman, T. F., Brice, K. A., Burniston, D. A., Cussion, S., Gatz, D. F., Harlin, K. and Schroeder, W. H.: 1996, *Atmos. Environ.* **30**, 3505-3527.

Hühnerfuss, H., Bester, K., Landgraff, O., Pohlmann, T. and Selke, K.: 1997, *Mar. Pollut. Bull.* **34**, 419-426.

Hurle, K., Giessl, H. and Kirchhoff, J.: 1987, *Schr.-Reihe Verein WaBoLu* **68**, 169-190.

Hüskes, R. and Levsen, K.: 1994, in Borrell, P. M., Borrell, P., Cvitas, T. and Seiler, W. (eds.), *Transport and transformation of pollutants in the troposphere. Proceedings of EUROTRAC symposium, Garmisch-Partenkirchen, Germany, 11-15 April, 1994*, SPB Academic Publishing bv, The Hague, p167-170.

Hüskes, R. and Levsen, K.: 1997, *Chemosphere* **35**, 3013-3024.

Jager, M. E., Bourbon, C. and Levsen, K.: 1998, *Int. J. Environ. Anal. Chem.* **70**,149-162.

Johnson, A. C., Haria, A. H., Bhardwaj, C. L., Williams, R. J. and Walker, A.: 1996, *Pestic. Sci.* **48**, 225-237.

Kallenborn, R., Oehme, M., Wynn-Williams, D. D., Schlabach, M. and Harris, J.: 1998, *Sci. Total Environ.* **220**, 167-180.

Kaufman, D. D. and Kearney, P. C.: 1970, *Residue Rev.* **32**, 235-265.

Kirknel, E. and Felding, G.: 1995, in Helweg, A. (ed.), *Pesticides in precipitation and surface water. TemaNord 1995:558*, Nordic Council of Ministers, Copenhagen, p45-54.

Kreuger, J.: 1995, in Helweg, A. (ed.), *Pesticides in precipitation and surface water. TemaNord 1995:558*, Nordic Council of Ministers, Copenhagen, p33-44 (in Swedish with English summary).

Kutz, F. W., Yobs, A. R. and Yang, S. C.: 1976, in Lee Jr, R. E. (ed.), *Air pollution from pesticides and agricultural processes*, CRC Press, Cleveland, OH, USA, p95-136.

Lane, D. A., Johnson, N. D., Hanley, M.-J. J., Schroeder, W. H. and Ord, D. T.: *Environ. Sci. Technol.* **26**, 126-133.

Leuenberger, C., Czuczwa, J., Tremp, J. and Giger, W.: 1988, *Chemosphere* **17**, 511-515.

Lode, O., Eklo, O. M., Holen, B., Svensen, A. and Johnsen, A. M.: 1995a, *Sci. Total Environ.* **160/161**, 421-431.

Lode, O., Eklo, O. M., Holen, B., Svensen, A. and Johnsen, A. M.: 1995b, in Helweg, A. (ed.), *Pesticides in precipitation and surface water. TemaNord 1995:558*, Nordic Council of Ministers, Copenhagen, p19-32.

Majewski, M. S. and Capel, P. D.: 1995, *Pesticides in the atmosphere*, Ann Arbor Press, Inc., Michigan.

Majewski, M. S., Foreman, W. T., Goolsby, D. A. and Nakagaki, N.: 1998, *Environ. Sci. Technol.* **32**, 3689-3698.

McConnell, L. L., LeNoir, J. S., Datta, S. and Seiber, J. N.: 1998, *Environ. Toxicol. Chem.* **17**, 1908-1916.

McConnell, L. L., Nelson, E., Rice, C. P., Baker, J. E., Johnson, W. E., Harman, J. A. and Bialek, K.: 1997, *Environ. Sci. Technol.* **31**, 1390-1398.

Millet, M., Wortham, H., Sanusi, A. and Mirabel, P.: 1996, *Arch. Environ. Contam. Toxicol.* **31**, 543-556.

Millet, M., Wortham, H., Sanusi, A. and Mirabel, P.: 1997, *Environ. Sci. Pollut. Res.* **4**, 172-180.

Mills, M. S. and Thurman, E. M.: 1994, *Environ. Sci. Technol.* **28**, 73-79.

Moore, W. S., Kjerfve, B. and Todd, J. F.: 1998, *J. Geophys. Res.* **103**, 7709-7717.

Muir, D. C. G. and Grift, N. P.: 1995, In: *Eighth international congress of pesticide chemistry; options 2000.* N. N. Ragsdale, P. C. Kearney and J. R. Plimmer (eds.), American Chemical Society, Washington D. C., 141-156.

Müller, M. D. and Buser, H.-R.: 1997, *Environ. Sci. Technol.* **31**, 1953-1959.

Nations, B. K. and Hallberg, G. R.: 1992, *J. Environ. Qual.* **21**, 486-492.

Neururer, H. and Womastek, R.: 1991, *Die Bodenkultur* **42**, 57-70.

Oberwalder, C., Giessl, H., Irion, L., Kirchhoff, J. and Hurle, K.: 1991, *Nachrichtenbl. Deut. Pflanzenschutzd.* **43**, 185-191.

Oberwalder, C., Kirchhoff, J. and Hurle, K.: 1992, *Z. PflKrankh. PflSchutz* **Suppl. XIII**, 363-376.

Oberwalder, C. and Hurle, K.: 1993, *8th EWRS Symposium "Quantitative approaches in weed and herbicide research and their practical application"*, Braunschweig, 391-398.

Oehme, M.: 1991, *Ambio* **20**, 293-297.

Oehme, M., Haugen, J.-E. and Schlabach, M.: 1996, *Environ. Sci. Technol.* **30**, 2294-2304.

Palm, W.-U., Elend, M, Krueger, H.-U. and Zetzsch, C.: 1997, *Environ. Sci. Technol.* **31**, 3389-3396.

Pearce, P. A., Brun, G. L. and Wittemann, J.: 1979, *Bull. Environ. Contam. Toxicol.* **23**, 503-508.

Pimentel, D., Tort, M., D'Anna, L., Krawic, A., Berger, J., Rossman, J., Mugo, F., Doon, N., Shriberg, M., Howard, E., Lee, S. and Talbot, J.: 1998, *BioSci.* **48**, 817-826.

Prospero, J. M. and Seba, D. B.: 1972, *Atmos. Environ.* **6**, 363-364.

Provincie Zuid-Holland: 1991, *Project Integratie Milieu Metingen: Het Westland (in Dutch).* Provincie Zuid-Holland, Dienst Water en Milieu, The Hague.

Provincie Zuid-Holland: 1994, *Pesticides in precipitation in Zuid-Holland (in Dutch).* Provincie Zuid-Holland, Dienst Water en Milieu, The Hague, 113p.

Quaghebeur, D.: 1995, *Water* **82**, 121-126 (in Dutch with English summary).

Rawn, D. F. K., Halldorson, T. H. J., Lawson, B. D. and Muir, D. C. G.: 1999a, *J. Environ. Qual.* **28**, 898-906.

Rawn, D. F., Halldorson, T. H. J. and Muir, D. C. G.: 1998, in: *Triazine herbicides: risk assessment.* L. G. Ballantine, J. E. McFarland and D. S. Hackett (eds.), ACS Symposium Series 683, American Chemical Society, Washington, D.C., 158-176.

Rawn, D. F. K., Halldorson, T. H. J. and Muir, D. C. G.: 1999c, *Environ. Sci. Technol.* submitted.

Rawn, D. F. K., Halldorson, T. H. J., Turner, W. N., Woychuk, R. N., Zakrevsky, J.-G. and Muir, D. C. G.: 1999b, *J. Environ. Qual.* **28**, 906-917.

Rice, C. P.: 1996, *Pesticide Outlook* **April**, 31-36.

Rice, C. P. and Chernyak, S. M.: 1997, *Chemosphere* **35**, 867-878.

Richards, R. P., Kramer, J. W., Baker, D. B. and Krieger, K. A.: 1987, *Nature* **327**, 129-131.

Risebrough, R. W., Huggett, R. J., Griffin, J. J. and Goldberg, E. D.: 1968, *Sci.* **159**, 1233-1236.

Robinson, E. and Fox, L. L.: 1978, *J. Air Pollut. Control Assoc.* **28**, 1015-1020.

Sandmann, E. R. I. C., De Beer, P. R. and Van Dyk, L. P.: 1991, *Chemosphere* **22**, 137-145.

Scharf, J. and Bächmann, K.: 1993, *Nachrichtenbl. Deut. Pflanzenschutzd.* **45**, 82-87.

Scharf, J., Wiesiollek, R. and Bächmann, K.: 1992, *Fresenius J. Anal. Chem.* **342**, 813-816.

Schössner, H. and Koch, A.: 1991, *Forum Städte-Hygiene* **42**, 109-112.

Schomburg, C. J., Glotfelty, D. E. and Seiber, J. N.: 1991, *Environ. Sci. Technol.* **25**, 155-160.

Seiber, J. N., Wilson, B. W. and McChesney, M. M.: 1993, *Environ. Sci. Technol.* **27**, 2236-2243.

Seiber, J. N. and Woodrow, J. E.: 1995, In: *Eighth international congress of pesticide chemistry; options 2000.* N. N. Ragsdale, P. C. Kearney and J. R. Plimmer (eds.), American Chemical Society, Washington D. C., 157-172.

Siebers, J., Gottschild, D. and Nolting, H.-G.: 1991, *Nachrichtenbl. Deut. Pflanzenschutzd.* **43**, 191-200.

Siebers, J., Gottschild, D. and Nolting, H.-G.: 1994, *Chemosphere* **28**, 1559-1570.

Spencer, W. F. and Cliath, M. M.: 1990, In: *Long range transport of pesticides.* D. A. Kurtz (ed.), Lewis Publishers, Inc., Chelsea, MI, 1-16.

Stähler, M.: 1993, *Nachrichtenbl. Deut. Pflanzenschutzd.* **45**, 32-36.

Suzuki, S.: 1996, *J. Pest. Sci.* **21**, 7-15.

Taylor, A. W. and Spencer, W. F.: 1990, in: *Pesticides in the soil environment: processes, impacts, and modeling.* H. H. Cheng (ed.), Soil Science Society of America, Inc., Madison, Wisconsin, USA, 213-269.

Tarrant, K. R. and Tatton, J. O'G.: 1968, *Nature* **219**, 725-727.

Trautner, F., Huber, K. and Niessner, R.: 1992, *J. Aerosol Sci.* **23**, suppl. 1, S999-S1002.

Trevisan, M., Montepiani, C., Ragozza, L., Bartoletti, C., Ioannilli, E. and Del Re, A. A. M.: 1993, *Environ. Pollut.* **80**, 31-39.

Tsai, W. T., Altwicker, E. R., Asman, W. A. H.: 1990, *Atm. Environ.* **24A**, 2485-2498.

Tsai, W., Cohen, Y., Sakugawa, H. and Kaplan, I. R.: 1991, *Environ. Sci. Technol.* **25**, 2012-2023.

Turnbull, A. B.: 1995, *An assessment of the fate and behaviour of selected pesticides in rural England. PhD Thesis.* University of Birmingham, Birmingham, UK.

Van Boom, L.: 1993, *Pesticides in precipitation and in surface water (in Dutch).* Heemraadschap Fleverwaard, Lelystad, 71p.

Van den Berg, F., Kubiak, R., Benjey, W. G., Majewski, M. S., Yates, S. R., Reeves, G. L., Smelt, J. H. and Van der Linden, A. M. A.: 1999, *Water Air Soil Pollut.* this issue.

Van Pul, W. A. J., De Leeuw, F. A. A. M., Van Jaarsveld, J. A., Van der Gaag, M. A., Sliggers, C. J.: 1998, *Chemosphere* **37**, 113-141.

Van Zoonen, P., Buijsman, E., Baumann, R. A., 't Hart, V., De Jong, A. P. J. M. and Den Engelsman, G.: 1991, *An investigation into the occurrence of some polar pesticides in rainwater (in Dutch)*, report 748704044, National Institute of Public Health and Environmental Protection (RIVM), Bilthoven, 16p.

Villeneuve, J.-P. and Cattini, C.: 1986, *Chemosphere* **15**, 115-120.

Waite, D. T., Grover, R., Wescott, N. D., Irvine, D. G., Kerr, L. A. and Sommerstad, H.: 1995, *Environ. Toxicol. Chem.* **14**, 1171-1175.

Waite, D. T., Gurprasad, N. P., Cessna, A. J. and Quiring, D. V.: 1998, *Chemosphere* **37**, 2251-2260.

Waite, D. T., Cessna, A. J., Gurprasad, N. P. and Banner, J.: 1999, *Atm. Environ.* **33**, 1513-1523.

Wania, F. D. and Mackay, D.: 1996, *Environ. Sci. Technol.* **30**, 390A-396A.

Wendlandt, E., Stabel, H.-H. and Wieland, K.: 1989, *Ber. Arbeitsgemeinschaft Wasserwerke Bodensee-Rhein (AWBR)* **21**, 125-142.

Wu, T. L.: 1981, *Water Air Soil Pollut.* **15**, 173-184.

Zabik, J. M. and Seiber, J. N.: 1993, *J. Environ. Qual.* **22**, 80-90.

Zhu, T., Desjardins, R. L., MacPherson, J. I., Patty, E. and St. Amour, G.: 1998, *Environ. Sci. Technol.* **32**, 1032-1038.

ECOTOXICOLOGICAL RISK ASSESSMENT OF PESTICIDES SUBJECT TO LONG-RANGE TRANSPORT

NICO M. VAN STRAALEN and CORNELIS A.M. VAN GESTEL

Vrije Universiteit, Department of Ecology and Ecotoxicology, De Boelelaan 1087, 1081 HV Amsterdam, The Netherlands
E-mail: straalen@bio.vu.nl

Abstract. Concern has arisen about the possible ecological effects of persistent pesticides that become airborne during or after application and are transported to regions far away from where they were applied. In this paper an ecotoxicological approach is outlined that may support assessments of products suspected of long-range transport capacity. It is proposed that the risk is estimated from a classical PNEC/PEC comparison for the surface layer of a remote area, where PEC is estimated from dose rate, emission factors, atmospheric residence time and persistence, while PNEC is estimated from ecotoxicological information collected as part of the registration procedure. According to this "null model", risk assessment of pesticides subject to long-range transport is not different from the usual risk assessment, provided that due attention is paid to losses occurring during transport and accumulation in remote areas with low temperature. A simplified equation is derived for estimating PEC from the recommended dose rate, which shows that the concentration in the remote area is higher than in the target area only if its residence time is at least two ordes of magnitude longer than the corresponding value in the target area. A review of ecotoxicity data for effects of volatile pesticides on arthropods indicates that effect levels in the air compartment are far above the concentrations of concern in long-range transport. Arguments supporting the view that remote areas, specifically the polar regions, are characterized by ecosystems that are more vulnerable than the ones on which the usual risk assessment is based, are reviewed. Although residues of organochlorines are of concern, there does not seem to be concrete epidemiological or experimental evidence about effects of modern pesticides on wildlife in remote areas. It is concluded that there is no reason to reject the "null model" at the moment, however, in view of the large uncertainty involved, it is proposed that the maximum acceptable ratio between PNEC and PEC be increased by an extra safety factor.

Keywords: ecotoxicology, longe-range transport, pesticides, residues, risk assessment

1. Introduction

The detection of persistent organochlorines in ecosystems far away from any industrial or agricultural sources has raised concern about the possible ecological effects of these compounds in ecosystems of a pristine nature. Pesticides that are banned from most industrialized countries are still measured in remote regions in significant quantities (*e.g.* Hargrave *et al.*, 1992). The profile of the different compounds reflects their historic use and the efficacy of banning measures (Calamari *et al.*, 1994). Although some of the

Water, Air, and Soil Pollution **115:** 71–81, 1999.

most obvious environmental disasters of hazardous persistent organochlorines have been overcome by banning certain plant protection products from the market, the question still remains whether the present registration procedures for pesticides are sufficiently conservative to avoid possible ecological effects of residues transported to remote areas. The reasons to be especially concerned about effects following long-range transport are:

- Once the product becomes airborne, it is out of control.
- Pesticide residues become distributed over a large area.
- Environmental conditions at remote ecosystems may promote long residence times.
- Organisms in remote ecosystems may be exposed to pesticide residues over long times.
- Remote ecosystems may be more vulnerable than agro-ecosystems.

The physical and chemical processes that are responsible for vertical and horizontal fluxes of pesticides in the atmosphere were reviewed by De Voogt and Jansson (1993). Particular attention has been devoted to the "cold trap" hypothesis, the nett transfer of persistent compounds from temperate to arctic regions due to temperature-dependent volatilization and condensation phenomena (Bidleman, 1988; Oehme, 1991). In addition to transport over a long-range, atmospheric pollution by pesticides may also be a problem on a regional scale (Seiber and Woodrow, 1995; De Jong *et al.*, 1995).

Ecotoxicological risk assessment usually boils down to a comparison of concentrations in the environment with no-effect concentrations for ecological targets (Van Leeuwen and Hermens, 1995). If there is a sufficiently large safety margin (*e.g.* a factor of 100) between environmental concentrations predicted from exposure analysis (PEC) and threshold concentrations predicted from ecotoxicological tests (PNEC), the assessment leads to the conclusion that the risk is acceptably low. The problem of potential effects of pesticides subject to long-range transport is particularly complicated because common registration procedures for pesticides were not designed to take possible effects in remote areas into account. Toxicity data for species that are endemic to arctic regions are not available, and the fate data relevant for the environmental conditions of these regions are only limited. As this situation is not likely to change soon, risk assessments for pesticides subject to long-range transport will have to be based on extrapolations from the information available for the area where the pesticide is applied. To allow such an extrapolation we formulate a "null model" comprising the following elements:

- Estimate the predicted environmental concentration in a remote area from the dose rate in the target area, using a simplified, generically applicable scenario for emission, transport and deposition.
- Assume that the no-effect concentration of the biological community in the remote area can adequately be estimated from the test data available for the target area.
- Compare PEC and PNEC using the ratio approach.

Basically, the "null model" assumes that risk assessment of chemicals subject to long-range transport is not different from ordinary risk assessment, provided that due attention is paid to the processes underlying loss and accumulation associated with transport and deposition. The problem is viewed as a source-sink relationship, where the sink is the surface layer of the remote area (specifically soil, but water can be dealt with in a similar way). It is also assumed that the concentrations in the air itself are not critical. The aim of this paper is to explain the approach and then to discuss it critically.

2. Estimation of environmental concentrations

A great variety of models has been proposed to estimate concentrations of chemicals in the environment away from emission sources (*e.g.* Van Jaarsveld, 1990; Mackay *et al.*, 1992). Usually these models are spatially explicit and they require a number of detailed data on properties of the chemical, metereological parameters and terrain data. Such models are highly flexible and can be adapted to many different situations. In the case of new chemicals and in pesticide registration procedures the required model input is, however, not always available and more simplistic, generally applicable, approaches can still be useful. Such a simplistic scenario for the processes associated with long-range transport of pesticides is illustrated in Figure 1. It is assumed that the expected concentration in a remote area can be estimated from the dose rate in the target area plus a series of parameters that may be derived from experimental or modelling studies. The quantities involved in our model are the following:

- $D =$ dose rate of the pesticide, expressed in mg m^{-2} y^{-1},
- $A =$ area on which the pesticide is applied, in m^2,
- $f_e =$ emission factor: fraction of the dose that becomes airborne during or after application (dimensionless),
- $\Delta t =$ transport time: average time for a mass of air to travel from the target area to the remote area, expressed in d,
- $\tau =$ atmospheric residence time of the pesticide, taking into account degradation and reactions, but not including deposition, expressed in d,
- $\alpha =$ ratio between the emission area (A) and the target area (αA): a dimensionless factor that accounts for dilution processes ($\alpha > 1$ if there is dilution),
- $d =$ depth of the surface layer on which the pesticide is deposited in the remote area, expressed in m,
- $\rho =$ bulk density of the surface layer in the remote area, in kg m^{-3},
- $\varepsilon =$ residence time of the pesticide in the surface layer of the remote area, taking degradation into account, expressed in y,
- $C =$ concentration in the surface layer of the remote area, expressed in mg/kg.

According to this simplified scenario the annual amount of a pesticide emitted from a target area of size A is f_eDA. Assuming exponential decay during transport, a fraction of $e^{-\Delta t/\tau}$ of the emission remains after transport. The amount which is not degraded in the atmosphere must fall out somewhere, since the atmosphere is not considered a place of accumulation. This amount ($f_eDAe^{-\Delta t/\tau}$) is assumed to deposit on an area which is a factor of α larger than the area from which emission took place. The deposition flux is then assumed to be distributed over a depth d and is expressed on a mass base by dividing by the mass per volume ratio, ρ. This will give the annual increase of the concentration if the pesticide would be completely persistent. It may, however, be assumed that the pesticide decays in the surface layer of the remote area and has a mean residence time ε there. The flux times the residence time will then give a predicted equilibrium concentration according to the formula:

$$C = \frac{\varepsilon\, f_e\, D\, e^{-\Delta t/\tau}}{\alpha\, d\, \rho} \tag{1}$$

The equation is scale-independent due to the fact that the absolute areas of application and deposition cancel each other and only their quotient, α, remains.

It may be interesting to explore under what conditions the concentration in the remote area will be larger than the concentration in the target area. Assuming that the pesticide has the same distributional depth in the two areas and the same density applies, the condition for C being larger than the concentration in the target area is:

$$\varepsilon > \left[\left(\frac{1-f_e}{f_e}\right)\alpha\; e^{\Delta t/\tau}\right]\varepsilon_0 \tag{2}$$

where ε_0 is the residence time in the target area. The inequality illustrates that the residence time in the remote area, ε, is a crucial parameter and that the concentration in the remote area will be higher than the concentration in the target area only if ε is considerably larger than ε_0. For example, if we take $\alpha = 1$, $f_e = 0.1$ and $\Delta t = \tau$ (these values may be considered as a worst case), the condition reads $\varepsilon > 24\varepsilon_0$. In other words environmental persistence of the pesticide must increase by one to two orders of magnitude after the transport, for the concentration to be similar to the concentration in the target area.

How can the parameters in equation (1) be estimated from experimental data? Residence time in the remote area, ε, can be derived from estimates for the half-life, DT_{50}, determined under different conditions ($\varepsilon = DT_{50}/\ln 2$). If the concern is transport to the polar regions, determinations of DT_{50} at different temperatures will allow an extrapolation to be made to the conditions prevailing in polar ecosystems. The emission factor, f_e, can be derived from models and observations on pesticide fate during spray drift and evaporation from treated land. It is highly dependent on substance properties such as volatility (see the other papers in this volume) and application technology (design of spray nozzles, tarping of treated fields). Dose rate per unit area, D, can be equalled to the rate recommended for crop protection. Atmospheric transport time, Δt, is highly variable, but in principle an average value can be assumed for the transport between specified areas on the globe (*e.g.* from Central Europe to the arctic). Atmospheric residence time, τ, is a property that is not usually determined in a registration procedure. Further research must be done to see whether it is possible to obtain realistic and reproducible estimates for this parameter in an experimental manner, or to derive it from basic properties of the substance (see other papers in this volume).

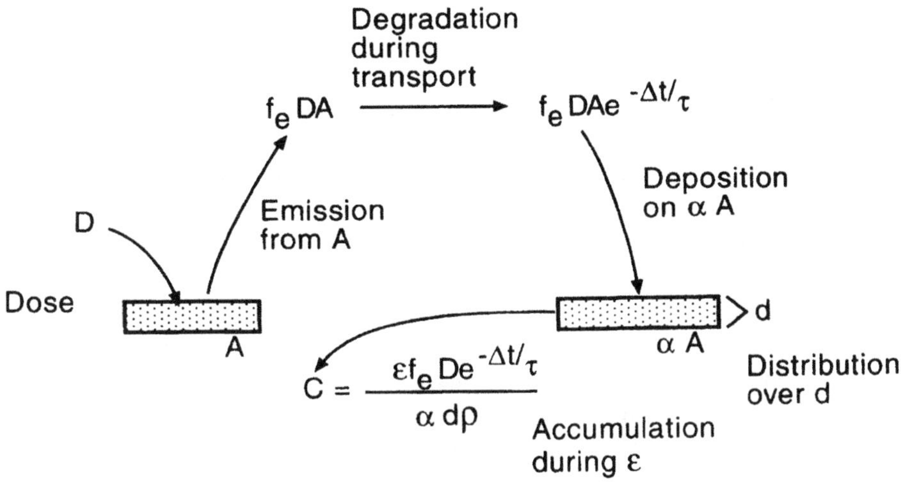

Fig. 1. Representation of a simplified scenario to estimate the equilibrium concentration, C, of a pesticide in a remote area from the annual dose rate (D) in a target area, with the aid of parameters for fraction emitted (f_e), atmospheric transport time (Δt), atmospheric residence time (τ), application/deposition area ratio (α), distribution depth (d), density (ρ), and residence time in the remote area (ε).

The surface ratio, α, is probably the most difficult of all the quantities involved in equation (1). It has to be taken into account that different source areas may contribute to the same remote area, but also that a single source area may contribute to different remote areas. A one-to-one relationship between source and sink areas cannot be made, therefore it is only on a large scale, assuming complete spatial averaging, that the factor α becomes useful. One might approach the problem by considering, for example, the complete agricultural area in the Northern hemisphere, relative to the complete arctic area on which pesticides might be deposited. Such a first guess must be refined by detailed spatially explicit models. A default (worst case) value may be $\alpha = 1$ (no dilution). Distributional depth, d, can be taken similar to the value used in common risk assessments (20 cm), although it may also be argued that in the absence of ploughing, pesticides will distribute over a small depth in natural ecosystems and a value of 5 cm seems to be more reasonable. For water surfaces, specific mixing processes have to be taken into account. Finally, assigning a value to ρ does not pose any problems.

The crude scenario illustrated in Figure 1 and formulated in equation (1) should not be considered as a real solution to the problem, it only expresses some basic mass balance rules. On the other hand, it shows that scenario-based predicted concentrations in remote areas can in principle be derived from quantities that are already required in pesticide registration procedures, or that demand only a limited extension of the requirements. More detailed residue measurements and spatially explicit modelling will be necessary to obtain a better underpinning for the quantities involved.

3. Estimation of no-effect concentrations

In the common registration procedures for pesticides, toxicological information is collected that may support the derivation of no-effect concentrations, to be compared with predicted environmental concentrations. For environmental targets, a variety of test protocols has been developed that will allow estimation of LC_{50}s or no-effect concentrations for organisms considered relevant for the agricultural ecosystem where the product is going to be applied. Overviews of ecotoxicological test procedures are provided in Calow (1993) and Løkke and Van Gestel (1998). Greig-Smith (1992) and Thomas (1995) summarize the situation in Europe with respect to the EU registration directive. Some of the more common test organisms are listed in Table I. These species are assumed to be jointly representative of the agricultural ecosystems where a product is going to be applied, although not all tests are done under all conditions. In addition to toxicity test results, data collected for bioaccumulation potential are also relevant for ecotoxicological evaluations, since these determine the risk of food-chain transfer and secondary poisoning (Noppert *et al.*, 1993; Jongbloed *et al.*, 1996).

TABLE I

Overview of common test species used to support the ecotoxicological evaluation of pesticides submitted for registration.

Water		
Algae	*Scenedesmus, Chlorella, Selenastrum*	Growth
Crustaceans	*Daphnia, Ceriodaphnia, Gammarus*	Survival, reproduction
Fish	*Salmo, Poecilia, Brachydanio*	Early life stage survival, adult survival, growth
Soil		
Plants	*Lactuca sativa, Avena sativa*	Seedling growth
Earthworms	*Eisenia fetida*	Survival, reproduction
Microorganisms	Undefined community	Nitrification, respiration
Birds		
Mallard duck	*Anas platyrhynchos*	Survival
Quail	*Colinus virginianus, Coturnix japonica*	Survival
Beneficial arthropods		
Honey bees	*Apis mellifera*	Survival
Parasitic wasps	*Encarsia formosa, Trichogramma cacoeciae*	Survival, parasitic capacity
Beetles	*Bembidion lampros, Poecilus cupreus*	Survival
Mites	*Phytoseiulus persimilis*	Survival

Ecotoxicity test results may be used to derive, in a quantitative way, maximum acceptable concentrations in water or soil (Van Leeuwen and Hermens, 1995). If there are sufficient data, a statistical approach may be applied, in which a frequency distribution of no-effect concentrations is used to estimate a concentration level below which the probability of finding a species more sensitive than that value is acceptably small (Smith and Cairns, 1993; Solomon, 1996). If there are only few data, a safety factor (0.1, 0.01 or 0.001) is usually applied to the lowest toxicological criterion to derive a PNEC. Chapman *et al.* (1998) provided a critical evaluation of the use of such safety (uncertainty) factors. Ecotoxicity test results may also be used as benchmark values in decision rules for identification of hazard, see, *e.g.*, Oomen (1986). In that case, the outcome of the assessment is not a PEC/PNEC, but a ratio between the proposed application rate and some toxicological criterion.

Data for risk assessment derived from laboratory tests may be verified in field studies or mesocosm studies. These studies are usually conducted as part of a higher tier in the evaluation, usually when laboratory tests have not given sufficient evidence that the product is safe for the targets of concern. Several reviews have been published in which laboratory ecotoxicity data of pesticides were compared to the results of field testing (Van Gestel, 1992; Heimbach, 1992; Brock and Budde, 1994). These reviews allow the conclusion that the direct effects of pesticides on specified targets can be predicted with reasonable accuracy if the exposure concentration in the field is known. Indirect effects, arising from trophic interactions within the affected community (herbivory, predation, competition), are difficult to predict, since they critically depend on the structure of the community. Van Straalen and Van Rijn (1998) summarized laboratory data for pesticide toxicity to soil invertebrates and from these data estimated the time needed for a soil-applied pesticide to decay until a safe concentration was reached. In 25 out of 32 cases, predicted "ecotoxicological recovery times" were longer than recovery times actually observed for soil invertebrate populations. On the basis of this and other studies cited above, the conclusion seems to be justified that no-effect concentrations of pesticides in the field can be predicted from laboratory toxicity data within reasonable limits.

We propose that, as a first approximation, the common procedures for deriving PNEC as described above are also applicable for evaluating residues in remote areas. The concentrations in air itself must be evaluated in another way. In contrast to risk assessment for water, soil, and higher animals exposed through the food-chain, ecological risk assessment for air is usually not considered explicitly. Standards for the air compartment are based on rodent inhalation studies and on human toxicology. The question may be asked whether volatile pesticides will directly affect ecological targets exposed through the air. This is specifically relevant for organisms that maintain an intensive gas exchange with the air (birds, bats, insects). Rademaker and Van Gestel (1993) compiled literature data on toxicity of volatile pesticides and other volatile compounds towards arthropods. Most of the studies reviewed were conducted as part of investigations into the efficacy of the product to target (pest) insects, such as fleas, wood-boring beetles and mites. An abstract of the data is given here as Table II. The

LC_{50} values from these studies are in the mg/m^3 range, which is considerably above the concentrations of concern to the problem of long-range transport (ng/m^3 and lower, see other papers in this volume). The wide safety margin implied indicates that direct exposure through the air compartment is not critical to the assessment of long-range transport potential of pesticides.

TABLE II

Toxicity of some volatile pesticides tested on arthropods. Data were extracted from a more extensive overview in Rademaker and Van Gestel (1993); the lowest value for each product is listed, plus the animal group for which that value was obtained.

Active ingredient	Lowest LC_{50} (mg/m^3)	Animals
Dichlorvos	0.03	Fleas
Dichloropropene	239	Spiders
Ethylene chlorobromide	3600	Beetles
Ethylene dibromide	900	Beetles
Lindane	0.12	Mites
Metallyl chloride	10,000	Beetles
Chloropicrine	1500	Beetles
Hydrocyan acid	400	Beetles
Methylisothiocyanate	2.9	Spiders

4. Discussion

In the introduction of this paper we have suggested that, as a first approximation, risk assessment of pesticides subject to long-range transport can be conducted as a PEC/PNEC comparison where PEC is derived from the recommended dose rate in the target area, plus parameters for transport and dilution, and PNEC is derived from toxicity results available in registration procedures using common test species. The question remains whether these assumptions can be justified. The representativeness of standard test species for remote ecosystems seems a particularly weak point. Maybe the species in remote areas, *e.g.* arctic ecosystems, are more vulnerable than estimated from standard tests?

Van Straalen (1994) argued that vulnerable ecological targets are characterized by three properties:

• Vulnerable species are exposed to a higher than average dose due to the fact that they live or feed in places where the pesticide accumulates. Pesticides, like many contaminants, are usually not homogeneously distributed over an ecosystem, but tend to accumulate in transition zones, *e.g.*, soil-air and plant-air surface layers. Species that take their food from these accumulation sites or live directly in them are exposed to high concentrations.

• Vulnerable species have an intrinsic susceptibility to the toxicant, due to badly developed detoxification mechanisms or sensitive biochemical targets. The capacity to concentrate toxicants to levels much higher than those in the environment is also an aspect related to intrinsic susceptibility, although high internal concentrations do not necessarily imply that the species is susceptible.

• Vulnerable species have a badly developed capacity to recover from toxic insults, due to properties of their life-cycle. The rate at which populations recover from pesticide effects depends largely on their reproductive and colonization potentials. Long-lived species with a low rate of reproduction will not recover easily.

Based on these considerations one may argue that slow growing plants, lichens, mosses, deer, seals, whales, piscivorous birds and birds of prey are among the vulnerable targets of concern when considering the potential effects of pesticide residues in remote areas. Certainly, these species are not covered by the present set of standard tests in registration procedures for new pesticides. On the other hand, is there any evidence that these species are threatened by new products that have appeared on the market recently?

Most of the concern about possible effects of pesticide residues after long-range transport stems from the detection of residues of organochlorines (lindane, DDT, dieldrin) in remote areas (Carlberg *et al.*, 1983; Oehme, 1991; Hargrave *et al.*, 1992; Calamari *et al.*, 1994; Letcher *et al.*, 1996). There is abundant evidence that these residues may significantly affect reproduction in predatory birds (Peakall, 1993), seals (Reijnders, 1986) and other vertebrates, such as mustelids (Leonards *et al.*, 1995). Measurements of biotransformation enzymes in representatives from arctic systems, such as polar bear (Letcher *et al.*, 1996) and beluga whale (White *et al.*, 1994) have indicated biochemical changes due to organochlorine residues. Regional effect studies consider off-target deposition of aircraft sprays, and spray drift on a scale of a few kilometers outside the area of application (Ernst *et al.*, 1991; Forsyth and Westcott, 1994; De Jong *et al.*, 1995). There are also suggestions that herbicides or their degradation products may be involved with forest decline around areas with intensive agricultural use (Rippen *et al.*, 1987). We did not, however, find any reports in the literature that document ecotoxicological effects of residues of modern pesticides in areas remote from agricultural sources.

Given the uncertainty illustrated above, and the fact that ecotoxicological risk assessment should consider both false negatives and false positives (Koeman, 1982), there does not seem to be concrete evidence to abandon the "null model" outlined in this paper. To deal with the uncertainties, one might, however, consider the application of an extra large safety margin (*e.g.* 1000 rather than 100) in the comparison of PEC and PNEC for pesticide residues prone to atmospheric transport.

Acknowledgements

The authors are indebted to participants of the Netherlands Health Council workshop on Atmospheric Transport of Pesticides (April 1998) for fruitful feed-back and discussions, and to two anonymous referees and the editors of this volume for comments on the manuscript and additional literature references.

References

Bidleman, T.F.: 1988, *Environ. Sci. Technol.* **22**, 361-367.

Brock, T.C.M. and Budde, B.J.: 1994, In: *Freshwater Field Tests for Hazard Assessment of Chemicals* (ed. Hill, I.R., Heimbach, F., Leeuwangh, P. and Matthiesen, P.), pp. 19-56, Lewis Publishers, Boca Raton.

Calamari, D., Tremeloda, P., Di Guardo, A. and Vighi, M.: 1994, *Environ. Sci. Technol.* **28**, 429-434.

Calow, P. (ed.): 1993, *Handbook of Ecotoxicology*. Blackwell Scientific Publications, London.

Carlberg, G.E., Ofstad, E.B., Drangsholt, H. and Steinness, E.: 1983, *Chemosphere* **12**, 341-356.

Chapman, P.M., Fairbrother, A. and Brown, D.: 1998, *Environ. Toxicol. Chem.* **17**, 99-108.

De Jong, F.M.W., Van der Voet, E., and Canters, K.J.: 1995, *Ecotox. Environ. Safety* **30**, 77-84.

De Voogt, P. and Jansson, B.: 1993, *Rev. Environ. Contam. Toxicol.* **132**, 1-27.

Ernst, W.R., Jonah, P., Doe, K., Julien, G. and Hennigar, P.: 1991, *Environ. Toxicol. Chem.* **10**, 103-114.

Forsyth, D.J. and Westcott, N.D.: 1994, *Environ. Toxicol. Chem.* **13**, 299-306.

Greig-Smith, P.W.: 1992, *Environ. Toxicol. Chem.* **11**, 1673-1689.

Hargrave, B.T., Harding, G.C., Vass, W.P., Erickson, P.E., Fowler, B.R., and Scott, V.: 1992, *Arch. Environ. Contam. Toxicol.* **22**, 41-54.

Heimbach, F.: 1992, *Soil Biol. Biochem.* **24**, 1749-1753.

Jongbloed, R.H., Traas, T.P. and Luttik, R.: 1996, *Ecotox. Environ. Safety* **34**, 279-306.

Koeman, J.H.: 1982, *Ecotox. Environ. Safety* **6**, 358-362.

Leonards, P.G., De Vries, T.H., Minnaard, W., Stuijfzand, S., De Voogt, P., Cofino, W.P., Van Straalen, N.M. and Van Hattum, B.: 1995, *Environ. Toxicol. Chem.* **14**, 639-652.

Letcher, R.J., Norstrom, R.J., Lin, S., Ramsay, M.A. and Bandiera, S.M.: 1996, *Toxicol. Appl. Pharmacol.* **137**, 127-140.

Løkke, H. and Van Gestel, C.A.M. (ed.): 1998, *Handbook of Soil Invertebrate Toxicity Tests*. John Wiley and Sons, Chichester.

Mackay, D., Paterson, S. and Shiu, W.Y.: 1992, *Chemosphere* **24**, 695-717.

Noppert, F., Dogger, J.W., Balk, F. and Smits, A.J.M.: 1993, In: *Integrated Soil and Sediment Research. A Basis for Proper Protection* (ed. Eijsackers, H.J.P. and Hamers, T.), pp. 303-307, Kluwer Academic Publishers, Dordrecht.

Oehme, M.: 1991, *Ambio* **20**, 293-297.

Oomen, P.A.: 1986, *Meded. Fac. Landbouww. Rijksuniversiteit Gent*, **51**, 1205-1213.

Peakall, D.B.: 1993, *Environ. Rev.* **1**, 13-20.

Rademaker, M.C.J. and Van Gestel, C.A.M.: 1993, *Verkennend onderzoek naar de effecten van niet-verzurende luchtverontreiniging op arthropoden*. Report D93003, Department of Ecology and Ecotoxicology, Vrije Universiteit, Amsterdam.

Reijnders, P.J.H.: 1986, *Nature* **324**, 456-457, 418.

Rippen, G., Zietz, E., Frank, R., Knacker, T. and Klöpffer, W.: 1987, *Environ. Technol. Lett.* **8**, 475-482.

Seiber, J.N. and Woodrow, J.E.: 1995, In: *Eighth International Congress of Pesticide Chemistry. Options 2000* (ed. Ragsdale, N.N., Kearney, P.C. and Plimmer, J.R.), pp. 157-172, American Chemical Society, Washington.

Smith, E.P. and Cairns, J. Jr: 1993, *Ecotoxicology* **2**, 203-219.

Solomon, K.R.: 1996, *Risk Analysis* **16**, 627-633.

Thomas, B.: 1995, In: *Eighth International Congress of Pesticide Chemistry. Options 2000* (ed. Ragsdale, N.N., Kearney, P.C. and Plimmer, J.R.), pp. 382-388, American Chemical Society, Washington.

Van Gestel, C.A.M.: 1992, *Ecotoxicol. Environ. Safety* **23**, 221-236.

Van Jaarsveld, J.A.: 1990, *An operational atmospheric transport model for priority substances*. RIVM report 222501002. National Institute of Public Health and the Environment, Bilthoven.

Van Leeuwen, C.J. and Hermens, J.L.M. (ed.): 1995, *Risk Assessment of Chemicals. An Introduction*. Kluwer Academic Publishers, Dordrecht.

Van Straalen, N.M.: 1994, *Neth. J. Zool.* **44**, 112-129.

Van Straalen, N.M. and Van Rijn, J.P.: 1998, *Rev. Environ. Contam. Toxicol.* **154**, 83-141.

White, R.D., Hahn, M.E., Lockhart, W.L. and Stegeman, J.J.: 1994, *Toxicol. Appl. Pharmacol.* **126**, 45-57.

MICROMETEOROLOGIC METHODS FOR MEASURING THE POST-APPLICATION VOLATILIZATION OF PESTICIDES

M. S. MAJEWSKI

U.S. Geological Survey, 6000 J Street, Placer Hall, Sacramento, CA 95819-6129, U.S.A.
(e-mail: majewski@usgs.gov)

Abstract. A wide variety of micrometeorological measurement methods can be used to estimate the postapplication volatilization of pesticides from treated fields. All these estimation methods require that the entire study area have the same surficial characteristics, including the area surrounding the actual study site, and that the pesticide under investigation be applied as quickly and as uniformly as possible before any measurements are made. Methods such as aerodynamic profile, energy balance, eddy correlation, and relaxed eddy accumulation require a large (typically 1 or more hectare) study area so that the flux measurements can be made in a well developed atmospheric boundary-layer and that steady-state conditions exist. The area surrounding the study plot should have similar surficial characteristics as the study plot with sufficient upwind extent so the wind speed and temperature gradients are fully developed. Mass balance methods such as integrated horizontal flux and trajectory simulations do not require a large source area, but the area surrounding the study plot should have similar surficial characteristics.

None of the micrometeorological techniques for estimating the postapplication volatilization fluxes of pesticides disturb the environment or the soil processes that influence the gas exchange from the surface to the atmosphere. They allow for continuous measurements and provide a temporally averaged flux value over a large area. If the behavior of volatilizing pesticides and the importance of the volatilization process in redistributing pesticides in the environment are to be fully understood, it is critical that we understand not only the processes that govern pesticide entry into the lower atmosphere, but also how much of the millions of kilograms of pesticides that are applied annually are introduced into, and redistributed by, the atmosphere. We also must be aware of the assumptions and limitations of the estimation techniques used, and adapt the field of pesticide volatilization flux measurements to advances in atmospheric science.

Keywords: pesticide volatilization flux, aerodynamic, energy balance, eddy correlation, theoretical profile shape

1. Introduction

Pesticides have many beneficial uses both in agriculture and in urban areas, and they are used extensively throughout the world. They have greatly increased crop production, lowered maintenance costs, and helped to control many public health hazards. About 500 million kilograms of pesticides are used each year in the United States (Aspelin *et al.*, 1992; Aspelin, 1994; Gianessi and Anderson, 1995) and about 255 million kilograms in Europe (NEFYTO, 1996) for controlling many different types of weeds, insects, and other pests in a wide variety of agricultural and nonagricultural settings. In the U.S., total use and the number of different chemicals applied have grown steadily since the early 1960s, when the first

Water, Air, and Soil Pollution **115:** 83–113 , 1999.

reliable records were kept. For example, national use of herbicides and insecticides on cropland and pasture grew from 90 million kilograms of active ingredient in 1964 to about 319 million kilograms in 1992 (Gianessi and Anderson, 1995). In addition to the many beneficial uses of pesticides, concerns about the potential adverse effects that pesticides can have on the environment and human health have been steadily growing.

Pesticides have been recognized as potential air pollutants since 1946 (Daines, 1952). Early in the history of agricultural pesticide use in the United States, off-target drift of the applied pesticides was a concern, and much effort has been expended studying the factors that affect drift and the best ways to control it (Akesson and Yates, 1964; Yates and Akesson, 1973; Grover *et al.*, 1978). On the other hand, mosquito abatement and other large-scale programs to eradicate such pests in the urban and agricultural environments in developed as well as third-world countries are examples of pesticide applications directly into the atmosphere with the intention of maximizing the coverage area using aerial drift.

Until the 1960s, atmospheric pesticide contamination was generally thought of as a local problem caused by spray drift. Long-range movement of pesticides was thought to be minimal, if any, because of their physical and chemical properties (low volatility and low solubility in water). The detection of DDT and other organochlorine compounds in fish and mammals in the Arctic (Cade *et al.*, 1968; Addison and Smith, 1974) and Antarctic (George and Frear, 1966; Sladen *et al.*, 1966; Peterle, 1969) changed this notion. Today, many different types of pesticides, like many types of other anthropogenically manufactured chemicals, are found dispersed throughout the world, even in remote areas where they have never been used. Pesticides have been detected in various atmospheric matrices (air, precipitation, and fog) all over the world in agricultural locations (Capel, 1991; Nations and Hallberg, 1992; Chevreuil and Garmouma, 1993; Seiber *et al.*, 1993; Trevisan *et al.*, 1993; Waite *et al.*, 1995; Baker *et al.*, 1996; Chevreuil *et al.*, 1996; Dörfler and Scheunert, 1997; Goolsby *et al.*, 1997; Hüskes and Levsen, 1997), in urban areas (Tabor, 1965; Baker *et al.*, 1996; Chevreuil *et al.*, 1996; Müller *et al.*, 1996; Majewski *et al.*, 1998), and even in remote, pristine areas (Bidleman *et al.*, 1981; Eisenreich *et al.*, 1981; Oehme and Mano, 1984; Rice and Evans, 1984; Bidleman *et al.*, 1989; Welch *et al.*, 1991; Trevisan *et al.*, 1993; McConnell *et al.*, 1997; Rice and Chernyak, 1997) where atmospheric transport and deposition are often the major or only way pesticides can get to these areas.

Because of the vast amount of pesticides used worldwide, and the fact that the atmosphere acts as a receptacle that transports and deposits these chemicals in areas where their presence was never intended, it is critical that we understand not only the processes that govern pesticide entry into the lower atmosphere, but also how much of the millions of kilograms that are applied annually are introduced into, and redistributed by, the atmosphere. Knowing the methods available to study the volatilization process in actual field situations will aid researchers to better understand the total environmental fate of these chemicals. This paper describes a variety of micrometeorological field volatilization flux estimation methods, and their underlying theory, uses, and limitations.

2. Pesticide Volatilization

Volatilization losses of a material from a surface to the atmosphere (F_O) can be estimated by measuring the vertical flux density (F_z) of the material through a plane located a discrete distance (z) above the surface source, but within the atmospheric surface layer. These measurements can be made directly using eddy correlation techniques, if sufficiently fast response sensors are available for the material in question, or by using variations of profile or mass balance measurement techniques.

In atmospheric theory, the transfer of any conservative entity, i.e. water vapor, heat, momentum, or chemical gases such as pesticides, from an evaporating surface to the atmosphere is governed primarily by the atmospheric turbulence generated by wind moving over the surface. These vertical fluxes can be expressed by a generalized flux gradient equation such as:

$$F_z = -K_z(\partial c/\partial z) \tag{1}$$

where K_z is the transfer coefficient of eddy diffusivity, an unknown function of measurement height z, that must be empirically determined, and $\partial c/\partial z$ is the gas phase concentration gradient of the material being studied. By convention, the negative sign indicates movement away from the surface. Equation 1 is appropriate to use when the gradients and fluxes are measured at a height sufficiently above the surface roughness elements.

A variety of micrometeorological techniques that are based on equation 1 have been developed to estimate volatilization fluxes of various environmental constituents from agricultural fields. Many of these techniques have been developed by the agricultural meteorology and atmospheric science communities for measuring fluxes of momentum, heat, water vapor, CO_2, and NH_3. Transferring these micrometeorological techniques for measuring fluxes of gases in relative abundance, such as CO_2 and NH_3, to those in very low concentrations, such as pesticides, requires making the assumption that the behavior of the pesticide molecules in the atmosphere is similar to that of all other atmospheric gases or any other conservative scalar property. This is not an unreasonable assumption, because once the molecules are outside the laminar surface boundary layer where molecular diffusion effects dominate, they are transported away from the surface by the turbulence generated by mechanical and thermal mixing of the air. The eddy transfer coefficient, K, in equation 1 depends on the turbulent flow of the atmosphere into which the pesticide gas is dissipated. K is a function of the meteorological conditions and not of any physical or chemical property of the pesticide, provided the airborne pesticide behaves as a molecular gas.

If the relation between the flux and the gradient for one conservative scalar property is known, it is known for all conservative scalar properties. This relation is referred to as *similarity theory*. It is the basis for most of the pesticide volatilization flux estimation methods. All fluxes characterized by equations similar to equation 1 have corresponding K_z values. For neutral atmospheric conditions, these K terms are considered to be equivalent (Pasquill, 1949; Pruitt and Lourence, 1966; Brutsaert, 1984; Sharma, 1985). In practice, it is assumed that all K's are equal all the time, and if one K is known, it can be used in place of another, unknown K value. However, the assumption that $K_m = K_w = K_h$, where K_m, K_w, and K_h are the coefficients of eddy diffusivity for momentum, water vapor, and heat, respectively, has been shown not to be the case during most of the day (Thom, 1975; Webb, 1975; Warhaft, 1976; Angus and Watts, 1984; Raupach and Legg, 1984). It has also been speculated that the K values for pesticides may differ from those of momentum, water vapor, and heat, and that each pesticide may have a unique value that is a function of it's molecular weight and structure (Glotfelty *et al.*, 1983). There are no definitive answers to these assumptions at this time (Glotfelty and Zoller, 1984; Hicks *et al.*, 1984; Horst and Slinn, 1984) and much more research is needed in this area. If we are to fully understand how the environment in general, and the atmosphere in particular, affects the volatilization behavior of pesticides, we must be aware of the assumptions and limitations of the flux estimation techniques used and be ready to adapt the discipline of pesticide volatilization flux measurements to the advances in atmospheric science.

2.1. METHOD REQUIREMENTS

The purpose of most pesticide field volatility flux experiments is to use existing atmospheric theories to estimate pesticide volatilization fluxes as accurately as possible. It is usually assumed that $K_w = K_h = K_p$, where K_p is the coefficient of eddy diffusivity for pesticide molecules in the gas phase. Equation 1 is the basis for many of the methods used to estimate the field volatilization fluxes for pesticides. Before any field measurements are made, however, two conditions must be met before all the volatilization flux estimation techniques can be used accurately. These include a spatially uniform surface area and a spatially uniform surface source strength, both of which are discussed in detail below.

A spatially uniform surface area means that the treated surface area must be spatially uniform in terms of surface characteristics such as roughness element type, height, and density, with the area surrounding the study plot having similar surface characteristics. Theoretically, equation 1 applies to gradients measured over an infinite plane source during thermally neutral ($\partial T/\partial z = 0$) atmospheric conditions. Neutral conditions usually occur only for brief periods near sunrise and sunset, or on cloudy days when a brisk wind is blowing over a moist surface. An infinite plane, however, can be approximated by placing the top measurement height at about 1 percent of the distance of the upwind edge of the field. This is known as the "1:100 rule of thumb," which states that an upwind distance (fetch) of at least 100 m is needed to readjust the internal boundary layer for each meter above the effective surface (Monteith, 1973; Businger, 1986). This minimum fetch-to-height ratio ensures that the boundary layer in which the fluxes are being determined has the same characteristic as the adjacent underlying surface and that the fluxes are constant with height. That is, the pesticide air concentration gradient is fully developed and nearly constant within the fully developed boundary layer (Panofsky and Townsend, 1964; Munro and Oke, 1975; Wesely *et al.*, 1982; Rosenberg *et al.*, 1983), at least within several meters of the surface. A large upwind fetch also ensures that the gradients of wind speed, temperature, and humidity are fully developed and log-linear within the fully developed boundary layer.

A spatially uniform source strength means that the surface source must be uniformly distributed. That is, the concentration of the test chemical must be uniformly distributed over the entire study area. A uniform source strength of sufficiently long upwind fetch ensures that the air is moving over a homogeneous exchange surface, that the vertical gaseous concentration profiles are in equilibrium with the local rate of exchange, and that horizontal concentration gradients are negligible ($\partial c/\partial x, \partial c/\partial y \approx 0$). Under these conditions, the vertical flux density of the volatilizing gas will be constant with height in the air layers close to the ground, and a one-dimensional (vertical) analysis can be made (Denmead, 1983). A spatially uniform surface with similar surrounding surfaces and a spatially uniform source strength ensures that steady-state conditions ($\partial c/\partial t = 0$) can exist during the measurement periods. Steady-state conditions are only approximated, however, as the concentration gradients of all measured species are continuously changing with time. Because these changes usually occur gradually, the length of the sampling period is critical in meeting this requirement. A long sampling period may result in conditions in which the measured gradients are significantly different at the end of the period than those at the beginning of the sampling period.

Measuring the volatilization flux of pesticides from an infinite plane source is clearly not possible. In most field experiments, the 1:100 rule is usually observed, but mass-balance estimation techniques are not constrained by it. With mass-balance techniques, small source areas can be used, but the requirement for a spatially uniform surface area still needs to be applied for the proper development of the wind, temperature, and humidity gradients.

If the above conditions are met, the gradients of wind speed, temperature, and pesticide air concentrations are typically log-linear with height. The log profiles generally occur in the dynamic sublayer, which begins at a height above 1.5 to 3.5 h_0 from the soil surface and extends for tens of meters, where h_0 is the height of a typical roughness element (Brutsaert, 1984). Nonsteady-state conditions are usually manifest as a nonlog-linear concentration profile. In small fields, edge effects may also contribute to nonsteady-state conditions, but volatilization fluxes can still be estimated from these plots if the appropriate measurement technique is used. In practice, a uniform surface source of pesticide is never achieved because it takes time to apply the material to the study area. A typical study field is on the order of 1 hectare or more in size and application times can vary from 1 hour or less, to the better part of a day, depending on the application method and the size of the study area. The best results that can be expected using these types of field methods is a temporally and spatially averaged volatilization flux value. These average numbers are very useful, however, especially when estimating atmospheric loading and potential exposure to airborne pesticides.

3. Flux Measurement Methods

All the volatilization flux estimation methods discussed below require a uniform source strength and a uniform surface area with the surrounding land having similar surface characteristics. These conditions are required even for those methods that use small source areas.

3.1. Aerodynamic profile method

One of the most frequently used field techniques for determining pesticide volatilization fluxes from treated fields is an ADP (aerodynamic profile) method. This method measures profiles of wind speed, air temperature, and pesticide air concentration, usually at the center of a treated field. The equation that is used to estimate the volatilization rate was originally derived by Thornthwaite and Holzman (1939), and later generalized by Holzman (1949). It is based on the flux-profile relation between momentum and water vapor. The equation, as originally derived, is applicable only during neutral atmospheric stability conditions. An empirically derived correction term must be added to compensate for the deviations in the logarithmic gradient profiles due to the thermal buoyancy effects on atmospheric mixing during non-neutral stability conditions. During unstable conditions ($\partial T/\partial z < 0$), thermal buoyancy increases mixing which produces smaller vertical gradients than would exist under neutral conditions (Tanner, 1968). During stable (or inversion) conditions ($\partial T/\partial z > 0$), atmospheric mixing due to thermal turbulence is dampened and an air parcel tends to return to its original level because of the density differences with elevations (Saugier and Ripley, 1978; Rosenberg *et al.*, 1983). Equation 2 is the Thornthwaite–Holzman equation modified for use in general atmospheric stability conditions.

$$\overline{F}_z = -\frac{k^2(\Delta\bar{c})(\Delta\bar{u})}{\psi\left[\ln\frac{(z_2 - d)}{(z_1 - d)}\right]^2} \tag{2}$$

where $\overline{F}_z$ is the estimated average flux through a plane at the geometric mean height of $(z_1 z_2)^{1/2}$, k is von Kármán's constant (dimensionless, ≈ 0.40), $\Delta\bar{c}$ (μg m^{-3}) and $\Delta\bar{u}$ (m s^{-1}) are the averaged air concentration and averaged horizontal wind speed differences, respectively, between heights z_1 and z_2 (m), d is the zero-plane displacement (m), and ψ (dimensionless) is an atmospheric stability correction term. The ψ expression compensates for the thermal stability effects on momentum (φ_m) and the material being measured, such as water vapor (φ_v) or pesticide (φ_p).

Equation 2 can also be derived using equation 3.

$$\overline{F}_z = -u_* c_* \tag{3}$$

where u_* and c_* are the scaling quantities of frictional velocity and the trace gas flux concentration, respectively. These scaling quantities are related to their respective gradients by:

$$\frac{\partial\bar{u}}{\partial z} = \frac{u_*}{k(z - d)}\varphi_m(\zeta) \tag{4}$$

and

$$\frac{\partial\bar{c}}{\partial z} = \frac{c_*}{k(z - d)}\varphi_c(\zeta) \tag{5}$$

A variety of the φ atmospheric stability correction expressions have been reported in the literature (Prandtl, 1932; Obukhov, 1946; Priestley, 1954; Panofsky *et al.*, 1960; Swinbank, 1968; Dyer and Hicks, 1970; Webb, 1970; Monin and Yaglom, 1971; Businger, 1973; Pruitt *et al.*, 1973; Dyer, 1974; Yaglom, 1977; Dyer and Bradley, 1982). These φ functions usually take the general form of $\varphi_n = \alpha(1 \pm \beta\zeta)^{\pm\delta}$, where α, β, and δ are constants derived by semi-empirical arguments or experimental data (Brutsaert, 1984), and ζ is expressed in terms of the dimensionless Monin–Obukhov stability parameter z/L:

$$z/L = \frac{kzgH}{\rho_a C_p T u_*^3} \tag{6}$$

where L is the Obukhov length (m), z is the measurement height (m), k is von Kármán's constant, **g** is the gravitational acceleration (m s^{-2}), ρ_a is the density of air (g m^{-3}), C_p is the specific heat of air at constant pressure, and T is the ambient temperature (K), or a gradient Richardson number (Ri, dimensionless, equation 7).

$$Ri = \frac{g(\partial T / \partial z)}{T(\partial \bar{u} / \partial z)^2} \tag{7}$$

z/L is a more precise indicator of thermal stability, but because it is a function of surface fluxes, it requires measurements of sensible heat flux (H) and the frictional velocity (u_*). Ri, on the other hand, can be evaluated using only gradient measurements of wind speed ($\partial u/\partial z$) and temperature ($\partial T/\partial z$), which can be easily measured. Ri, however, does vary with height (Brutsaert, 1984).

The Monin–Obukhov and Richardson stability parameters are dimensionless expressions that relate the consumption of turbulent energy by buoyant forces to the rate of production of turbulent energy by wind shear. Large negative values are representative of predominantly convective conditions and are associated with strong vertical motions that diffuse gases and fine particles both laterally and vertically. As the Richardson number approaches zero, the ratio between buoyancy and mechanical turbulence decreases. Depending on the wind velocity, however, the turbulence intensity can still be quite high. Large positive values are representative of conditions in which vertical motions are dampened, and diffusion is at a minimum (Yates and Akesson, 1973).

These φ functions have been well characterized for unstable conditions, but are less well understood for stable conditions (Brutsaert, 1984; Raupach and Legg, 1984). During unstable conditions, Ri is considered to be nearly equivalent to z/L. In spite of all the research that has been done to mathematically describe atmospheric stability, there are no universally accepted expressions for the φ terms. There is also no consensus on the value of von Kármán's constant (Swinbank,

1968; Dyer and Hicks, 1970; Businger *et al.*, 1971; Pruitt *et al.*, 1973; Dyer and Bradley, 1982; Raupach and Legg, 1984), although k is generally accepted as 0.40 ± 0.04 (Raupach and Legg, 1984; Tennekes, 1984). In spite of the uncertainty associated with the description of atmospheric stability, the affect on the estimated flux value is usually small.

The advantages of the ADP method are that the theory has been well documented and the method has been used for years. The instrumentation is relatively simple and readily available. The profile measurements, usually between 4 and 8 sampling points within 2 m of the surface, provide a very detailed description of the meteorological parameters and the air concentration profiles. If one point along the profile is lost or significantly deviates from the overall trend, the analyst can still generate a regression line through the remaining data points and feel confident that the resulting flux value is representative of what occurred in the environment.

Disadvantages of this method include the fetch and land area conditions that require relatively large quantities of a chemical and the preparation of the application surface and surrounding area. The method is also very labor intensive. The large field needs to be prepared, the pesticide applied, the measurement masts put in place, and 4 or more samples changed at regular intervals. A typical ADP field experiment, running 24 h a day for 1 to 2 weeks or more, can generate hundreds of samples for analysis. Field surface parameters, such as roughness height (z_0) and zero-plane displacement height (d), need to be considered if the method is used over rough surfaces or a crop. Although z_0 and d are obtained in a straightforward manner over bare ground, for plant canopies an estimation can be difficult in some cases (Paw U and Meyers, 1989). A high degree of accuracy in the profile measurements is needed because the concentration differences between levels are often very small. In addition, the ADP method may fail or give spurious results during conditions of low or no wind. This can be a serious limitation during summer days with very calm and very hot midday conditions, when considerable volatilization may be occurring from moist fields.

3.2. INTEGRATED HORIZONTAL FLUX METHOD

The IHF (Integrated Horizontal Flux) technique for estimating volatilization fluxes is a mass balance-profile method similar to the ADP method in that profile measurements of wind speed and pesticide air concentration are taken. The IHF method differs from the ADP method in that it estimates surficial emission rates by measuring the time averaged movement of material across a plane perpendicular to the wind at some distance downwind of the leading edge of the source (Denmead *et al.*, 1977). The transport rate of gaseous material from the surface source across a unit area of the plane at height z is the product of the averaged horizontal wind speed ($\overline{u}_z$) and the averaged air concentration ($\overline{c}_z$, mass per unit volume). The time averaged horizontal flux density at any level z ($\overline{F}_z$) can be expressed as:

$$\overline{F}_z = \overline{u_z c_z} = \overline{u}_z \, \overline{c}_z + \overline{u_z' c_z'} \qquad (8)$$

where $\overline{u_z c_z}$ is the mean rate of transport (the mean horizontal advective flux) over a period of time at level z. $\overline{u}_z \, \overline{c}_z$ represents the transport due to advection and $\overline{u_z' c_z'}$ represents transport due to horizontal diffusion (the mean horizontal turbulent flux). A prime denotes an instantaneous departure from the average value. This diffusion term is very small and is usually neglected in the overall flux estimation. The total mean flux across the surface of the plane can be estimated by integrating equation 8 from the surface to an infinite height. The result is equation 9, the total mean horizontal flux across the face of unit width in the plane.

$$\overline{F}_z = \frac{1}{X} \int_0^{\infty} \overline{u}_z \, \overline{c}_z \, dz \qquad (9)$$

Equation 9 is also the net rate of emission per unit width of the surface source, where X is the upwind distance to the leading edge of the source. Because there is no horizontal flux at the surface or at an infinite height, the integral limits can be restricted to the height of the affected air layer. Denmead (1983) defined the height of the affected air layer as $0.1X$ and the lower limit can be set at z_o because there is no horizontal motion below this height (Wilson *et al.*, 1982). At least 5 sampling heights are recommended to obtain well defined profiles of wind speed and air concentration and the integral can be evaluated using the trapezoidal rule (Denmead, 1983).

The IHF method was first developed and applied to estimate ammonia losses from surface applied nitrogen fertilizers (Denmead *et al.*, 1977; Denmead *et al.*, 1982; Wilson *et al.*, 1982; Wilson *et al.*, 1983; McInnes *et al.*, 1985; McInnes *et al.*, 1986) and sewage sludge (Beauchamp *et al.*, 1978). It was later applied to estimate volatilization fluxes of pesticides (Majewski *et al.*, 1990) and fumigants (Yates *et al.*, 1996; Yates *et al.*, 1997) from treated fields.

The IHF method relies on measurements of both pesticide air concentration and wind speed at multiple heights but does not provide any information above the highest sampling point or below the lowest without extrapolation. The total error associated with the method depends on the quality of the analytical method used and the graphical integration procedure (Denmead *et al.*, 1977). Errors in the latter depend on how well the wind speed and air concentration profiles are defined and if the height of the profiles encompass the full height of the affected air layer. An integration error of up to 10% was estimated on the basis of a four-point profile between 31 and 224 cm. Beauchamp, *et al.* (1978) estimated the fraction of the total flux that occurred above the integration of the four sampling heights between 10 and 150 cm to be about 38% when measuring ammonia volatilization from sewage sludge.

This method requires the same uniform source area requirements as the ADP method. A long upwind source fetch is not critical, but the actual distance to the upwind source edge must be known. This implies that wind direction must also be monitored if a circular source plot is not used. If a small source area is used, the area surrounding the source must be similarly surfaced with a sufficient upwind fetch for the proper development of the wind speed, temperature, and humidity gradients. One advantage with the IHF method is that it is independent of any *K*-similarity considerations and does not require atmospheric stability corrections. As with any method that relies on measurements of the horizontal wind velocity, this method will break down during periods of low to no wind speed where well defined gradients are not formed.

3.3. Theoretical profile shape

Another mass balance micrometeorological method that has been used to estimate pesticide volatilization fluxes was developed by Wilson *et al.* (1981a; 1981b; 1981c). This method estimates the rate of gaseous mass transfer from a small circular source to the atmosphere on the basis of numerical simulation predictions of particle trajectories in nonhomogeneous turbulence. This TPS (Theoretical-Profile-Shape) method is a two-dimensional trajectory simulation model that has been shown to be in good agreement with observed vertical profiles of horizontal flux at the center of a circular source (Wilson *et al.*, 1981a; 1981b; 1981c; Wilson *et al.*, 1982). This method shows that it is possible to infer surface fluxes from measurements of the horizontal flux at a single height and, if the measurement height is properly chosen, is nearly independent of atmospheric stability effects.

The TPS method differs from the other methods discussed in that profile measurements of meteorological parameters and air concentrations are not necessary. Since the fluxes are not related to gradients, the assumptions about the equivalence of eddy diffusivity transfer coefficients are avoided (Wilson *et al.*, 1982). The TPS method only requires a circular plot source of 50 m radius, or less, and a single measurement of horizontal wind speed and pesticide air concentration taken above the center of the source plot.

For a given source radius, the appropriate measurement height (ZINST) is a function of the surface roughness height (z_0) and the Monin–Obukhov stability length (L). In plotting the model predictions of normalized horizontal flux ($\overline{uc/F_O}$, dimensionless, the ratio of the horizontal to vertical flux) with height for various atmospheric stability conditions (very stable, very unstable, and neutral) at the center of a circular source of radius R, there is a height at which the predicted flux values intersect. At this point, the normalized horizontal flux is nearly independent of atmospheric stability considerations, and volatilization flux measurements can be made throughout the day.

The emission rate from a circular source of radius R and surface roughness height (z_o) is estimated using equation 10.

$$\overline{F_z} = \frac{(\bar{u}\ \bar{c}\ )^{measured}}{\overline{uc/F_O}} \tag{10}$$

where $(\bar{u}\ \bar{c})^{measured}$ is the product of the measured averaged wind speed and air concentration at ZINST. The TPS method has been used to estimate the rate of gaseous mass transfer of ammonia from a surface source to the atmosphere (Wilson *et al.*, 1982; Wilson *et al.*, 1983; Freney *et al.*, 1985; Brunke *et al.*, 1988; Sherlock *et al.*, 1989), various pesticides (Majewski *et al.*, 1989; Majewski *et al.*, 1990; Majewski *et al.*, 1991; Jenkins *et al.*, 1993; Whang *et al.*, 1993), and the fumigant methyl bromide (CH_3Br) (Yates, 1993; Yates *et al.*, 1996; Yates *et al.*, 1997). Yates *et al.* (1996) have modified the model and applied it to predicting off-site transport and concentrations of volatilized pesticides downwind of agricultural fields.

The relatively small source area requirements for the TPS method reduces field preparation time as well as the amount of the chemical needed. The area surrounding the circular source plot, however, needs to be similar in surface characteristics (roughness, moisture content, etc.) and of sufficient upwind fetch so that the wind speed, temperature, and humidity gradients are fully developed and uniform over the source plot. The single point measurement of wind speed and air concentration reduces the number of samples generated and, thus, the number of analyses that need to be done. If, however, a sample is lost or in question, the data for the whole period is lost. In addition, there is no way of determining the error associated in sampling. To avoid this, replicate samples can be taken at ZINST. This would increase the analytical work load, but it would still be far less than the standard five to eight sample profile measurements, but multiple measurements detract from the purpose of the method.

One disadvantage of the TPS method, as with any other method that relies on rotating cup anemometers for horizontal wind speed measurements, is that during intervals of very light or no wind, flux determinations can be uncertain. Some knowledge of the surface roughness is needed for accurate ZINST placement, and the TPS method is limited to bare surfaces or crops that do not exceed 10 cm in height (Wilson *et al.*, 1982).

Wilson's method is not the only trajectory simulation method that has been used to estimate volatilization fluxes. McInnes*et al.* (1985; 1986) used a method proposed by Philip (1959) to estimate ammonia losses from fertilizer applied to mowed wheat straw field. Philip's method, called here the TPSP method, is a solution for local advection, which is based on a two-dimensional diffusion equation. The solution is easy to calculate for any size plot, and the measurements can be made at any suitable height. The TPSP method, however, requires an estimate of the actual atmospheric stability. McInnes*et al.* (1985) briefly discussed the theory, and uses the IHF method as a reference to compare the results of the TPS and TPSP methods. Their results show that the two trajectory simulation methods are comparable, but that the Wilson method has more scatter. One advantage of the TPSP method is that the sampling height can be lower than the ZINST. Sampling closer to the source has the advantage of higher atmospheric concentrations and less chance of sample contamination by background concentrations. No reported use of the TPSP method to estimate pesticide volatilization flux was found in the literature.

3.4. Energy balance

The EB (Energy Balance, also known as the Bowen-ratio Energy Balance) method is an indirect way to estimate pesticide volatilization fluxes. This method accurately accounts for the energy sources and sinks in the experimental field and usually measures temperature, vapor pressure, and pesticide air concentrations at two levels. The direct energy fluxes and the appropriate gradient measurements are used to calculate eddy diffusivity transfer coefficients (K) that are then used to estimate the volatilization flux of the applied pesticide using equation 1. The EB method is based on the energy budget described by equation 11.

$$Rn + H + LE + G + P + M = 0 \tag{11}$$

where Rn is net radiation, H is sensible heat flux, LE is latent heat flux, G is soil heat flux, and P and M represent photosynthesis and miscellaneous energy exchanges, respectively. All the terms in equation 11 have the units of watts per square meter. LE is the major energy consumer when water is present, and Rn is the major energy supplier. In some instances, however, H can also be a significant energy source (Jensen, 1973). Both the P and M terms are usually negligible and are often ignored.

The energy balance method has been extensively used to study evapotranspiration in a variety of crops (Pruitt and Lourence, 1966; Lourence and Pruitt,

1971; Sinclair *et al.*, 1975; Garratt, 1984; Ashktorab *et al.*, 1989). Before the advent of fast response sensors with the capabilities of directly measuring environmental fluxes, the Bowen-ratio (Bowen, 1927) had been widely used to calculate H and LE and their corresponding K values. The Bowen-ratio (ß) is the ratio of sensible and latent heat fluxes and is proportional to the ratio of air temperature ($\partial T/\partial z$) and water vapor pressure ($\partial q/\partial z$) gradients as described in equation 12.

$$\beta = \frac{H}{LE} = \frac{\rho C_p K_h(\partial \overline{T}/\partial z)}{\rho L\varepsilon K_w(\partial \overline{q}/\partial z)} \approx \frac{C_p VP}{0.622 L} (\Delta\overline{T}/\Delta\overline{q}) \quad (12)$$

where ρ is the density of air (g m^{-3}), C_p is the specific heat of air at constant pressure (J kg^{-1}), L is the latent heat of vaporization of water (J kg^{-1}), ε is the ratio of the molecular weight of water to that of dry air (0.622, dimensionless), $\partial q/\partial z$ is the specific humidity (vapor pressure) gradient, and VP is the atmospheric pressure (hPa). This equation assumes that $K_h = K_w$. H and LE can be calculated by substituting equation 12 into 11 and rearranging into equations 13 and 14.

$$H = -\beta(Rn + G)/(1 + \beta) \quad (13)$$

$$LE = -(Rn + G)/(1 + \beta) \quad (14)$$

Pesticide fluxes can then be calculated using equation 1 and the K_h or K_w value (assuming K_w and $K_h = K_p$) determined from the calculated H and the measured temperature gradient or LE and the measured vapor pressure gradient.

The field size and source considerations are the same for the EB method as for the ADP method. That is, a large upwind source fetch and uniform source strength is required. The EB method is relatively simple to use, but needs sophisticated instrumentation. The setup is readily portable and automated. These combined measurement units are usually custom made, but their commercial availability is increasing. The EB method works during low wind situations because the horizontal wind speed is not a critical measurement. The method calculates pesticide fluxes from direct measurements of other fluxes and gradients and does not require any additional atmospheric stability correction terms. Surface roughness and displacement height are also not considered in the flux determinations. Balancing the energy equation also serves as a check on the accuracy of the calculated K value and the resulting pesticide flux estimation.

When ß < 1, equations 13 and 14 are relatively insensitive to errors in ß (Tanner, 1989). When ß approaches −1, that is, when H and LE are of similar magnitude but of opposite sign, usually at night, equations 13 and 14 are undefined and the Bowen-ratio method cannot be used. With thoroughly wet soil, most of the available heat goes into evaporating water. In this situation, ß will be very small and any associated inaccuracies will have little effect on the flux determination. In dryer conditions, ß may not be small and any imprecision in the measurements will affect the flux values to a much greater degree (Lourence and Pruitt, 1971). The Bowen-ratio method typically takes measurements at only two levels, and it is usually assumed that all the gradients are log-linear with height. Any error in the gradient measurements will directly affect the accuracy of ß.

One of the appealing features of the EB method is that it requires sampling at only two levels per period to estimate a flux. This, however, can lead to errors if the method is used in non-ideal situations where all of the measured gradients may not be uniform with height. It is not uncommon for one or two measurements in a multilevel pesticide air concentration profile, such as that used in the ADP method, to be off the general trend of the observed gradient. If this deviation occurred at one of the EB sampling levels, the resulting flux would not be representative of what was actually occurring. The analyst would have no idea that this was happening and would report a flux value that was higher or lower than the actual value.

3.5. Eddy correlation

Eddy Correlation (EC) offers a direct way to measure environmental fluxes if the appropriate fast-response sensor is available. EC techniques measure the instantaneous deviation from the mean vertical values of wind speed and the scalar being measured. In atmospheric turbulent flow, all entities exhibit short-term fluctuations about their mean value. The instantaneous value (w) consists of a time-averaged mean component ($\overline{w}$) and a fluctuating component (w', the instantaneous deviation from the mean), as shown in equation 15.

$$w = \overline{w} + w' \tag{15}$$

The mean vertical flux at height z of the material being measured can be expressed as the product of the mean vertical wind speed and the mean concentration of the material. Because each one of these properties can be broken down into mean and fluctuating components, the resulting equation is very complex. It can, however, be greatly simplified by using the Reynolds rules of averaging, and by assuming that the air density is constant in the lower atmosphere and that the measurements are being made over a uniform, horizontal surface for a sufficiently long period of time so the mean vertical velocity is negligible ($\overline{w} = 0$). The resulting expression is in the form of equation 16.

$$\overline{F_z} = -\rho_a \overline{w'v'} \tag{16}$$

where the overbar denotes the time average of the instantaneous covariance of w and v. On the basis of equation 16, the vertical heat flux (H) can be expressed as:

$$H = -\rho_a C_p \overline{w'T'} \tag{17}$$

where $\overline{w'}$ and $\overline{T'}$ are the instantaneous deviations from the mean vertical wind velocity and air temperature, respectively.

Sufficiently fast response sensors are available to use the EC technique to measure momentum, heat, water vapor, and CO_2 fluxes (Dyer, 1961; Wesely *et al.*, 1970; Desjardins, 1972; McBean and Miyake, 1972; Desjardins, 1974; Desjardins, 1977a; Desjardins, 1977b; Desjardins *et al.*, 1984). EC is frequently used to measure H and LE for the EB method. There is, however, no instrument currently available with the sensitivity and reaction time (frequencies on the order of 10 to 20 Hz) necessary to directly measure the volatilization flux for any pesticide.

The EC method, like the EB method described above, is an indirect way of estimating pesticide volatilization fluxes using a direct measurement of an environmental parameter. In pesticide field volatilization experiments, EC is often used to measure heat flux with a sonic anemometer and a fine wire thermocouple. With the measured temperature gradient, and the assumption that $K_h = K_p$, pesticide flux can be calculated using equation 1.

The eddy correlation method is based on a simple theory. It has the advantage of measuring fluxes of environmental parameters directly, and corrections for atmospheric stability and surface roughness parameters (z_o and d) are not needed. The instrumentation has the capability of measuring even fine-scale turbulence, and flux measurements can be made during periods of very light winds. The field surface area and source requirements, however, are the same as for the ADP and EB methods.

Disadvantages of the EC method include the complexity of the instrumentation and the need for fast-response sensors, especially in the region of measurement where these types experiments are carried out (within 2 or 3 meters of the surface). Severe measurement errors can occur if the sonic anemometer is not aligned perpendicular with respect to the horizontal wind streamlines (Kaimal, 1969; McBean, 1972). If the anemometer is not properly aligned, fluctuations in the horizontal wind will appear as fluctuations in the measured w, and, therefore, $\overline{w} \neq 0$. Equation 17 represents surface fluxes only when the mean component of equation 15 is zero (i.e., $\overline{w} = 0$). In situations in which the sensor is not properly aligned, or the terrain is irregular and causes an apparent mean vertical velocity, the computation of the flux covariance can be compensated for by using a three-dimensional anemometer and rotating the coordinate computations (Baldocchi *et al.*, 1988; Tanner, 1989).

When using the EC technique to correctly measure vertical turbulent fluxes, other factors that should be considered are the sensor time response, separation distance between sensors, measurement height above the surface, sampling period length, sensor tilt and orientation, sensor noise, and sampling rates. The sampling period must be chosen to ensure that the spectrum of eddies that contribute to the transfer process are taken into account. The appropriate frequency response of the instrument and the sampling rate need to be determined to ensure that the highest frequency eddies that contribute to the flux will be detected. The proper height at which to place the instrument for a given wind speed range can be calculated (Baldocchi *et al.*, 1988) for an instrument with a known frequency response. If vertical flux measurements of heat, water vapor, or CO_2 are being made, the response of both instruments must be matched and sufficiently fast to detect the smallest eddies contributing to the flux. These requirements become harder to meet closer to the ground as the frequency of these fluctuations also increases closer to the ground. An excellent extended discussion on the sensor and sampling period requirements and considerations using eddy correlation is given by Baldocchi *et al.* (1988) and Tanner (1989).

3.6. Relaxed eddy accumulation technique

The EC technique has long been recognized as a more direct approach for measuring atmospheric fluxes of sensible heat, water vapor and carbon dioxide (Dyer, 1961; Hicks, 1970; Miyake and McBean, 1970; Desjardins *et al.*, 1984; Verma *et al.*, 1986). Because of the lack of fast response sensors for agrochemicals it has not been possible to use this technique to estimate pesticide fluxes. Desjardins (1972) suggested an alternative sampling approach in which air is collected at a rate proportional to the magnitude of the vertical wind velocity. This technique is called *eddy accumulation* (EA). The air sampling part of the instrument consists of two sampling ports: one inlet concentrates gases or particles associated with updrafts, and the other concentrates gases or particles associated with downdrafts. This allows sufficient material to be collected and detected by conventional analytical techniques. Despite the intrinsic appeal, this technique has not yet been successfully employed in the field because of problems related to biased vertical wind velocity measurements, limited accuracy of volume and chemical analysis measurements, and difficulty in controlling the proportional sampling valves (Dyer, 1961; Speer *et al.*, 1985; Buckley *et al.*, 1988).

Businger and Oncley (1990) suggested a relaxation of the EA technique where the air is sampled at a constant flow rate but conditional to the vertical wind direction. This relaxed eddy accumulation (REA) technique (Businger and Oncley, 1990; MacPherson and Desjardins, 1991) greatly simplifies the valving design, and pesticide volatilization fluxes can be estimated on the basis of the concentration difference between the upward and downward moving air masses. The average vertical fluxes at height z ($\overline{F_z}$) can be calculated using equation 18.

$$\overline{F_z} = \overline{w'c'} = As_w(\overline{c^+} - \overline{c^-}) \tag{18}$$

where A is a dimensionless, empirically derived coefficient, s_w (m s^{-1}) is the standard deviation of the vertical wind speed measured with a sonic anemometer, and $\overline{c^+}$ and $\overline{c^-}$ (μg m^{-3}) are the mean pesticide concentrations associated with the upward and downward moving air masses, respectively. Simulations of the REA technique made with a series of vertical wind speed, water vapor, and air temperature fluctuation measurements using fast response sensors showed that the A coefficient is nearly constant and equal to 0.59 (Businger and Oncley, 1990; MacPherson and Desjardins, 1991).

One technical factor that makes it difficult to sample air with the same precision using a valve system in place of a fast response sensor is the lag between the vertical wind speed (w) signal and the valve control caused by the valve response time and the data logger processing time (35 to 85 ms). This can cause an underestimation of $\overline{c^+} - \overline{c^-}$ by diverting a fraction of the up- and downdrafts to the wrong collectors. Under similar atmospheric turbulence conditions, the degree of $\overline{c^+} - \overline{c^-}$ underestimation is a characteristic of the sampling system and can be accounted for by adjusting the empirically determined A coefficient. The determination of the appropriate adjusted A value (A_s) can be achieved by the simultaneous measurement of latent heat flux (L) using the EC and REA techniques and the assumption that the pesticide and water vapor sources had similar distributions. The mean up and downdraft water vapor concentration measurements ($\overline{q^+}$ and $\overline{q^-}$, g m^{-3}) are associated with the same filtered w signal that also controls the pesticide sampling solenoid valve. A_s integrates the effects of the system design and the similarity assumption is made that the coefficient can be used for both the different scalar properties, water vapor and herbicide vapor. The impact of an error in $\overline{c^+} - \overline{c^-}$ in equation 18 is linear and can be controlled by the A_s coefficient and the analytical method (Majewski *et al.*, 1993).

The REA system for measuring the volatilization fluxes of pesticides is still an experimental system. It is appealing in that only two air concentration measurements are needed to estimate fluxes, and it is independent of atmospheric stability considerations. The instrument and data handling requirements, as with the EA method, are very sophisticated, and the field and source requirements are the same as with the ADP method.

REA sampling has recently been applied to the measurement of pesticide fluxes on a regional scale by attaching the sampling system to an aircraft (Zhu *et al.*, 1998). This application of the REA method may prove to be a useful way to measure concentrations and fluxes at a regional scale and improve our understanding of the long-range transport of pesticides and other semivolatile chemicals.

4. The Field Experiment

A typical micrometeorological flux field experiment requires a great deal of logistical planning before sampling ever begins. Sampling and measurement equipment needs to be cleaned, tested, and calibrated. Trapping efficiency studies need to be done on the selected trapping matrix for the study compounds. The field needs to be located and the surface prepared, if necessary. Those methods that use a circular field require additional plot preparation by simulating a circular shape using rectangles of various length and a width of the spray boom. Each spray swath also needs to be marked so the applicator knows where to start and stop applying the material. The flux measurement equipment (air sampling and meteorological masts) is usually set up in the middle of the field and sampling begun soon after the application is completed. The sampling masts can also be placed at an edge of the field if it is downwind of the predominant wind direction. Typical measurements include air concentration, wind speed and direction, air temperature, relative humidity, barometric pressure, solar radiation, precipitation, and soil concentration, temperature, and moisture content. The duration of the sampling periods are variable and depend on the objectives of the experiment, trapping efficiency of the air sampling matrix, and sensitivity of the analytical method. In general, the more information the analyst has, the better he or she will be able to evaluate the environmental conditions that most influence the volatilization flux behavior during each sampling period.

4.1. VOLATILIZATION FLUX RESULTS

Volatilization flux estimates using micrometeorological profile methods require accurate measurements of wind speed, temperature, and air concentration. Volatilization fluxes can then be calculated using the discrete measured values at the selected heights above the surface. If the experimental field has the appropriate fetch, and the air concentration, wind speed, and temperature gradients are fully developed and linear with respect to the logarithm of height above the surface, linear regression of the best fit line through each measured profile can be used to determine the required values to use in the flux calculations. Linear regression of the measured profiles smooths the scatter in the measured data. Because only one measurement is usually taken per height per sampling period, and a variety of misfortunes can affect any measurement at any time during the sampling or analytical process, regression analysis also allows those data points that are lost or in question to be estimated from interpolation.

In reality, the measured profiles are not always linear with height. Significant deviations from linearity typically occur at the highest or lowest sampling point. This nonlinearity is due to an insufficient development of the overlying boundary layer, insufficient up-wind fetch, or interactions with the surface. Regression analysis of the profile measurements can still be done, provided the values used in the flux calculations fall within a linear portion of the profile.

Volatilization of the applied chemical is usually a first-order process, and can begin immediately upon application, depending on the formulation, application method, and nature of the surface. A number of factors influence the volatilization process from soil, water, and plant surfaces, but it is primarily dependent on the chemical's effective vapor pressure at the surface and environmental conditions that control the rate of movement away from the surface. The application method and formulation type, and surficial interactions such as sorption, temperature, and moisture content, can all affect the effective vapor pressure of the compound. Solar energy, atmospheric turbulence, and surface roughness can affect the movement of the chemical away from the surface. A number of reviews have discussed these influential processes in detail (Spencer, 1987; Taylor and Glotfelty, 1988; Glotfelty and Schomburg, 1989; Spencer and Cliath, 1990; Taylor and Spencer, 1990).

Field studies have used a variety of micrometeorological methods to estimate the volatilization fluxes for a number of pesticides from a variety of field situations and several in-depth reviews of these studies are also available (Taylor and Glotfelty, 1988; Taylor and Spencer, 1990; Grover, 1991; Majewski and Capel, 1995). Many of the flux estimation methods described above can be used during the same field experiment and several studies have compared the results from different combinations of methods. The ADP method is usually used as the standard to which the other methods are compared. This does not mean that the ADP is the most accurate method; most likely, it is done because the ADP method has been in use for the longest time.

Majewski *et al.*, (1990) compared the results from the ADP, IHF, TPS, EB, and EC methods in a side-by-side comparison using a 2.5 ha rectangular plot and duplicate 20 m radius (0.13 ha) circular plots. They estimated the volatilization fluxes of chlorpyrifos (*O,O*-diethyl *O*-3,5,6-trichloro-2-pyridyl phosphorothioate), diazinon (*O,O*-diethel *O*-2-isopropyl-6-methylpyrimidin-4-yl phosphorothioate), lindane (γ-1,2,3,4,5,6-hexachlorocyclohexane), and nitrapyrin (2-chloro-6-trichloromethylpyridine) during daylight hours only for four consecutive days. During periods of high winds (greater than 4 m s^{-1} at 1 m above the surface), the ADP fluxes were generally the highest, while during periods of low winds (less than 2 m s^{-1} at 1 m above the surface), the EC method generally had the highest fluxes. The ADP fluxes were almost always greater than those of the IHF method, and the EB and EC fluxes showed less variability than the TPS and IHF methods when compared to the ADP values. The ADP fluxes were also estimated using two different atmospheric stability correction expressions. This comparison showed a consistent 17% difference between flux values and is a good example of how sensitive the ADP method is to the various ψ descriptions. A statistical comparison using a paired t test found no significant difference at the 95% confidence interval between methods for each of the compounds studied. No mass balance accounting of the applied mass was attempted, however.

Yates *et al.*, (1996) did a similar comparison of the ADP, TPS, and IHF methods using methyl bromide applied to a 3.5 ha rectangular field. The experiment involved injecting the chemical into the soil at a depth of 0.25 m and immediately covering the soil with a plastic film. Yates*et al.*, (1996) adapted the TPS method for use on a noncircular plot and also calculated the ADP results using the discrete measured values (ADP-D) and those from regression analysis of the air concentration profiles. The average difference in estimated volatilization losses between the methods was about 5% with IHF > ADP > ADP-D > TPS. An average of 103% (range 88% to 112%) of the applied mass was accounted for with about 65% of the losses being attributed to volatilization.

This experiment showed that the TPS method is not limited to circular shaped fields. They developed a regression relationship for the appropriate ZINST level or $\overline{uc}/F_0$ value as a function of the average wind direction per time period and resulting upwind fetch distance that allowed them to use the TPS method on the irregular shaped field. The determination of these relations, however, required a large number of simulations as well as multilevel air concentration and wind speed measurements which detracts, somewhat, from the simplicity of the method.

In a similar experiment, Yates*et al.*, (1997) used the same 3.5 ha rectangular field, but applied the methyl bromide deeper into the soil (0.68 m) and did not cover the soil with a plastic film. They again estimated volatilization fluxes using the ADP, ASP-D, TPS, and IHF methods. The results between the methods were much more variable this time, with an average method variation of about 48% with ADP-D > ADP > TPS > IHF. The estimated volatilization losses were between 2% and 5% of the applied mass, but a mass balance accounting suggested that 21% of the applied material should have been lost by volatilization. The discrepancy between the measured and mass balance estimated volatilization losses was attributed to those losses during the application process and before the first sampling period. These unaccounted losses during application are common for all field volatility experiments. Measuring these losses quantitatively would be very difficult, if not impossible, since they are very localized and transient in nature (Yates *et al.*, 1997).

In a field experiment that compared the volatilization losses of dacthal (dimethyl tetrachloroterephthalate) using the ADP and TPS methods from a 50 m radius (0.79 ha) circular plot, Majewski *et al.*, (1991) found that the TPS method gave averaged volatilization flux results about 20% higher than those of the ADP method. They reported that 45% of the applied material was lost by volatilization, but an accurate mass balance accounting was not done because the air sampling was not continuous over the 21-day experiment and no analysis for transformation products was done.

Using a larger 150 m radius (7.1 ha) circular plot, Majewski *et al.*, (1993) compared the volatilization losses of triallate (*S*-[2,3,3-trichloro-2-propenyl] bis [1-methylethyl] carbamothioate) and trifluralin (α,α,α-trifluoro-2,6-dinitro-*N,N*-dipropyl-*p*-toluidine) using the ADP and REA methods. For both triallate and trifluralin, the ADP flux results were higher than those of the REA method by 27% and 13%, respectively. The total mass lost by volatilization was estimated at 95% and 65% for the ADP and REA methods, respectively. These loss estimates were adjusted for the actual area contributing to the measured fluxes using the method of Schuepp *et al.*, (1989).

The actual area contributing to the measured fluxes, sometimes called the *footprint* or *effective fetch*, can be estimated as a simple function of roughness length and measurement height (Gash, 1986). More detailed studies incorporating the effects of atmospheric stability conditions (LeClerc and Thurtell, 1990; Schmid and Oke, 1990; Horst and Weil, 1992) have shown that the 1 to 100 fetch to height ratio may underestimate the fetch requirements, especially during stable conditions. It is clear from these studies that the source area contributing to the measured volatilization fluxes can and does change during the course of the experiment, quite possibly from one sampling period to the next. This change in source area needs to be accounted for if an accurate mass balance is to be done.

A variety of factors contribute to the uncertainty in volatilization flux results including environmental conditions, estimation method, and study design. The above field studies show that the various volatilization flux estimation methods used in the same field experiment can give results that are very similar or very different. The results can also be quite variable from one field experiment to the next. While the mass balance accounting for the methyl bromide experiments (Yates *et al.*, 1996; Yates *et al.*, 1997) was good, results from other field experiments using the ADP method (Glotfelty *et al.*, 1984; Seiber and McChesney, 1987; Grover *et al.*, 1988) suggest that the ADP measurements may be low by approximately 50%. These low mass balance estimates may result from attributing the volatilization losses to the whole field and not just from the footprint area contributing to the measured volatilization fluxes, unaccounted for transformation products, or bound residues. It is not certain that any currently used method gives an accurate description of what is actually occurring in the field. Continuous volatilization flux measurements over an extended time period and an accurate mass balance accounting of the applied material are the most reliable validation technique.

4.2. Effect of Sampling Time Length on Flux Results

Field volatilization flux experiments using short sampling periods of one to two hours or less can generate hundreds of samples over the course of an experiment, and they are very labor intensive. This sampling frequency produces a very detailed picture of the volatilization process and is extremely valuable in investigating how various meteorological and field conditions influence the volatilization process. Not every field investigation, however, is interested in such fine detail. Often, the required result is in terms of how much of the applied material was lost through volatilization over a given time period. One approach for producing this result is to use longer sampling periods to estimate the volatilization losses. Longer sampling periods can be used provided the collection efficiency of the sampling matrix remains within acceptable limits. When long sampling periods are used, the requirement that steady-state conditions exist during the sampling period are, almost certainly, violated, especially during the first few periods after application when the fluxes are usually very high and decrease rapidly. How do long sampling times affect the estimated cumulative volatilization losses?

Majewski (1996) addressed this question using data collected over a ten-day field experiment that estimated the volatilization flux of methyl bromide using the ADP method (Majewski *et al.*, 1995). The measured data (continuous 2 and 4 h averages of air concentration, wind speed, and air temperature) were combined as time weighted averages for 8, 12, and 24 h sampling periods. Volatilization fluxes were then calculated for each of the extended time periods.

The results of this simulated increase in sampling time showed that the fluxes estimated using the extended time-weighted averages (8, 12, and 24 h) of measured air concentration and meteorological data were generally higher than the simple average of the original 2 and 4 h flux periods. The cumulative volatilization losses over the ten-day experiment also increased by approximately 25%. Majewski *et al.*, (1991) did a similar, but less rigorous manipulation with dacthal fluxes and found the simple 24 h average of the original per period fluxes were about 10% higher than the fluxes recalculated using the time-weighted 24 h measurement averages, the opposite of the methyl bromide results. The differences in both cases were attributed to neutralizing the effects of the atmospheric stability terms over the longer time periods.

These two studies show that cumulative volatilization losses estimated from long sampling periods are different from those estimated from shorter sampling periods. Environmental conditions and the physical–chemical properties of the pesticides may contribute to the observed differences, but no definite conclusions can be made that are based on the data from these two experiments. It is clear, however, that more research is needed if we are to fully understand how the length of the sampling time affects the cumulative volatilization loss estimates.

4.3. ERROR ASSOCIATED WITH FLUX ESTIMATES

Various studies have compared volatilization flux estimation methods, and all of them have found differences in the results between the methods. In some cases, the variability between the methods was small, but in others, the variability was quite large. Because most of these flux estimation methods rely on a single measurement of air concentration, wind speed, and temperature at each height, an uncertainty, or error cannot be assigned to each measurement. For those flux estimation methods that use profile measurements, and where linear regression can be done on the profile data, an error analysis for each value used in the flux value can be done using a technique described by Majewski (1996). This error analysis calculates the standard deviation associated with each measured value and estimates the percent uncertainty associated with the resulting fluxes as well as a maximum and minimum range. An assumption that T, ΔT, and Δz are measured accurately and that their contributions to the overall error is negligible can be made to simplify the calculations. Using data from a methyl bromide field study (Majewski *et al.*, 1995), Majewski (1996) estimated the average percent uncertainty associated with the ADP volatilization flux measurements to be 50%.

The results from this type of error analysis indicate that the variation in flux values between the various techniques described above are within the error estimated for the profile measurements. This type of error analysis has been done for only one field experiment, though, and no definite conclusions can be made as to how these errors compare with those of other field experiments. The accuracy of the postapplication volatilization flux methods will be resolved only by doing very careful mass balance experiments combined with estimating total volatilization losses associated only with the footprint area contributing to the estimated flux and not the entire field, and then, estimating the error associated with each flux value.

5. Summary

For many pesticides used in agriculture, as well as in the urban environment, volatilization is a continuous process that begins at the time of application and continues until the applied residues are depleted. In some cases, such as with very volatile pesticides like butylate (*S*-ethyldi-isobutylthiocarbamate), EPTC (*S*-ethyl dipropylthiocarbamate), molinate (*S*-ethyl *N*,*N*-hexamethylenethiocarbamate), and trifluralin, much of what is applied can volatilize within a few hours or days after application. For other less volatile, environmentally stable compounds, such as the organochlorine insecticides DDT (1,1,1-trichloro-2,2-bis[4-chlorophenyl] ethane), chlordane (1,2,4,5,6,7,8,8 - octachloro - 2,3,3a,4,7,7a - hexahydro - 4,7-methanoindene), and aldrin ([1R,4S,4aS,5S,8R,8aR]-1,2,3,4,10,10-hexachloro-1,4,4a,5,8,8a-hexahydro-1,4:5,8-dimethanonaphthalene), residues remain in the soil today and are still volatilizing into the atmosphere even though their wide-spread use was banned in the United States decades ago.

All of the micrometeorological techniques for estimating the volatilization fluxes of applied pesticides described above have both advantages and disadvantages associated with them. None of these methods disturb the environment or the soil processes that influence the gas exchange from the surface to the atmosphere. They allow for continuous measurements and provide a temporally and spatially averaged flux measurement from a large source area.

Disadvantages of using these methods include the need for a large experimental area with a commensurate amount of the study chemical. If the study area is near other treated fields, application drift and postapplication volatilization and drift from the other fields must be taken into account. Very accurate measurements must be made of environmental parameters such as wind speed and air temperature, as well as the gas phase pesticide concentrations in air. Often, the differences between heights for the various measurements are small, requiring the analytical techniques to be very sensitive and accurate.

Those methods that rely on rotating cup anemometers for wind speed measurements will all fail or give spurious results during very calm periods. Those methods that use sophisticated and very sensitive instrumentation, such as the EC and REA methods, may work well during very calm periods, but the electronic components may be adversely affected during inclement weather. Other methods that rely on only one or two measurements to estimate fluxes, such as the TPS, EB, EC, and REA methods, are appealing in their simplicity and low number of samples generated, but if one of the measured values is lost or in question, the data for the whole period is lost. In addition, those methods that rely on measurements at only two levels are assuming that the gradients are always behaving in a log-linear manner and that the measured data points are always reflective of the trend in the gradient. This is not always the case. The fluxes estimated using an errant data point will be incorrect, and in most cases, the analyst will not even be aware of the error. In contrast, profile techniques such as the ADP and IHF methods generate many samples per period, but if the data from one sampling level is lost or in question, it can be estimated using regression analysis of the remaining data.

Field experiments using the flux measurement methods discussed above have measured the volatilization flux of a variety of pesticides under a variety of environmental conditions since the 1960s. The accurate evaluation of the amount of the applied chemical lost to the atmosphere by volatilization is not a simple task. All the flux methods estimate the movement of the test compound away from the applied surface (soil, water, or plant). They do not deal with the processes that control the volatilization from the surface or account for atmospheric reactions. A volatilization field study requires a very detailed mass balance accounting of the compound applied as well as the major transformation products produced in the atmosphere and on or in the applied surface. An underlying question that has troubled many researchers is the accuracy of the measured flux values. Uncertainties are associated with every aspect of a field flux experiment starting with the actual amount of the study compound applied and how evenly it is distributed over the surface. Uncertainties are associated with the meteorological measurements, the sample collection method, and the analytical method. The flux estimation technique for profile measurements may have an average uncertainty of about 50%.

If the behavior of volatilizing pesticides, and the importance of the volatilization process in redistributing pesticides in the environment are to be fully understood, it is critical that we understand not only the processes that govern pesticide entry into the lower atmosphere, but how much of the millions of kilograms of pesticides that are applied annually are introduced into, and redistributed by, the 9atmosphere, and redeposited in areas where they were never intended. We must also be aware of the assumptions, uncertainties, and limitations of the estimation techniques used, and adapt the field of pesticide volatilization flux measurements to advances in atmospheric science.

Acknowledgments

Portions of this material were presented in April 1998 at a workshop on the Atmospheric Transport of Pesticides, in Driebergen, the Netherlands, organized by the Health Council of the Netherlands. This work was supported, in part, by the U.S. Geological Survey's National Water Quality Assessment Program. The use of brand, firm, or trade names in this report is for identification purposes only and does not imply endorsement by the U.S. Geological Survey.

References

Addison, R. F. and Smith, T. G.: 1974, *Oikos* **25**, 335.

Akesson, N. B. and Yates, W. E.: 1964, in Smith, R. F. and Mittler, T. E., (eds.), *Annual Review of Entomology*, Annual Reviews, Inc., Palo Alto, Calif., pp. 285-318.

Angus, D. E. and Watts, P. J.: 1984, *Agricultural Water Management*, Elsevier Science Publishing, Amsterdam.

Ashktorab, H., Pruitt, W. O., Paw U, K. T., and George, W. V.: 1989, *Agric. Forest Meteorol.* **46**, 259.

Aspelin, A. L.: 1994, *Pesticide Industry Sales and Usage. 1992 and 1993 Market Estimates*, 773-K-94-001, U.S. Environmental Protection Agency, Washington, D.C.

Aspelin, A. L., Grube, A. H., and Torla, R.: 1992, *Pesticide Industry Sales and Usage. 1990 and 1991 Market Estimates*, 773-K-92-001, U.S. Environmental Protection Agency, Washington, D.C.

Baker, L. W., Fitzell, D. L., Seiber, J. N., Parker, T. R., Shibamoto, T., Poore, M. W., Longley, K. E., Tomlin, R. P., Propper, R., and Duncan, D. W.: 1996, *Environ. Sci. Technol.* **30**, 1365.

Baldocchi, D. D., Hicks, B. B., and Meyers, T. P.: 1988, *Ecology* **69**, 1331.

Beauchamp, E. G., Kidd, G. E., and Thurtell, G.: 1978, *J. Environ. Qual.* **7**, 141.

Bidleman, T. F., Christensen, E. J., Billings, W. N., and Leonard, R.: 1981, *J. Marine Res.* **39**, 443.

Bidleman, T. F., Patton, G. W., Walla, M. D., Hargrave, B. T., Vass, W. P., Erickson, P., Fowler, B., Scott, V., and Gregor, D. J.: 1989, *Arctic* **42**, 307.

Bowen, I. S.: 1927, *Phys. Rev.* **27**, 779.

Brunke, R. R., Alvo, P., Schuepp, P., and Gordon, R.: 1988, *J. Environ. Qual.* **17**, 431.

Brutsaert, W.: 1984, *Evaporation into the Atmosphere*, D. Reidel Publishing, Dordrecht, Holland.

Buckley, D. J., Desjardins, R. L., LaLonde, J. L. M., and Brunke, R.: 1988, *Comput. Electron. Agric.* **2**, 243.

Businger, J. A.: 1973, in D. A. Haugen (ed.), *Workshop on Micrometeorology*, American Meteorological Society, Boston, pp. 67-100.

Businger, J. A.: 1986, *J. Climate Appl. Meteorol.* **25**, 1100.

Businger, J. A. and Oncley, S. P.: 1990, *J. Atmos. Ocean. Tech.* **7**, 349.

Businger, J. A., Wyngaard, J. C., Izumi, I., and Bradley, E. F.: 1971, *J. Atmos. Sci.* **28**, 181.

Cade, T. J., White, C. M., and Haugh, J. R.: 1968, *The Condor* **70**, 170.

Capel, P. D.: 1991, in *U.S. Geological Survey Toxic Substances Hydrology Program – Proceedings of the Technical Meeting*, Monterey, Calif., U.S. Geological Survey, pp. 334-337.

Chevreuil, M. and Garmouma, M.: 1993, *Chemosphere* **27**, 1605.

Chevreuil, M., Garmouma, M., Teil, M. J., and Chesterikoff, A.: 1996, *Sci. Total Environ.* **182**, 25.

Daines, R. H.: 1952, in L. C. McCabe (ed.), *Air pollution, Proceedings of the United States Technical Conference on Air Pollution*, McGraw-Hill Book Co., Inc., New York, pp. 140-143.

Denmead, O. T.: 1983, in Freney, J. R. and Simpson, J. R., (eds.), *Gaseous Loss of N_2 from Plant-Soil Systems*, Martinus Nijhoff/Dr Junk Publishers, The Hague, Netherlands, pp. 133-157.

Denmead, O. T., Freney, J. R., and Simpson, J. R.: 1982, *J. Environ. Qual.* **11**, 568.

Denmead, O. T., Simpson, J. R., and Freney, J. R.: 1977, *Soil Sci. Soc. Amer. J.* **41**, 1001.

Desjardins, R. L.: 1972, *A Study of Carbon-dioxide and Sensible Heat Fluxes Using the Eddy Correlation Technique*, Cornell University, Ph. D. dissertation.

Desjardins, R. L.: 1974, *Int. J. Biometeorol. Agric. Forest Meteorol.* **18**, 76.

Desjardins, R. L.: 1977a, *Boundary-Layer Meteorol.* **11**, 147.

Desjardins, R. L.: 1977b, *J. Appl. Meteorol.* **16**, 2498.

Desjardins, R. L., Buckley, D. J., and St. Amour, G.: 1984, *Agric. Forest Meteorol.* **32**, 257.

Dörfler, U. and Scheunert, I.: 1997, *Chemosphere* **35**, 77.

Dyer, A. J.: 1961, *Q. J. Roy. Meteorol. Soc.* **87**, 401.

Dyer, A. J.: 1974, *Boundary-Layer Meteorol.* **7**, 363.

Dyer, A. J. and Bradley, E. F.: 1982, *Boundary-Layer Meteorol.* **22**, 3.

Dyer, A. J. and Hicks, B. B.: 1970, *Q. J. Roy. Meteorol. Soc.* **96**, 715.

Eisenreich, S. J., Looney, B. B., and Thornton, J. D.: 1981, *Environ. Sci. Technol.* **15**, 30.

Freney, J. R., Leuning, R., Simpson, J. R., O.T., D., and Muirhead, W. A.: 1985, *Soil Sci. Soc. Amer. J.* **49**, 1049.

Garratt, J. R.: 1984, *Agric. Water Manag.* **8**, 99.

Gash, J. H. C.: 1986, *Boundary-Layer Meteorol.* **35**, 409.

George, J. L. and Frear, E. H.: 1966, *J. Appl. Ecol.* **3**, 155.

Gianessi, L. P. and Anderson, J. E.: 1995, *Pesticide Use in U.S. Crop Production*, National Center for Food and Agriculture Policy, Washington, D.C.

Glotfelty, D. E. and Schomburg, C. J.: 1989, in *Reactions and Movement of Organic Chemicals in Soils*, Soil Science Society of America and American Society of Agronomy, SSSA Special Publication 22, Madison, Wis., pp. 181-206.

Glotfelty, D. E., Taylor, A. W., Turner, B. C., and Zoller, W. H.: 1984, *J. Agric. Food Chem.* **32**, 638.

Glotfelty, D. E., Taylor, A. W., and Zoller, W. H.: 1983, *Science* **219**, 843.

Glotfelty, D. E. and Zoller, W. H.: 1984, *Science* **224**, 905.

Goolsby, D. A., Thurman, E. M., Pomes, M. L., Meyers, M. T., and Battaglin, W. A.: 1997, *Environ. Sci. Technol.* **31**, 1325.

Grover, R.: 1991, in Grover, R. and Cessna, A. J., (eds.), *Environmental Chemistry of Herbicides*, CRC Press, Boca Raton, Fla., pp. 89-118.

Grover, R., Kerr, L. A., Maybank, J., and Yoshida, K.: 1978, *Can. J. Plant Sci.* **58**, 611.

Grover, R., Smith, A. E., Shewchuk, S. R., Cessna, A. J., and Hunter, J. H.: 1988, *J. Environ. Qual.* **17**, 543.

Hicks, B. B.: 1970, *J. Appl. Meteorol.* **9**, 386.

Hicks, B. B., Rao, K. S., Durham, J. L., Gifford, F. A., Lenschow, D. H., and Reible, D. D.: 1984, *Science* **224**, 905.

Holzman, B.: 1949, *Ann. New York Acad. Sci.* **44**, 13.

Horst, T. W. and Slinn, W. G. N.: 1984, *Atmos. Environ.* **18**, 1339.

Horst, T. W. and Weil, J. C.: 1992, *Boundary-Layer Meteorol.* **59**, 279.

Hüskes, R. and Levsen, K.: 1997, *Chemosphere* **35**, 3013.

Jenkins, J. J., Curtis, A. S., and Cooper, R. J.: 1993, in Racke, K. D. and Leslie, A. R., (eds.), *Pesticides in the Urban Environments, Fate and Significance*, American Chemical Society, Washington, D.C., pp. 228-242.

Jensen, M. E.: 1973, *Consumptive Use of Water and Irrigation Water Requirements*, American Society of Civil Engineers, New York.

Kaimal, J. C.: 1969, *Radio Sci.* **4**, 1147.

LeClerc, M. Y. and Thurtell, G. W.: 1990, *Boundary-Layer Meteorol.* **52**, 247.

Lourence, F. J. and Pruitt, W. O.: 1971, *Agron. J.* **63**, 827.

MacPherson, J. L. and Desjardins, R. L.: 1991, in *Seventh AMS Symposium on Meteorological Observations and Instrumentation*, New Orleans, La., pp. 6-11.

Majewski, M. S.: 1996, in Seiber, J. N., Knuteson, J. A., Woodrow, J. E., Wolfe, N. L., Yates, M., and Yates, S. R., (eds.), *Fumigants: Environmental Fate, Exposure, and Analysis*, American Chemical Society, Washington, D.C., pp. 135-153.

Majewski, M. S. and Capel, P. D.: 1995, *Pesticides in the Atmosphere: Distribution, Trends, and Governing Factors*, Ann Arbor Press, Inc., Chelsea, Mich..

Majewski, M. S., Desjardins, R. L., Rochette, P., Pattey, E., Seiber, J. N., and Glotfelty, D. E.: 1993, *Environ. Sci. Technol.* **27**, 121.

Majewski, M. S., Foreman, W. T., Goolsby, D. A., and Nakagaki, N.: 1998, *Environ. Sci. Technol.* **32**, 3689.

Majewski, M. S., Glotfelty, D. E., Paw U, K. T., and Seiber, J. N.: 1990, *Environ. Sci. Technol.* **24**, 1490.

Majewski, M. S., Glotfelty, D. E., and Seiber, J. N.: 1989, *Atmos. Environ.* **23**, 929.
Majewski, M. S., McChesney, M. M., and Seiber, J. N.: 1991, *Environ. Toxicol. Chem.* **10**, 301.
Majewski, M. S., McChesney, M. M., Woodrow, J. W., Prueger, J. H., and Seiber, J. N.: 1995, *J. Environ. Qual.* **24**, 742.
McBean, G. A.: 1972, *J. Appl. Meteor.* **11**, 1078.
McBean, G. A. and Miyake, M.: 1972, *Q. J. Roy. Meteorol. Soc.* **98**, 383.
McConnell, L. L., Nelson, E., Rice, C. P., Baker, J. E., Harman, J. A., and Bialek, K.: 1997, *Environ. Sci. Technol.* **31**, 1390.
McInnes, K. J., Ferguson, R. B., Kissel, D. E., and Kanemasu, E. T.: 1986, *Soil Sci. Soc. Amer. J.* **50**, 969.
McInnes, K. J., Kissel, D. E., and Kanemasu, E. T.: 1985, *Agron. J.* **77**, 884.
Miyake, M. and McBean, G. A.: 1970, *Boundary-Layer Meteorol.* **1**, 88.
Monin, A. S. and Yaglom, A. M.: 1971, *Statistical Fluid Mechanics: Mechanics of Turbulence*, The MIT Press, Cambridge, Mass..
Monteith, J. L.: 1973, *Principles of Environmental Physics*, Elsevier Science Publishing, New York.
Müller, S. R., Zweifel, H. R., Kinnison, D. J., Jacobsen, J. A., Meier, M. A., Ulrich, M. M., and Schwarzenbach, R. P.: 1996, *Environ. Toxicol. Chem.* **15**, 1470.
Munro, D. S. and Oke, T. R.: 1975, *Boundary-Layer Meteorol.* **9**, 53.
Nations, B. K. and Hallberg, G. R.: 1992, *J. Environ. Qual.* **21**, 486.
NEFYTO: 1996, *Landbouw en Chemische Gewasbescherming In cijfers (Gegevens over 1995)*, Vol. 8, Nederlandse Stichting voor Fytofarmacie (NEFYTO), The Hague, The Netherlands.
Obukhov, A. M.: 1946, *Trudy Instit. Teoret. Geofiz: AN-S.S.S.R., No. 1 (English Translation: (1971), Boundary-Layer Meteorol. 2, 7).*
Oehme, M. and Mano, S.: 1984, *Fresenius Z. Anal. Chem.* **319**, 141.
Panofsky, H. A., Blackaday, A. K., and McVehel, G. E.: 1960, *Q. J. Roy. Meteorol. Soc.* **86**, 390.
Panofsky, H. A. and Townsend, A. A.: 1964, *Q. J. Roy. Meteorol. Soc.* **90**, 147.
Pasquill, F.: 1949, *Proc. Roy. Soc. Lond.* **A198**, 116.
Paw U, K. T. and Meyers, T. P.: 1989, *Agric. Forest Meteorol.* **47**, 259.
Peterle, T. J.: 1969, *Nature* **224**, 620.
Philip, J. R.: 1959, *J. Meterol.* **16**, 535.
Prandtl, L.: 1932, *Beitr. Phys. Fr. Atmosph.* **19**, 188.
Priestley, C. H. B.: 1954, *Australian J. Phys.* **6**, 290.
Pruitt, W. O. and Lourence, F. J.: 1966, in Brooks, F. A. and Pruitt, W. O. (eds.) *Investigation of Energy, Momentum and Mass Transfer Near the Ground*, University of California, Davis, Calif., pp. 37-63.
Pruitt, W. O., Morgan, D. L., and Lourence, F. J.: 1973, *Q. J. Roy. Meteorol. Soc.* **99**, 370.
Raupach, M. R. and Legg, B. J.: 1984, in M. L. Sharma (ed.), *Evapotranspiration from Plant Communities*, Elsevier Science Publishing, Amsterdam, pp. 119-132.
Rice, C. P. and Chernyak, S. M.: 1997, *Chemosphere* **35**, 867.
Rice, C. P. and Evans, M. S.: 1984, in Nriago, J. O. and Simmons, M. S., (eds.), *Toxic Contaminants in the Great Lakes*, John Wiley and Sons, New York, pp. 163-194.
Rosenberg, N. J., Blad, B. L., and Verma, S. B.: 1983, *Microclimate: The Biological Environment*, John Wiley and Sons, New York.
Saugier, B. and Ripley, E. A.: 1978, *Q. J. Roy. Meteorol. Soc.* **104**, 257.
Schmid, H. P. and Oke, T. R.: 1990, *Q. J. Roy. Meteorol. Soc.* **116**, 965.
Schuepp, P. H., Leclerc, M. Y., MacPherson, J. I., and Desjardins, R. L.: 1989, *Boundary-Layer Meteorol.* **50**, 355.
Seiber, J. N. and McChesney, M. M.: 1987, *Measurement and Computer Model Simulation of the Volatilization Flux of Molinate and Methyl Parathion from a Flooded Rice Field*, California Department of Food and Agriculture, Sacramento, Calif.

Seiber, J. N., Wilson, B. W., and McChesney, M. M.: 1993, *Environ. Sci. Technol.* **27**, 2236.

Sharma, M. L.: 1985, *Advances in Irrigation*, Academic Press, New York, pp. 213.

Sherlock, R. R., Freney, J. R., Smith, N. P., and Cameron, K. C.: 1989, *Fert. Res.* **21**, 61.

Sinclair, T. R., Allen, L. H., and Lemon, E. R.: 1975, *Boundary-Layer Meteorol.* **8**, 129.

Sladen, W. J. L., Menzie, C. M., and Reichel, W. L.: 1966, *Nature* **210**, 670.

Speer, R. E., Peterson, K. A., Ellestad, T. G., and Durham, J. L.: 1985, *J. Geophys. Res.* **90**, 2119.

Spencer, W. F.: 1987, in Biggar, J. W. and Seiber, J. N., (eds.), *Fate of Pesticides in the Environment*, University of California Publication 3320, pp. 61-68.

Spencer, W. F. and Cliath, M. M.: 1990, in D. A. Kurtz (ed.), *Long Range Transport of Pesticides*, Lewis Publishing, Chelsea, Mich., pp. 1-16.

Swinbank, W. C.: 1968, *Q. J. Roy. Meteorol. Soc.* **93**, 460.

Tabor, E. C.: 1965, *J. Air Pollut. Control Assoc.* **15**, 415.

Tanner, B. D.: 1989, in *Specialty Conference of the Irrigation and Drainage Division*, American Society of Civil Engineers, ASCE, Lincoln, Neb.

Tanner, C. B.: 1968, in T. T. Kozplwski (ed.), *Water Deficits and Plant Growth*, Academic Press, New York, pp. 73-106.

Taylor, A. W. and Glotfelty, D. E.: 1988, in R. Grover (ed.), *Environmental Chemistry of Herbicides*, CRC Press, Inc., Boca Raton, Fla., pp. 89-129.

Taylor, A. W. and Spencer, W. F.: 1990, in H. H. Cheng (ed.), *Pesticides in the Soil Environment*, Soil Science Society of America, Madison, Wis., pp. 213-269.

Tennekes, H.: 1984, in Nieuwstadt, F. T. M. and van Dop, H., (eds.), *Atmospheric Turbulence and Air Pollution Modelling*, D. Reidel Publishing, Dordrecht, Holland, 37-68.

Thom, A. S.: 1975, in J. L. Monteith (ed.), *Vegetation and the Atmosphere*, Academic Press, London, pp. 57-109.

Thornthwaite, C. W. and Holzman, B.: 1939, *Mon. Weather Rev.* **67**, 4.

Trevisan, M., Montepiani, C., Ragozza, L., Bartoletti, C., Ioannilli, E., and Del Re, A. A. M.: 1993, *Environ. Pollut.* **80**, 31.

Verma, S. B., Baldocchi, D. D., Anderson, D. E., Matt, D. R., and Clement, R. J.: 1986, *Boundary-Layer Meteorol.* **36**, 71.

Waite, D. T., Grover, R., Westcott, N. D., Irvine, D. G., Kerr, L. A., and Sommerstad, H.: 1995, *Environ. Toxicol. Chem.* **14**, 1171.

Warhaft, Z.: 1976, *Q. J. Roy. Meteorol. Soc.* **102**, 703.

Webb, E. K.: 1970, *Q. J. Roy. Meteorol. Soc.* **96**, 67.

Webb, E. K.: 1975, in Chapman, T. G. and Dunin, F. X., (eds.), *Predictions in Catchment Hydrology*, Australian Academy of Sciences, Canberra, pp. 203-236.

Welch, H. E., Muir, E. C. G., Billeck, B. N., Lockhart, W. L., Brunskill, G. J., and Kling, H. J.: 1991, *Environ. Sci. Technol.* **25**, 280.

Wesely, M. L., Eastman, J. A., Stedman, D. H., and Yalvac, E. D.: 1982, *Atmos. Environ.* **16**, 779.

Wesely, M. L., Thurtell, G. W., and Tanner, C. B.: 1970, *J. Appl. Meteorol.* **9**, 45.

Whang, J. M., Schomburg, C. J., Glotfelty, D. E., and Taylor, A. W.: 1993, *J. Environ. Qual.* **22**, 173.

Wilson, J. D., Catchpoole, V. R., Denmead, O. T., and Thurtell, G. W.: 1983, *Agric. Meteorol.* **29**, 183.

Wilson, J. D., Thurtell, G. E. W., and Kidd, G. E.: 1981a, *Boundary-Layer Meteorol.* **21**, 295.

Wilson, J. D., Thurtell, G. E. W., and Kidd, G. E.: 1981b, *Boundary-Layer Meteorol.* **21**, 423.

Wilson, J. D., Thurtell, G. E. W., and Kidd, G. E.: 1981c, *Boundary-Layer Meteorol.* **21**, 443.

Wilson, J. D., Thurtell, G. E. W., Kidd, G. E., and Beauchamp, E. G.: 1982, *Atmos. Environ.* **16**, 1861.

Yaglom, A. M.: 1977, *Boundary-Layer Meteorol.* **11**, 89.

Yates, S. R.: 1993, *J. Environ. Qual.* **22**, 481.

Yates, S. R., Ernst, F. F., Gan, J., Gao, F., and Yates, M. V.: 1996, *J. Environ. Qual.* **25**, 192.

Yates, S. R., Wang, D., Ernst, F. F., and Gan, J.: 1997, *Environ. Sci. Technol.* **31**, 1136.

Yates, W. E. and Akesson, N. B.: 1973, in W. van Valkenburg (ed.), *Pesticide Formulations*, M. Dekker, New York, pp. 275-341.

Zhu, T., Desjardins, R. L., MacPherson, J. L., Pattey, E., and St. Amour, G.: 1998, *Environ. Sci. Technol.* **32**, 1032.

ATMOSPHERIC TRANSPORT AND AIR-SURFACE EXCHANGE OF PESTICIDES*

Terry F. Bidleman

Atmospheric Environment Service, 4905 Dufferin St., Downsview, Ontario M3H 5T4, Canada.
E-mail. terry.bidleman@ec.gc.ca

Abstract. Atmospheric transport and exchange of pesticides with soil, vegetation, water and atmospheric particles are discussed, with an emphasis on applying physicochemical properties of the compound to describe environmental partitioning. The octanol-air partition coefficient is promoted as a unifying property for describing volatilization of pesticides from soil and sorption to aerosols. Present-day sources of organochlorine (OC) pesticides to the atmosphere are continued usage in certain countries and volatilization from contaminated soils where they were used in the past. Models are available to predict volatilization from soil; however, their implementation is hampered by lack of soil residue data on a regional scale. The need to differentiate "new" and "old" sources is increasing, as countries negotiate international controls on persistent organic pollutants (POPs). A new technique, based on the analysis of individual pesticide enantiomers, is proposed to follow emission of chiral OC pesticides from soil and water. Air monitoring programs in the Arctic show the ubiquitous presence of OC pesticides, PCBs and other POPs, and recently a few "modern" pesticides have been identified in fog and surface seawater. Atmospheric loadings of POPs to oceans and large lakes take place mainly by air-water gas exchange. In the case of OC pesticides and PCBs, aquatic systems are often near air-water equilibrium or even oversaturated. Measurement of water/air fugacity ratios suggests revolatilization of PCBs and several OC pesticides in the Great Lakes and, for α-hexachlorocyclohexane (α-HCH), in the Arctic Ocean. Outgassing of α-HCH in large lakes and arctic waters has been confirmed by enantiomeric tracer studies. The potential for pesticides to be atmospherically transported depends on their ability to be mobilized into air and the removal processes that take place enroute: wet and dry deposition of gases and particles and chemical reactions in the atmosphere. Measurement of reaction rate constants for pesticides in the gas and particle phase at a range of environmental temperatures is a critical research need. The transport distance of a chemical is related to its overall environmental persistence, determined by the partitioning among different compartments (water, sediment, soil, air), degradation rates in each compartment and mode of emission (into water, soil, air). Several pesticides found in the arctic environment have predicted lifetimes in the gas phase of only a few days in temperate climates, pointing out the need for monitoring and evaluation of persistence in cold regions.

Key words: atmospheric transport, enantiomers, gas exchange, octanol-air partition coefficient, particle-gas partitioning, persistent organic pollutants, POPs, pesticides, soil-air exchange.

---

* Figures 2 and 6 are reprinted from *Environmental Pollution* **102**: 43–49, 1998, by T.F. Bidleman et al., and Figure 3 from *Environmental Pollution* **102**: 51–62, 1998, by C.J. Halsall et al., with permission from Elsevier Science.

Water, Air, and Soil Pollution **115:** 115–166, 1999.

1. Introduction

Intercontinental transport of persistent organic pollutants (POPs) is not a new concept. Contamination of the polar regions by DDT residues was recognized in the late 1960s and early 1970s (Bowes and Jonkel, 1975; Peterle, 1969, Sladen et al., 1966; Tatton and Ruzicka, 1967). At about the same time, DDT and dieldrin were identified on Sahara dust carried from Africa to Barbados by the northeast trade winds (Prospero and Seba, 1972; Risebrough et al., 1968; Seba and Prospero, 1971). These findings sparked interest in the atmospheric transport of contaminants to the oceans. Between the early 1970s to mid-1980s the U.S. National Science Foundation sponsored two large-scale investigations of atmospheric transport, the International Decade of Ocean Exploration (IDOE) and the Sea-Air Exchange (SEAREX) program. Baseline measurements of organochlorine (OC) pesticides and polychlorinated biphenyls (PCBs) in marine air were made in the North Atlantic, North Pacific and South Pacific oceans. Results were synthesized by Duce et al. (1991) to provide estimates of loadings to the world's oceans typical of the 1970s and early 1980s. Atmospheric processes accounted for 80-99% of total loadings. Although rivers directly impact coastal areas, their contributions were judged as minor on a global scale.

Today the world is moving away from OC pesticides to less persistent compounds, seeking to avoid the pitfalls that led to global contamination in the past. In July, 1998 international negotiations were started in Montreal to eliminate the "dirty dozen" POPs: DDT, hexachlorobenzene (HCB), chlordane, heptachlor, toxaphene, aldrin, dieldrin, endrin, mirex, PCBs, and chlorinated dioxins and furans (information on UNEP activities concerning POPs can be found on the website http://irptc.unep.ch/pops/). To design new pesticides and manage those already in use, it is essential to understand the factors that cause them to migrate from application sites and be transported over long distances. The general pattern of POPs transport is already well known. Nearly twenty-five years ago, Goldberg (1975) coined "global distillation" to describe the tendency of pesticides to evaporate from temperate and tropical regions and condense in cold climates. "Global chromatography" (Risebrough, 1990), and the "grasshopper effect" (Wania and Mackay, 1996) have been used to describe the tendency of POPs to undergo cycles of deposition and re-emission during transport. These concepts have been quantified through global-scale modelling by Wania and Mackay (1995,1996). Their model predicts that cycles of deposition and revolatilization will lead to fractionation of the POPs during transport from warm to cold regions, with the more volatile ones migrating most rapidly and the less volatile ones lagging behind. Such behavior depends largely on the physicochemical properties of the chemicals themselves, especially the partition coefficients between air and environmental stationary phases: soil, vegetation, water, snow and atmospheric particles.

This paper begins with a discussion of pesticide physicochemical properties, since they are central to modelling environmental transport and fate. Following this is a section on atmospheric sources of pesticides with an emphasis on vapor-phase emissions from soil and the use of chemical markers as source indicators. Subsequent sections discuss the occurrence of pesticides in the remote atmosphere, case studies of air-water gas exchange in large lakes and marine regions and factors which control the lifetime of pesticides in air. The occurrence of pesticides in rural air of temperate regions is not emphasized in this article, but an extensive review for the U.S. has been published by Majewski and Capel (1995). Two relatively new concepts are promoted. The octanol-air partition coefficient is recommended as a descriptor for volatilization of pesticides from soil and partitioning of vapor-phase POPs to atmospheric particles. A new approach is presented for identifying emission of "old" (i.e., OC) pesticide residues from soil and water, based on the analysis of individual enantiomers of chiral compounds.

2. Physicochemical Properties of Pesticides

The properties that are commonly used to predict environmental partitioning and mobility of pesticides are saturation vapor pressure (P), water solubility (S_w), acid dissociation constant for ionizable compounds (pK_a), and the partition coefficients between octanol-water (K_{ow}), soil organic carbon-water (K_{oc}), air-water (Henry's law constant, H) and octanol-air (K_{oa}). The bioconcentration factor (BCF) can be derived since it is closely correlated with K_{ow} (Mackay, 1982). Models of pesticide fate also require estimates of the degradation rate constants in air, water, sediments and soil, and metabolism or depuration rate constants for biota. A database of physicochemical properties and environmental degradation rates for pesticides has been compiled by Mackay et al. (1997), who also provide discussions of properties determination methods, illustrative level I, II and III fugacity models of environmental distribution and fate, and quantitative structure-property relationships. The fugacity-based Equilibrium Criteria (EQC) model is designed to assess the fate of organic chemicals from their physicochemical properties (Mackay et al., 1996a,b,c). Properties for a large number of pesticides have also been compiled by Majewski and Capel (1995).

In practice, several problems arise in the use of physicochemical data taken from the literature for environmental modelling. Properties for a specific compound often differ greatly among reporting laboratories, and the user is forced to pick and choose to obtain "reliable" values. Experience and judgement are valuable in this exercise! Properties are often reported at a single temperature, whereas vapor pressure, Henry's law constant, water solubility and octanol-air partition coefficient vary strongly with temperature (Wania et al.,

1998a,b). The required values for vapor pressure and water solubility are for the liquid-phase, rather than the solid-phase compound (Mackay, 1991; Mackay et al., 1997). This presents a dilemma for substances that are solids at ambient temperature. The liquid/solid fugacity ratio (F) is:

$$F = P_L/P_S = S_{w(L)}/S_{w(S)} = \ln (\Delta H_f/RT_m)(T_m - T)/T \quad (1)$$

Subscripts L and S refer to the liquid and solid phase, T_m and T are the melting point and ambient temperatures (K), ΔH_f is the enthalpy of fusion (J/mol), and R is the gas constant (8.314 Pa m^3/mol K = 8.314 J/mol K). The term $\Delta H_f/T_m$ is the entropy of fusion, ΔS_f (J/mol K). Eq 1 can be used to estimate liquid-phase properties from solid-phase values. For rigid organic molecules, ΔS_f is approximately 56 J/mol K (Mackay, 1991; Mackay et al., 1997), but varies among compounds (Hinckley et al., 1990). Values have been measured for some POPs (Dannenfelser et al., 1993; Donnelly et al., 1990; Hinckley et al., 1990; Mackay et al., 1997), however for many it is necessary to assume 56 J/mol K. The errors in doing so can be fairly substantial, especially for high melting point compounds. For example, the values of P_L at 25°C, obtained by using actual vs. default values of ΔS_f, are 0.13 vs. 0.30 Pa for HCB (T_m = 503 K) and 0.010 vs. 0.022 Pa for dieldrin (T_m = 449 K) (Hinckley et al., 1990), and the error increases at lower ambient temperatures.

Alternate ways of estimating the liquid-phase vapor pressure for compounds that are solids at ambient temperature are extrapolation of vapor pressure data measured above the melting point and capillary gas chromatography (GC). Both approaches involve extrapolation uncertainties. GC is a reliable technique for non-polar compounds, but is less accurate for polar pesticides such as organophosphates (Hinckley et al., 1990). The GC method for polar compounds can be improved by using a vapor pressure standard from the same chemical class (Kim et al., 1984).

Mackay (1991) suggested that solubility in octanol is a suitable surrogate for organic phases such as lipids and organic carbon, thus leading the way to the definition of a new partition coefficient between octanol-air (K_{oa}). K_{oa}, or its equivalent K_{ow}/H, has been recommended for describing partitioning between air and plant waxes (Bacci et al., 1990; Kömp and McLachlan, 1997a,b; Paterson et al., 1991), soils (Harner et al., 1998a; Harner and Mackay, 1995) and aerosols (Finizio et al., 1997; Harner and Bidleman, 1998a; Pankow, 1998). The octanol-air partition coefficient (K_{oa}) can be estimated from the ratio of K_{ow}/H, however there is an inconsistency in the definition of the phases. Measurements of K_{ow} are carried out with water-saturated octanol and octanol-saturated water, whereas K_{oa} refers to partitioning of the gaseous pesticide into pure octanol. Also, as pointed out above, there is often the problem of selecting K_{ow} and H from the range of literature values. For these reasons, it is preferable to determine K_{oa} independently (Harner and Bidleman, 1996; Harner and Mackay, 1995). Values of K_{oa} as a function of temperature have been recently measured by generator column methods for chlorobenzenes and p,p'-DDT (Harner and Mackay, 1995), PCBs (Harner and Mackay, 1995; Harner and Bidleman, 1996; Kömp and McLachlan, 1997c), and polychloronaphthalenes (PCNs) and polycyclic aromatic hydrocarbons (PAHs) (Harner and Bidleman, 1998b).

3. Sources of Pesticides to the Atmosphere

Agricultural pesticides are released into the atmosphere by spray drift, post-application volatilization and wind erosion of soil. A great number of physical and chemical factors influence emissions by these different mechanisms: application methods, formulations, type of spray cloud, tillage practices, erosion conditions, solar energy input, atmospheric stability, soil moisture and temperature. Covering these factors is beyond the scope of this article, and the reader is referred to a book by Majewski and Capel (1995). Volatilization can remove a large fraction of pesticide initially applied to the field (Whang et al., 1993). Willis et al. (1983) followed losses of toxaphene and DDT after spray application to cotton plants and found that 53-58% volatilized within a month. Measured volatilization fluxes of trifluralin and triallate accounted for 54-74% of total losses from soil within five days after application (Majewski et al., 1993; Majewski and Capel, 1995).

Pesticides are also emitted from urban usage, e.g. lawns, parks, and gardens, and buildings treated for pest control (Coupe et al., 1998; Majewski and Capel, 1995). Halsall et al. (1995) estimated that release of PCBs from buildings contributed significantly to the observed PCB burden in urban air, and Bidleman et al. (1998a) suggested that this may also be true for chlordane which was applied to foundations as a termiticide.

Emissions from soil and water that were contaminated with pesticides in the past may be a significant contributor to atmospheric burdens, especially for OC compounds (Bidleman and Falconer, 1999). This is discussed further in sections 3.1.5 and 3.1.6.

3.1. AIR-SURFACE EXCHANGE OF VAPOR-PHASE COMPOUNDS

3.1.1. *Soil and Vegetation*

Woodrow et al. (1997) view pesticide saturation vapor pressure as the underlying driving force for volatilization, while other factors, such as soil adsorption, depth of soil incorporation and water solubility are operators that result in a reduced or effective vapor pressure under a particular set of conditions. The authors compiled results of pesticide flux studies from treated soil, plant foliage, water and inert surfaces (glass, plastic) to derive relationships between flux (mass/area x time) and the physicochemical properties of the compound. The fluxes were determined within 12 - 24 h after pesticide application, and so represent the "worst case" situations. Flux values were estimated in some cases by analysis of the medium to determine pesticide residues remaining at some time after application. In others, fluxes were measured by micrometeorological method. A log-log plot of the flux was made vs. R_{inc}, a combination of the pesticide application rate (AR), and the physicochemical properties of the pesticide which control volatilization: vapor pressure (P), water solubility (S_w) and K_{oc}.

$$R_{inc} = AR(P/K_{oc}S_w) \qquad (2)$$

An excellent correlation ($r^2 = 0.93$) was obtained for fifteen substances with a range of polarities: OCs, hydrocarbon oils, organophosphates, triazines and other herbicides with different functional groups. The relationship was subsequently modified for soil-incorporated pesticides by including a depth of incorporation term (d):

$$R_{inc} = (AR/d)(P/K_{oc}S_w) \qquad (3)$$

Pesticide flux from plants and inert surfaces was well correlated with vapor pressure, while emissions from water bodies were better explained by the Henry's law constant. The SCREEN-2 dispersion model of the U.S. Environmental Protection Agency was used to predict downwind concentrations of pesticides at 1.5 m height, using the correlation relationships (eq 2 or 3) to estimate fluxes from the source. The predicted downwind concentrations agreed well with those measured near treated fields.

Karickhoff (1981) showed that the relationship $K_{oc} = 0.41\ K_{ow}$ holds for a large number of non-polar and polar pesticides. Substituting into eq 3:

$$R_{inc} = (AR/0.41d)(P/K_{ow}S_w) \qquad (4)$$

The combination of physicochemical properties in eq 3 and 4 is the reciprocal of the partition coefficient of the pesticide between octanol and air (K_{oa}):

$$(1/RT)(P/K_{ow}S_w) = C_{air}/C_{oct} = 1/K_{oa} \qquad (5)$$

The soil/air partition coefficient (Hippelein and McLachlan; 1998) and the volatilization half life of POPs in soil (Cousins and Jones, 1998) were highly correlated with K_{oa} or its equivalent, K_{ow}/H. Thus $1/K_{oa}$ could be used as an alternative to the combination of properties in eq 2 - 4 for estimating the potential for pesticides to volatilize from soils.

Uptake of POPs by plants (Calamari et al., 1991, 1994; France et al., 1997; Eriksson et al., 1989; Jensen et al., 1992; Ockenden et al., 1998a; Tremolada et al., 1996) and tree bark (Simonich and Hites, 1995) has been used to monitor atmospheric contamination. Plants provide a link between the atmosphere and the terrestrial food chain (Lorber et al., 1994; McLachlan, 1996; Thomas et al., 1998a,b; Welsch-Paulsch and McLachlan, 1998). Partitioning of tetrachlorodibenzo-p-dioxin (TCDD) to and degradation within vegetation is important in governing the travel distance in the atmosphere (Bennett et al., 1998). The forest canopy can be expected to play a key role in the environmental fate of POPs by decreasing atmospheric half lives and transferring POPs to the forest soil (McLachlan and Horstmann, 1998).

Vapor exchange of pesticides (Woodrow et al., 1997) and PCDD/Fs (Wagrowski and Hites, 1998) with plants has been correlated with vapor pressure. Others (Bacci et al., 1990; Paterson et al., 1991; Thomas et al., 1998b,c) found that plant/air partition coefficients (K_{pa}) could be described by the ratio of $K_{ow}/H = K_{oa}$. A log-log plot of K_{pa} vs K_{ow}/H for a laboratory study of the accumulation of semivolatile compounds by azalea leaves had a slope of 1.14 (Bacci et al., 1990). However plots of log K_{pa} vs log K_{oa} from field studies often have slopes that are different from unity. Slopes ranged from 0.32 - 0.47 for air-pasture transfer of PCBs (Thomas et al., 1998b) and 0.35 - 0.53 for PCB partitioning to pine needles and lichens in Norway (Ockenden et al., 1998a). In a comparison of PCB partitioning from air to six plant species, Kömp and McLachlan (1997a) found that log-log plots of K_{pa} vs. K_{oa} were highly correlated for a particular plant, but the slopes and intercepts of the plots showed considerable variability among plant species, ranging from 0.57 - 1.15. The authors suggested that the lipophilicity of the contaminant storage compartment of plants is different from octanol and varies widely. Goss and Schwartzenbach (1998) state that these plots will have slopes of unity only if there is a high chemical similarity between octanol and the sorbing phase.

Dowdy and McKone (1998) compared a molecular connectivity index (MCI) descriptor, corrected for the presence of polar substituent groups, vs. K_{oa} for describing plant/air partitioning. Both approaches were somewhat successful, but the test was limited by the rather small database for K_{pa} values. The authors concluded that a hybrid quantitative structure-activity relationship (QSAR), based on the combination of K_{oa} with MCI should be considered as more data on K_{pa} values become available, since MCI should be better able to account for metabolic processes and other transformations.

Kinetic factors are important in controlling air-plant exchange (Hung and Mackay, 1998; Paterson et al., 1991). Uptake and loss of gaseous POPs by plants is a complex process and air-vegetation equilibrium is not always established, particularly for less volatile POPs (Kömp and McLachlan, 1997b). In laboratory experiments, azalea leaves approached steady state with POPs vapors on a time scale of a few hundred hours (Bacci et al., 1990). Pine needles seem to achieve steady state with airborne POPs very slowly or not at all (Kylin and Sjödin, 1998; Tremolada et al., 1996), and PCDD/Fs with log $K_{oa} > 9$ were also out of equilibrium with grass and corn plants (McLachlan, 1996). On the other hand, Thomas et al. (1998b) found that uptake of PCBs by pasture grass was nearly the same regardless of whether the pasture growth (exposure time) had been 2, 6 or 12 weeks.

3.1.2. *Water*

Gas exchange of POPs across lake and sea surfaces has been described by many authors (Ballschmiter, 1992; Bidleman and McConnell; 1995; Eisenreich et al., 1997; Mackay, 1991; Pacyna et al., 1998). Wania et al. (1998b) reviewed atmospheric deposition and gas exchange processes for POPs, with special attention to some important factors that are not often included in such reviews: temperature effects on physicochemical properties and mass transfer coefficients, sorption of POPs to dissolved organic matter, vertical transport in the water column, and the influence of the sea surface microlayer.

The net gas flux (F_{net}, ng/m^2 s) can be estimated from the two-film model in which mass transfer is limited by diffusive exchange across air and water films at the interface.

$$F_{net} = 10^9 M D_{aw} (f_w - f_a) \quad (6)$$

$$D_{aw} = K_{OL}/H \quad (7)$$

$$1/K_{OL} = 1/k_w + RT/Hk_a \quad (8)$$

The terms f_a and f_w are the fugacities (partial pressures) of the compound in air and water. Fugacities are related to the gaseous and dissolved concentrations (C_a and C_w, ng/m^3), Henry's law constant as a function of water temperature (Pa m^3/mol), air temperature (T_a, K) and molecular mass (M) (Bidleman and McConnell, 1995; Mackay, 1991):

$$f_a = 10^{-9} C_a R T_a / M \quad (9)$$

$$f_w = 10^{-9} C_w H / M \quad (10)$$

D_{aw} (mol/m^2 s Pa), the mass transfer coefficient used in fugacity-based flux equations, is calculated from the overall mass transfer coefficient K_{OL} (m/s) and the Henry's law constant, H by eq 7. K_{OL} is related to the individual mass transfer coefficients for the air and water films (k_a and k_w) by eq 8. These individual mass transfer coefficients are functions of wind speed and the molecular diffusivity or Schmidt number of the compounds in air and water. Relationships based on empirical observations or theoretical models are summarized by the above authors. Many field experiments have been done in recent years to determine k_w in lakes and oceans, using SF_6, ^3He or Rn as tracers (summarized by Eisenreich et al., 1997), and these have considerably reduced the uncertainty in the relationship of k_w to wind speed. Mass transfer coefficients vary in a non-linear manner with wind speed, and this effect must be taken into account when integrating flux values over time. Livingstone and Imboden (1993) suggested using a Weibull distribution of wind speed to estimate the variability due to time-dependent wind speeds.

The direction of net gas exchange is determined by the water/air fugacity ratio (combining eq 9 and 10):

$$f_w/f_a = C_wH/C_aRT_a \qquad (11)$$

Fugacity ratios of <1 and >1 imply net deposition and volatilization, while a ratio = 1 indicates air-water equilibrium. Thus gas exchange with water can be a source or sink for atmospheric pesticides, as discussed later. It is impossible to make realistic estimates of gas exchange without knowing the concentrations of POPs in both air and surface water. It is also important to recognize that even though the net flux at equilibrium is zero, the individual deposition and volatilization fluxes can be quite high and dominate the mass loading of POPs to surface water. For this reason, it is desireable to present the individual rather than net flux values (Murphy, 1995), although net fluxes have also been used successfully (Hornbuckle and Eisenreich, 1995). Individual fluxes can be calculated simply by separating the water and air fugacity parts of eq 6. Because of the bidirectional nature of gas exchange, instantaneous values of water/air fugacity ratios vary with short-term changes in atmospheric concentrations brought about by on- and off-shore breezes (deWulf et al., 1998) and "urban plumes" (Simcik et al., 1997).

3.1.3. *Snow and Ice*

Wania et al. (1998b) point out that traditional descriptions of wet deposition have ignored vapor-phase scavenging by snow. Recently that view has changed because of new information on the capacity of air-water and air-ice interfaces to sorb chemicals (Hoff et al., 1993a, 1995; Wania et al., 1998b,c, 1999). The theory and experimental evidence for sorption of vapor-phase organic compounds by ice and snow surfaces, field studies of hydrophobic contaminant deposition by snow scavenging, and diagenesis in snowpacks were reviewed by Franz et al. (1997) and Wania et al. (1998b). Both articles identified research needs and recommended future studies. Wania et al. (1998b) concluded that snow and ice is a major vector for POPs transport and is a more effective scavenger of airborne POPs than rain. Estimates of wet deposition of HCHs to the Arctic were greatly increased by the inclusion of vapor-phase sorption to snow (Barrie et al., 1997). The ability of snow to retain POPs changes greatly after deposition and metamorphosis of the snowpack is expected to be a significant factor in the remobilization of POPs (Barrie et al., 1997; Gregor, 1991). Measurements of the specific surface area of snow (Hoff et al., 1998a) and formulation of a model to describe the behavior of POPs in an aging snowpack (Wania, 1997) are important recent developments.

3.1.4. *Modelling Pesticide Emission from Soils*

Models of pesticide volatilization from soils are often based on a series of classic papers by Jury et al. (1983, 1984a,b,c), who described volatilization in terms of properties of the chemical, the soil, and evaporation of water. Key properties of the pesticide are the Henry's law constant (H) and the partition coefficient between soil organic carbon and water (K_{oc}). The model considers pesticide transport in the soil by diffusion in the liquid and vapor phases and convective transport by water movement. Release to the atmosphere takes place by vapor diffusion through a boundary layer of air near the soil. Jury et al. divided pesticides into three classes, depending on whether their soil-to-air transfer was dominated by resistance in the soil layer (Category I), the air layer (Category III) or both (Category II). Category I compounds are those with relatively high Henry's law constants and include OC pesticides, volatile fumigants such as methyl bromide, and some organophosphates. Members of Category III have low Henry's law constants, exemplified by atrazine and 2,4-D. After incorporation of the pesticide into the soil, the soil-to-air flux of Category I compounds decreases with time, while the flux of the Category III and some of the II substances increases with time when evaporation of moisture takes place. The behavior of the latter compounds is caused by the "wick effect", whereby pesticide is carried to the soil surface by the water flux that accompanies evaporation. The Jury model was not intended for predicting fluxes under field conditions, but rather to group pesticides according to their behavior in environmental screening tests.

A recent model developed by the Canadian Global Emissions Interpretation Centre (Scholtz et al., 1997) has the capability of estimating emissions from soils and vegetation on field and regional scales. The soil volatilization model differs from the Jury model by having multilayer soil and air compartments and coupling to hourly meteorological conditions to account for variability in soil temperature, soil moisture content and evaporation, which are main drivers of pesticide volatilization. Simulations were done for three different modes of pesticide application (spray, in-furrow and soil incorporated), and also for old residues that remained in the soil over winter and were released by tilling in spring. Modelled volatilization rates for 2,4-D, lindane and chlordane indicated high diurnal variability, up to two orders of magnitude. The major factor influencing these short-term fluctuations was water flux. Emissions of trifluralin and triallate applied as a spray to bare soil were measured by the relaxed eddy accumulation technique (Majewski et al., 1993) and compared to results of model simulations. The model runs closely tracked the shape and magnitude of the observed peak in volatilization flux following the application. The measured cumulative loss of the two pesticides over 3-5 days was 23 - 30% lower than model predictions. The canopy volatilization model assumes a series of resistances in movement of the pesticide from the plant cuticle to water and air. Transfer of pesticide from the leaf surface to the atmosphere is controlled by the quasi-laminar boundary layer resistance in the air.

The soil and canopy models were used to prepare a gridded pesticide emissions inventory for Canada and the United States, each grid being 127 x 127 km. The emissions inventory took into account the time, application rate and mode of application of the pesticides, gridded soil texture, and several geophysical, climatological and crop parameters. Seasonal and annual emission factors, presented as the fraction of the application rate, were generated for nine pesticides that are still used in two countries, as well as for several banned OC pesticides.

A Dynamic Exchange of Pollutants between Air and Soil Surface (DEPASS) model has been developed to describe air-soil exchange under actual meteorological conditions (van Jaarsveld, 1996). The model contains multilayered soil and atmospheric compartments and is driven by observed meteorological parameters (wind speed, temperature, radiation and precipitation). The atmospheric part of the model includes wet and dry removal of particulate and gaseous species. The model was run with lindane as representative of a pesticide that is predominently in the gas phase and relatively water soluble. Starting with lindane in the air but not in the soil, the initial flux was downward and the top layer of the soil reached saturation after about 24 h. Thereafter, lindane exhibited strong diurnal variations in flux which were due largely to temperature and evaporation, as in the Scholtz model. The model was also run for lindane incorporated to a depth of 10 cm in the soil. Again, strong diurnal variations in flux were predicted. The magnitude of the flux and shape of the volatilization curve over 2 months' simulation was quite similar to that obtained from the Scholtz model. Depending on the organic carbon fraction of the soil, 10 - 30% of the incorporated lindane was predicted to volatilize within a year.

A fugacity model for soils was developed and applied to emission of PCBs from U.K. soils (Harner et al.; 1995). Treatment of the volatilization mechanism was according to the Jury model. An updated version has recently been described which considers multiple soil layers and uses K_{oa} as the key property which governs volatilization (Harner et al., 1998a). The soil model is coupled to a simple atmospheric box model, which considers input and loss through advection, volatilization, wet and dry deposition. An initial test of the model gave a seasonal cycle of predicted toxaphene concentrations in Alabama within a factor of two of observations.

3.1.5. *The "Old" vs. "New" Source Problem for Organochlorine Pesticides*
Most OC pesticides have been banned for years to decades in Canada, the U.S. and European countries, but are still routinely found in air and precipitation from the North American Great Lakes region (Cortes et al., 1998; Hillery et al., 1998; Hoff et al., 1996) and the Arctic (Halsall et al., 1998a; Oehme et al., 1996). What are the sources of these pesticides today? Are they atmospherically transported from countries where they are still applied, or "ghosts of the past" – recycled from contaminated soils, water bodies and vegetation? Answers to the old vs. new source question are particularly relevant today, as the world's nations work toward an agreement to eliminate persistent pesticides.

Field measurements support the hypothesis that agricultural soils containing "old" OC residues are still capable of releasing them to the atmosphere. DDT residues in soil and air samples collected at 20 and 80 cm above the soil were measured in 1994 at an experimental farm in California where DDT was last applied twenty-three years previously (Spencer et al., 1996). The ΣDDT in the farm soil was 3516 ng/g dry wt., and concentrations in the overlying air averaged 9.2 and 16.9 ng/m^3 in February and September. These are 2 - 3 orders of magnitude higher than atmospheric levels measured at stations of the Integrated Atmospheric Deposition Network (IADN) on the Great Lakes (Cortes et al., 1998; Hillary et al., 1998; Hoff et al., 1996). In British Columbia, Finizio et al. (1998) found that concentrations of HCHs, chlordanes, and DDTs were highest in air samples collected 5 cm above a farm soil containing these residues, and decreased by a factor of 2-3 over a height of 140 cm.

Toxaphene was heavily used in the southern U.S. until the early 1980s, and high atmospheric concentrations have been historically found in the region (Rice et al., 1986; Bidleman et al., 1998a). In 1994-95, about ten years after deregistration of toxaphene, the average concentration of toxaphene in ambient air of South Carolina was still 180 pg/m^3 (Bidleman et al., 1998a), about 5 - 30 times higher than values reported near the Great Lakes during the early and mid-1990s (Jantunen et al., 1998; McConnell et al., 1998a; Shoeib et al., 1999). Subsequent investigations found that substantial residues of toxaphene remain in the soils of Alabama cotton fields, and a fugacity model predicted emissions of toxaphene from the soil (Harner et al., 1998a,b). Episodes of high toxaphene concentrations in the Great Lakes region have been associated with air trajectories arriving from the southern U.S. (Hoff et al., 1992b, 1993b; McConnell et al., 1998a). Atmospheric deposition was the dominant source of toxaphene to the Great Lakes in the past (Swackhamer et al., 1999; Voldner and Schroeder, 1989) and transport from old sources appears to be continuing today.

Emissions modelling (e.g., Scholtz et al., 1997; Harner et al., 1998a; van Jaarsveld, 1996) should be able to provide estimates of pesticide volatilization from old reservoirs on a regional scale, but at the present time this approach is limited by the meagre database for OCs in soils and vegetation, both in agricultural and background areas. Table I presents some data on OC pesticides in soils with an emphasis on measurements from the late 1980s and 1990s. The list is not intended to be comprehensive, but rather to illustrate the wide range of concentrations. Even in a particular region, pesticide residues in agricultural soils vary by several orders of magnitude. In a recent survey of forty farms in the midwestern United States, concentrations of ΣDDT in the top 15 cm of the soil ranged from <0.5 - 11800 ng/g dry wt., with arithmetic and geometric means of 395 and 9.6 ng/g (Aigner et al., 1998). Similar variability was found for soils in Alabama (Harner et al., 1998b). Because of this inhomogeneity, difficulties arise in assessing the magnitude of the soil reservoir and in selecting representative soil concentrations for input to regional volatilization models. DDT is also held in the soils of forests that were sprayed for insect control, as indicated by a survey of residues in Maine forests (Dimond and Owen, 1996) (Table 1). Concentrations of ΣDDT in 1993 were about 25% of those measured in 1967. About 60% of the ΣDDT in 1993 was accounted for by the metabolites DDE and DDD, compared to 7% in 1967. Other forests that were sprayed with DDT in the past include those in Washington and Oregon (Orgill et al., 1976) and eastern Germany (Eriksson et al., 19879 Jensen et al., 1992). No modelling nor experimental estimates have been made of the potential for these residues to be re-emitted.

3.1.6. *Chemical Markers for Differentiating "Old" and "New" OC Pesticides*

Ratios of the parent compounds p,p'-DDT and o,p'-DDT to their DDE metabolites have been used to infer pesticide sources and the age of the residues. Rapaport et al. (1985) found that ΣDDT residues in the upper slices of peat cores from the Great Lakes region and eastern Canada contained a high proportion of parent DDTs, leading to the hypothesis that "new" DDT continued to be atmospherically transported from Mexico, Central America and Caribbean countries after its 1972 ban in the U.S. Measurements of airborne pesticides in Belize, Central America during 1995-96 showed that concentrations of DDT, aldrin and dieldrin were typically 10 - 50 times higher than Great Lakes values (Alegria et al., 1999). Ratios of DDT/DDE in Belize air averaged 1.2 in winter and 6.3 in summer (Alegria et al., 1999), compared to 0.6 - 0.9 at IADN stations on the Great Lakes (Cortes et al., 1998; Sweet et al., 1996) and 0.3 - 0.4 in the Arctic (Halsall et al., 1998a). These observations suggest that the DDT residues in background air of the Northern Hemisphere are likely coming from a mixture of old and new sources. A problem with DDT/DDE ratios is that they are quite variable in soils. The DDT/DDE ratio in most farms surveyed in the midwestern U.S. was 0.5 - 2.0, but occasionally ≥ 5 (Aigner et al., 1998). Some soils in the southwestern U.S. contain a high proportion parent DDT due to their poor metabolic capability (Hitch and Day, 1992).

The proportion of γ-HCH/α-HCH has been suggested as a marker for atmospheric transport of lindane (pure γ-HCH) vs. the technical HCH product that contains about 70% α-HCH, 15% γ-HCH and other isomers. Canada, the U.S. and European countries use lindane. Large quantities of technical HCH were used in Asia during the 1970s and 80s, and at a reduced rate into the 1990s (Li et al., 1998a,b). The atmospheric signal consists of lindane superimposed on a background of technical HCH, and the γ-HCH/α-HCH ratio varies depending on the proportion of lindane transport. Episodic transport of lindane from central Europe to Svalbard (Oehme, 1991; Oehme et al., 1996) and southern Norway (Haugen et al., 1998) can be identified by the greater proportion of γ-HCH. A springtime peak in the γ-HCH/α-HCH ratio was noted in Québec which accompanied tillage of the soil and planting of lindane-treated seed (Poissant and Koprivnjak, 1996). Spring maxima in this ratio were also found in southern Ontario (Hoff et al., 1992a) and at arctic stations (Halsall et al., 1998a). A difficulty with using the γ-HCH/α-HCH ratio is that the two isomers are removed from the atmosphere at different rates during transport. This has been attributed to possible isomerization of γ- to α-HCH (Oehme, 1991) (although there is no evidence that this takes place in the gas phase) and to preferential deposition of γ-HCH by air-sea gas exchange (Iwata et al., 1993a,b).

A new approach to investigating pesticide emission from old sources uses the enantiomers of chiral pesticides as tracers (Aigner et al., 1998; Bidleman et al., 1998b; Bidleman and Falconer, 1999; Finizio et al., 1998). Most chiral pesticides are manufactured as a 1:1 (racemic) mixture of enantiomers, although a few are sold in single-enantiomer form. The proportion of enantiomers is not changed by abiotic reactions (hydrolysis, photolysis) nor transport processes (leaching, volatilization, atmospheric deposition). However enzymatic reactions in soil, water and biota are frequently enantioselective, leading to non-racemic residues of the parent compound and its metabolites. When these residues volatilize from soil and water, they carry the distinctive enantiomer signatures which can be used to characterize the source. In recent years it has become possible to determine the individual enantiomers of pesticides using capillary gas chromatography (GC) on chiral cyclodextrin stationary phases. Analytical methods and applications to OCs in biota have been reviewed by Vetter and Schurig (1997).

Enantioselective breakdown of OC pesticides has been reported in agricultural soils of the midwest "Cornbelt" in the United States (Aigner et al., 1998) and the Fraser Valley of British Columbia, Canada (Falconer et al., 1997;

TABLE I

Organochlorine pesticides in soils, ng/g dry wt[a].

Location	Year of Sampling	Category	ΣDDTs	ΣHCHs	ΣChlordanes	Hept. Epoxide	Dieldrin	Toxaphene	Reference[e]
U.S. (Maine)	1967	Forest (s)[b]	990 - 3872						1
	1967	Forest (ns)	86 -110						
	1973	Forest (s)	330 - 5900						
	1976	Forest (s)	280 - 4640						
	1976	Forest (ns)	20 - 70						
	1983	Forest (s)	808 - 3426						
	1993	Forest (s)	270 -1898						
	1993	Forest (ns)	nd[c] -11						
U.S. (Alabama)	1995	Agric.	nd - 2500	nd - 3.2	nd - 5.3	nd - 1.0	nd - 23	2 - 2420	2
U.S. (Ohio)	1996	Agric.	nd - 11800	nd - 0.3	nd - 752	nd - 121	nd - 4250		3
U.S. (Indiana)	1996	Agric.	nd - 214	0.07 - 1.2	0.4 - 364	nd - 40	nd - 69		3
U.S. (Illinois)	1996	Agric.	1 - 40	nd - 0.15	0.13 - 6.3	0.3 - 12	nd - 13		3
U.S. (Calif.)	1994	Agric.	1900 - 4100						4
	1985	Agric.	1 - 26700						5
Canada (Ontario)	1991	Agric.	1 - 70000	nd - 2.1	nd - 6.5	nd - 2.7	0.6 - 39		6

TABLE I (cont.)
Organochlorine pesticides in soils, ng/g dry wt[a].

Location	Year of Sampling	Category	ΣDDTs	ΣHCHs	ΣChlordanes	Hept. Epoxide	Dieldrin	Toxaphene	Reference[e]
Canada (Brit. (Columbia)	1991	Agric.	10 - 7162	nd - 899	nd - 1800	nd - 336	nd - 1280		7
Canada (other provinces)	1989	Agric.	1 - 433	0.4 - 1.4	0.16 - 2.9	nd - 3.1	nd - 4.1		6
Netherlands	1973	Agric.	3500 - 59 000	320			770 - 7300		8
	1989	Agric.	2090 - 35 000	10			430 - 4830		
Vietnam	1990	Agric.(rp)[d]	10 - 1300	0.2 - 55					9
	1990	Forest	0.7 - 44	0.1 - 0.2					
	1991	Agric.(rp)	1.0 - 290	0.09 - 2.3					10
	1991	Non-agric.	0.25 - 38	0.09 - 2.1					
Thailand	1988	Agric.(rp)	0.5 - 98	0.07 - 1.6					9
Taiwan	1990	Agric. (rp)	5.6 - 78	1.3 - 1.9					9
	1990	Roadside	5.0 - 31	0.85 - 4.8					
India	1989	Agric.(rp)	0.9 - 4.3	1.1 - 1100					11
	1989	Agric (rp)	0.85 - 2200	0.42 - 280					12
	1988	Agric.	2.0 - 4400	2.9 - 86 000					
	1989	Roadside	16 - 190	0.55 - 27					

TABLE I (cont.)

Organochlorine pesticides in soils, ng/g dry wt[a].

Location	Year of Sampling	Category	ΣDDTs	ΣHCHs	ΣChlordanes	Hept. Epoxide	Dieldrin	Toxaphene	Reference[e]
New Zealand	1977	Pasture	2500 - 4500						13
	1983	Pasture	1200 - 3100						
	1989	Pasture	1200 - 2300						
Russia	1992	Agric.	0.34 - 28	0.043 - 16	nd - 0.009				14
China	1980	Agric.	4 - 9000	31 - 3628					15,16

a) One or more sites, depending on study. ΣDDT may include p,p'-DDT, o,p'-DDT, p,p'-DDE and p,p'-DDD; ΣHCHs may include α-HCH and γ-HCH (lindane); Σchlordane = cis- and trans-chlordane.

b) s = forests sprayed with DDT between 1960-67, ns = not sprayed.

c) nd = not detected.

d) rp = rice paddy

e) 1 = Dimond and Owen, 1996; 2 = Harner et al., 1998b.; 3 = Aigner et al., 1998; 4 = Spencer et al., 1996; 5 = Mischke et al., 1985; 6 = Webber and Wang, 1995; 7 = Szeto and Price, 1991; 8 = Martijn et al., 1993; 9 = Thao et al., 1993a; 10 = Thao et al., 1993b; 11 = Ramesh et al., 1991; 12 = Kawano et al., 1992; 13 = Boul et al., 1994; 14 = Iwata et al., 1995; 15 = Li et al., 1998a; 16 = Li et al., 1999.

Finizio et al., 1998). Compounds investigated were o,p'-DDT, chlordanes, heptachlor, α-HCH and the metabolites HEPX and oxychlordane. Preferential degradation of (+)trans-chlordane and (-)cis-chlordane in Cornbelt soils is shown in Figure 1. Average enantiomer ratios (ER = (+)/(-) enantiomer) of trans- and cis-chlordane in soils of 23 farms were 0.70 $\pm$ 0.12 and 1.21 $\pm$ 0.14 respectively (Aigner et al., 1998). Air samples taken within a few tens of cm above farm fields showed that the enantiomer signatures in the soil were expressed in the overlying air (Falconer et al., 1998; Finizio et al., 1998; Leone, 1998).

Ambient air samples from Great Lakes IADN stations contained non-racemic chlordanes with ER values lying between racemic (ER = 1.00) and soil values (Bidleman and Falconer, 1999; Bidleman et al., 1998b; Falconer et al., 1998; Leone, 1998; Ulrich and Hites, 1998; Wiberg et al., 1997). Chlordanes in Great Lakes air therefore appear to be derived from a mixture of non-racemic and racemic sources. The non-racemic contribution is probably from regional soils (see above) and/or revolatilization of chlordanes from the lakes themselves. Racemic sources include atmospheric transport of fresh chlordane and release of chlordane from building foundations that were treated for termite control. To investigate this further, a survey of chlordane in the air of homes was carried out (Bidleman et al., 1998b; Falconer et al., 1998; Wiberg et al., 1997). Indoor concentrations of chlordane were elevated by factors of 100 - 1000 over ambient levels and the ER values were racemic.

Concentrations and ERs of HEPX and α-HCH were determined in air samples collected from 5 - 140 cm above soil at a British Columbia farm (Finizio et al., 1998). Concentrations of both OCs were highest near the soil surface and decreased with height. ER values of these OCs in the soil were 1.35 (α-HCH) and 1.38 (HEPX). The ER of α-HCH in air decreased from 1.35 at 5 cm to 1.19 at 140 cm, suggested some mixing of soil-derived α-HCH with advected α-HCH having a more nearly racemic composition. By contrast, the ER profile of HEPX varied from only 1.41 - 1.45 over the same heights, indicating that emission from the farm soil (or other soils with similar HEPX residues) supplied most of the HEPX to the air. Surveys of HEPX in ambient air from the Great Lakes region and the southern U.S. found that the HEPX was non-racemic in all cases (Bidleman et al., 1998c; Ulrich and Hites, 1998). Thus the source of HEPX to the atmosphere appears to be enantioselective epoxidation of heptachlor in soil followed by volatilization of HEPX, rather than achiral photolysis of heptachlor.

3.1.7. *Air-Surface Exchange as a Buffer of Atmospheric Concentrations*

Re-emission of POPs from soil, vegetation and water bodies exerts a major control on atmospheric concentrations. Seasonal (Cortes et al., 1998; Halsall et al., 1998; Hillery et al., 1997; Hoff et al., 1998b; Wania et al., 1998a) and even diurnal (Hornbuckle and Eisenreich, 1996; Lee et al., 1998; Wallace and Hites, 1996) cycles of PCBs and pesticides in ambient air occur in response to changes

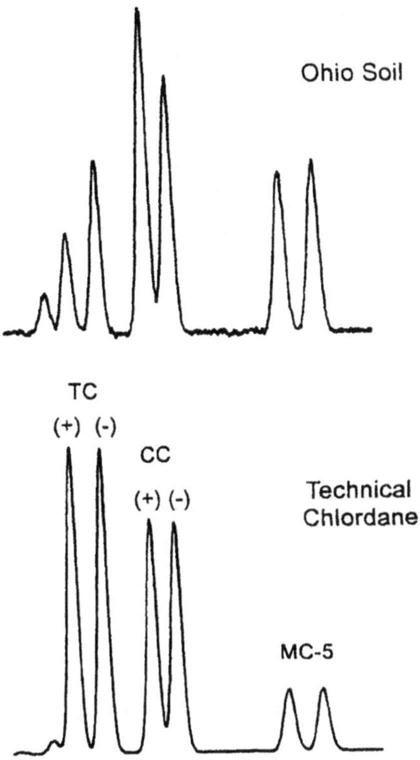

Figure 1. Gas chromatographic separation of the enantiomers of trans-chlordane (TC), cis-chlordane (CC) and another octachlordane MC-5 in a technical chlordane standard and an extract of soil from an Ohio farm. Data from Aigner et al. (1998).

in air temperature and, in some cases, relative humidity (Hornbuckle and Eisenreich, 1996). The thermodynamics of air-surface exchange and the relative contribution of local exchange vs. advective transport in governing ambient concentrations have been discussed by Hoff et al., 1998b, Pankow, 1993, and Wania et al., 1998a. The partial pressure (P) of a semivolatile compound in air is related to the enthalpy of air-surface exchange (ΔH_{ex}) by a form of the Clausius-Clapeyron equation:

$$\text{Ln } P = -\Delta H_{ex}/RT + b_{ex} \qquad (12)$$

In the case where air-surface exchange is the sole control of the partial pressure in air, ΔH_{ex} is expected to be bounded on the upper end by the enthalpy of vaporization (ΔH_{vap}) or octanol-air partitioning (ΔH_{oa}) for exchange with soil and vegetation, and on the lower end by the Henry's law constant (ΔH_{aw}) for exchange with water surfaces. It is possible for ΔH_{ex} to be greater than these physical partitioning enthalpies if stronger interactions take place, for example adsorption to mineral surfaces and particles from combustion sources (Goss and Eisenreich, 1997; Goss and Schwartzenbach, 1998; Pankow, 1993).

Values of ΔH_{ex} derived from field measurements are often lower than enthalpies for physical partitioning because partial pressures are controlled by both local air-surface exchange and advective transport (Hoff et al., 1998b; Wania et al., 1998a). The former depends on local air temperature, whereas the transport component is related to climatic and application conditions in the source region, which may be distant from where air samples are collected. Hoff et al. (1998b) examined slopes of eq 12 for trans-chlordane sampled on the shoreline of Lake Ontario over a temperature range of -17 to 26°C. The slope were flattest in the cold temperature range and steepened when points from the warm sampling days were included. Similar results were found for PCB-52, a tetrachlorobiphenyl, sampled at a terrestrial site in southern Ontario (Hoff et al., 1992b, 1998b). Models of this "hockey-stick" behavior (Hoff et al., 1998b; Wania et al., 1998a) indicate that emission of POPs to the atmosphere from air-surface exchange increases with temperature; thus the slope of eq 12 plots can be used as a rough indicator of exchange vs. advective contributions to the total atmospheric levels.

Hoff et al. (1998b) also found that the slopes for trans-chlordane were flatter when air trajectories passed over Lake Ontario and steeper when transport was from over land. Both slopes were lower than expected from air-surface exchange alone, and reflect contributions from advective transport and exchange controlled by vapor pressure or K_{oa} (land, vegetation, steeper slope) and Henry's law (water, flatter slope).

Results from three years of air monitoring in the Arctic indicate that atmospheric concentrations of chlorinated pesticides are controlled largely by long-range transport, but air-surface exchange also plays some part. Eq 12 plots for the relatively volatile HCHs displayed no significant correlation with temperature, whereas the less volatile ones (chlordane, dieldrin, endosulfan and the metabolites oxychlordane and HEPX) were correlated at $p < 0.01$, although with ΔH_{ex} values lower than in temperate latitudes (Halsall et al., 1998; Hoff et al., 1998b).

One goal of air monitoring programs is to determine temporal trends. Such attempts are confounded by the often larger seasonal and diurnal variations in atmospheric concentrations with temperature. Pesticide (Cortes et al., 1998) and PCB (Hillery et al., 1997) concentrations were measured from 1991-95 at the Great Lakes IADN stations. The data were treated as a function of both temperature and time using an equation of the form:

$$\ln P = a_o + a_1(1/T) + a_2(\text{time}) \qquad (13)$$

and the parameters a_o, a_1 and a_2 were determined by multiple linear regression. With this approach, the authors were able to estimate half-lives for the decline of airborne PCBs and pesticides in the range of 3-12 years. These half-lives should be looked upon not as loss rates in the atmosphere *per se*, but rather as rates of disappearance in the reservoirs that supply these compounds to the atmosphere. "Virtual elimination" dates were estimated for the pesticides, based on the time when atmospheric concentrations were predicted to fall below instrumental detection limits using current IADN sampling and analytical methods. Such dates ranged from 2010 for p,p'-DDT to 2060 for HCB.

4. Airborne Pesticides in Remote Regions

Pesticides and other POPs can be carried to remote regions in a relatively short time, typically days, by episodic movement of air parcels (Ballschmiter, 1992). Transport of lindane from central Europe to Norway (Haugen et al., 1998) and the Arctic (Harner et al., 1999; Oehme et al., 1996) is an example. Continual deposition and re-evaporation of the chemical ("grasshoppering") also leads to migration to polar regions, as described by the global fractionation - cold condensation model of Wania and Mackay (1995, 1996). Ockenden et al. (1998b) obtained experimental evidence for global fractionation by deploying semipermeable membrane devices (SPMDs) as passive atmospheric monitors along a latitudinal gradient from the south of the U.K to the north of Norway (50 - 70°N). PCBs collected by the SPMDs showed a changing profile in which the more volatile congeners was enriched in the more northerly samples.

Atmospheric monitoring of OC pesticides and metabolites has been carried out at the land-based stations in the Arctic shown in Figure 2 (Barrie et al., 1997; Fellin et al., 1996; Halsall et al., 1998; Oehme et al., 1996). These measurements have been supplemented by shipboard expeditions to the Arctic Ocean and its regional seas (Harner et al., 1999; Macdonald et al., 1998; Jantunen and Bidleman, 1995, 1996) (Figure 2), temperate and tropical oceans (Iwata et al., 1993a; Schreitmüller and Ballschmiter, 1995), and the Southern Ocean and Antarctica (Bidleman et al., 1993; Kallenborn et al., 1998; Iwata et al., 1993a; Larsson et al., 1992). Maps showing the global distribution of HCHs, DDTs, chlordanes and PCBs in air and surface water are presented by Iwata et al. (1993a).

The baseline for the remote Northern Hemisphere is perhaps given best by the annual mean concentrations of OC pesticides at four circumpolar arctic stations (Figure 3, Halsall et al., 1998). Compounds in highest abundance (10-100 pg/m^3) are the HCHs, while the lowest (0.2 - 0.5 pg/m^3) are the DDTs. Pesticides having intermediate concentrations (0.5 - 5 pg/m^3) are chlordanes, endosulfans, methoxychlor and toxaphene. Stable metabolites found in arctic air

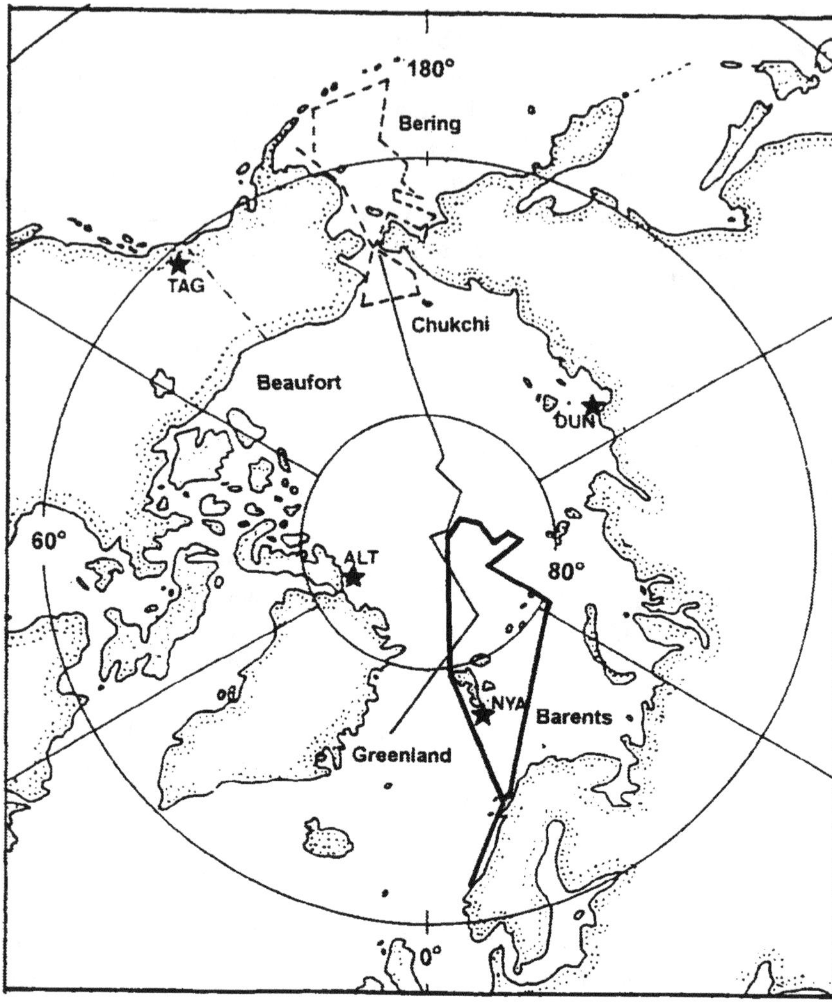

Figure 2. Land-based stations for measurement of POPs in arctic air, and expeditions for air-sea gas exchange measurements. ALT = Alert (Canada), TAG = Tagish (Canada), DUN = Dunai Island (Canada - Russia), NYA = Ny Ålesund (Norway). Cruises in the Arctic Ocean and regional seas: ----- 1993, —— 1994, ▬▬ 1996.

were HEPX, oxychlordane, pentachloroanisole and the DDEs. Other POPs identified were chloroveratroles, which are methylation products of chloroguaiacols, PCBs (Oehme et al., 1996; Stern et al., 1997), PAHs (Halsall et al., 1997) and PCNs (Harner et al., 1998c).

Time trends in the Arctic are difficult to establish because monitoring programs are fairly recent. Early measurements from Svalbard in 1981-84 detected seasonal changes in concentrations of HCHs and chlordanes, but the

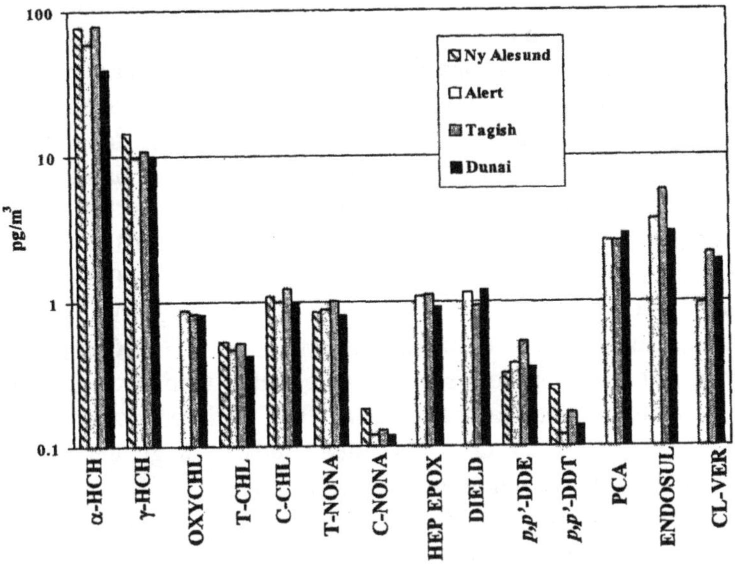

Figure 3. Organochlorine compounds in arctic air, measured at land-based monitoring stations. HCH=hexachlorocyclohexane, T-CHL,C-CHL=trans-,cis-chlordane, T-NONA=trans-nonachlor, OXYCHL=oxychlordane, HEP EPOX=heptachlor epoxide, DIELD=dieldrin, PCA = pentachloroanisole, ENDOSUL=endosulfan, CL-VER=chloroveratroles. From Halsall et al. (1998).

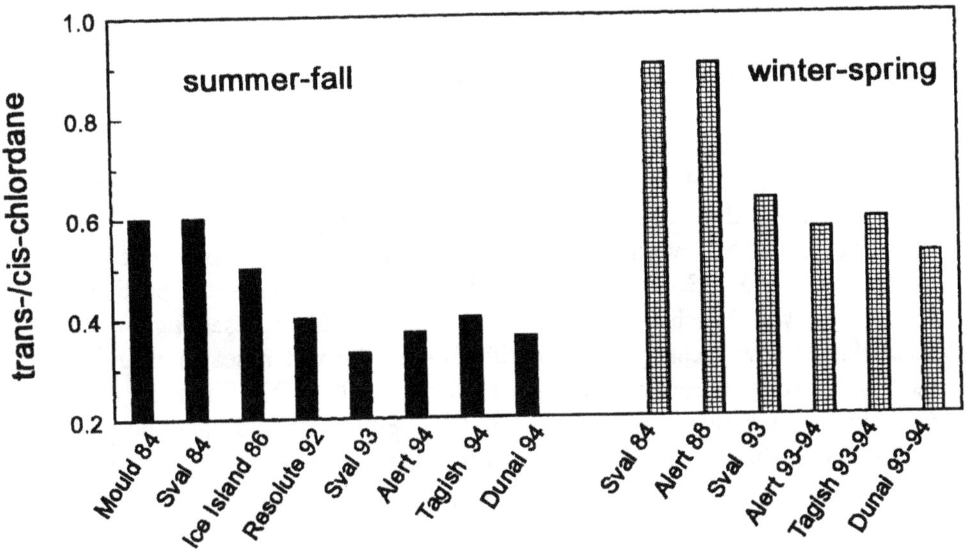

Figure 4. Ratios of trans-chlordane/cis-chlordane (TC/CC) in arctic air over a 10-year period. Data from Halsall et al., 1998; Oehme et al., 1996; Oehme, 1991; Bidleman et al., 1995b; Patton et al., 1989, 1991; Hoff and Chan, 1986. Stations are described in the original papers.

frequency of measurements was rather low and average concentrations were likely to be biased by transport episodes (Oehme et al., 1996). A greater frequency of monitoring at Svalbard (Oehme et al., 1996) and stations in the Canadian and Russian Arctic (Halsall et al., 1998) in the 1990s has enabled the seasonality of pesticide concentrations to be specified with greater confidence.

Clear seasonal variations in the ratio of trans-chlordane/cis-chlordane (TC/CC) were identified at all arctic stations (Halsall et al., 1998; Oehme et al., 1996). Average TC/CC ratios were 0.5 - 0.6 in winter, and 0.3 - 0.4 in summer. The winter-summer variability might be related to differences in the photochemical reactivity of the two isomers (Halsall et al., 1998; Oehme et al., 1996). Measurements in the 1980s also showed higher TC/CC ratios in winter than summer, but values in both seasons were higher than in the 1990s (Figure 4). Thus, in addition to the seasonal cycle of TC/CC, there appears to be a long-term trend toward a lower ratio. This might indicate a greater proportion of "recycled" chlordane in the atmosphere in recent years.

A strong downward trend in the concentration of α-HCH has been seen at arctic stations over the last two decades (Bidleman et al.,1995a; Li et al., 1998b). Summertime concentrations of α-HCH were 800 pg/m^3 in 1980 and $\leq$ 100 pg/m^3 in 1992-94. A decline of 10 pg/m^3 per year in α-HCH was also seen at the Lista air monitoring station in southern Norway between 1991-95 (Haugen et al., 1998). Li et al. (1998b) associated stepwise periods of α-HCH decline with the years when the People's Republic of China banned production and usage of technical HCH (1983), the former Soviet Union stopped using technical HCH (1990) and India eliminated its use on food crops (1990).

Transport of pesticides to remote environments is not restricted to OCs. A great number of modern agrochemicals are emitted into the atmosphere (Majewski and Capel, 1995). Currently used pesticides have been found in air along a transect of the Mississippi River from the southern to northern U.S. (Coupe et al., 1998; Majewski et al., 1998) and in precipitation from the central and northeastern U.S. (Goolsby et al., 1997). It is likely that some portion of these residues will reach remote areas. Compounds representing several chemical classes have recently been identified in air, fog water, seawater and the sea surface microlayer from arctic regions (Barrie et al., 1997; Chernyak et al., 1996, Rice and Chernyak, 1997) (Table II). Endosulfan, organophosphate pesticides and their oxon transformation products, and the fungicide chlorothalonil were found in air, snow and pine needles from the high Sierra mountains, apparently transported from California's Central Valley (Aston and Seiber, 1996; McConnell et al., 1998b; Zabik and Seiber, 1993). Chlorpyrifos and chlorothalonil were identified in tadpoles from a mountain lake in the Sierras (Datta et al., 1998). Increased concentrations of PCBs and pesticides in snow with altitude were found in the mountains of western Canada. Deposition increased with altitude, due to higher concentrations in snow and also to greater snowfall (Blais et al., 1998). The pesticides included endosulfan and lindane, which are currently used in Canada. Rawn et al. (1998) and Muir and Grift (1995) detected atrazine in air, precipitation and lake water in remote northwest Ontario lakes and Goolsby et al. (1997) found atrazine in precipitation collected on Isle Royale in Lake Superior.

The appearance of a chemical in the remote environment has been suggested as a screening criterion for long-range transport potential (Cowen-Ellsberry et al., 1999). For this reason, it is essential to monitor the polar regions and other remote areas for the appearance of modern agrochemicals and presently used industrial chemicals. Pesticides which are considered to be of low persistence in temperate climates are apt to be much more resistant to breakdown in cold regions. Based on hydrolysis rate constants published by Ngabe et al. (1993), the half-lives of α- and γ-HCHs in seawater at pH 8.1 are 0.4 - 0.5 y at 25°, but increase to 60 - 100 y at 0°. Half-lives of organophosphate pesticides in fresh water at warm (21 - 22°) and cold (4 - 6°) temperatures were reported by Frank et al. (1991) and Lartiges and Garrigues (1995). Half-lives of fifteen compounds at cold temperatures were 1.9 to 17 (average 6.5) times longer than those at warm temperatures.

5. Case Studies of Air-Water Gas Exchange for Pesticides

Loadings of pesticides into lakes and oceans take place by precipitation scavenging and dry particle deposition, and by gas exchange across the surface. Comparative loading estimates indicate that gas exchange dominates in many cases. This is true for OC pesticides and PCBs in the open ocean (Duce et al., 1991) and the Great Lakes (Hoff et al., 1996), PAHs and PCBs in Chesapeake Bay (Gustafson and Dickhut, 1997a; Nelson et al., 1998) and the insecticide chlorpyrifos in Chesapeake Bay (McConnell et al., 1997).

5.1. GREAT LAKES

The Canada - U.S. IADN program on the Great Lakes carries out continuous monitoring of air and precipitation for PCBs, OC pesticides, PAHs and trace metals. Less frequent measurements are made in Great Lakes water under separate programs. A synthesis of data through 1994 was made by Hoff et al. (1996) and Hillery et al. (1998) to estimate the relative loadings from various atmospheric processes. Precipitation and dry deposition of particles was most important for the metals and PAHs which are largely particulate, such as the benzopyrenes and benzofluoranthenes, while gas exchange dominated for PCBs, pesticides and 3-4 ring PAHs. On an annual basis, volatilization exceeded gas absorption for PCBs, HCB, DDE and dieldrin (Figure 5).

TABLE II.
"Modern" pesticides identified in the arctic environment

Compound	Use	Found In	Reference
chlorpyrifos	insecticide	seawater, ice, fog	Chernyak et al., 1996; Rice and Chernyak, 1997
terbufos	insecticide	seawater, ice, fog	Chernyak et al., 1996; Rice and Chernyak, 1997
endosulfan	insecticide	seawater, air	Chernyak et al., 1996; Rice and Chernyak, 1997 Jantunen and Bidleman, 1998 Hargrave et al., 1997 Bidleman et al., 1995b
chlorothalonil	fungicide	seawater, fog	Chernyak et al., 1996; Rice and Chernyak, 1997
metolachlor	herbicide	fog	Chernyak et al., 1996; Rice and Chernyak, 1997
atrazine	herbicide	air, ice	Chernyak et al., 1996; Rice and Chernyak, 1997
trifluralin	herbicide	seawater, fog air	Chernyak et al., 1996; Rice and Chernyak, 1997 Barrie et al., 1997

Other investigations in the Great Lakes for PCBs (Eisenreich et al., 1997; Hillery et al., 1997; Honrath et al., 1997; Hornbuckle et al., 1994,1995) and HCHs (Ridal et al., 1996, 1997) clearly indicated cycles of net deposition and net volatilization which depended on changes in atmospheric concentration and surface water temperature. This agrees with earlier assessments (reviewed by Bidleman and McConnell, 1995; Eisenreich et al., 1997 and Wania et al., 1998b). Volatilization from Lake Superior was estimated to remove 26,500 kg of PCBs from the lake between 1980-92, decreasing the concentration in the water column from 1.1 ng/L to 0.18 ng/L over this time (Eisenreich et al., 1997; Jeremiason et al., 1994). Mackay and Bentzen (1997) and Hillery et al. (1998) noted that OC compounds in the Great Lakes are close to being in air-water equilibrium and that further declines in water concentrations will not take place until atmospheric burdens are reduced.

Is there evidence, other than fugacity calculations, that volatilization of POPs is actually taking place? In the case of PCBs, Hornbuckle et al. (1993) found that the concentrations and congener distributions measured over the water of

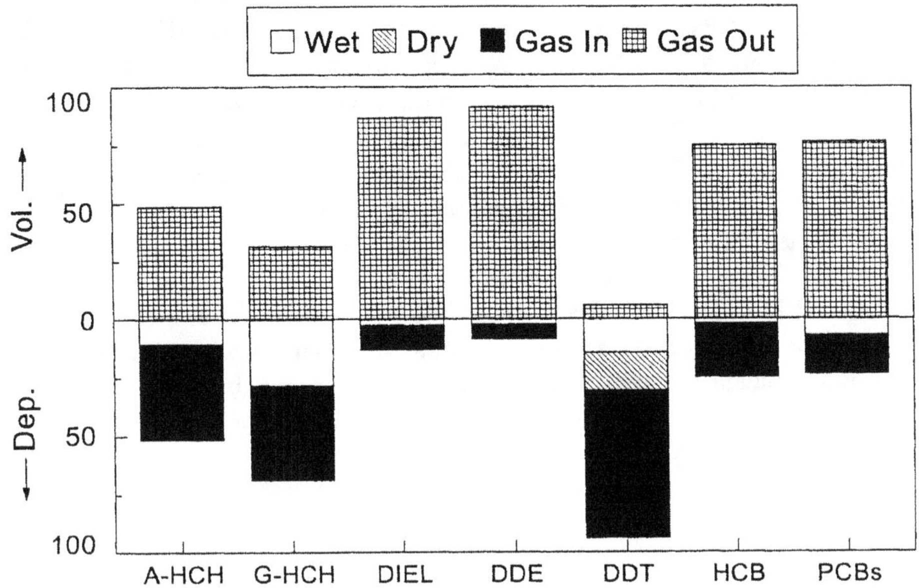

Figure 5. Atmospheric budget of OCs for Lake Ontario. The total mass of material crossing the air-lake interface (total loading) is apportioned into inputs from precipitation (Wet), dry particle deposition (Dry) and gas exchange (Gas In), and output from gas exchange (Gas Out). Data from Hoff et al.; 1996.

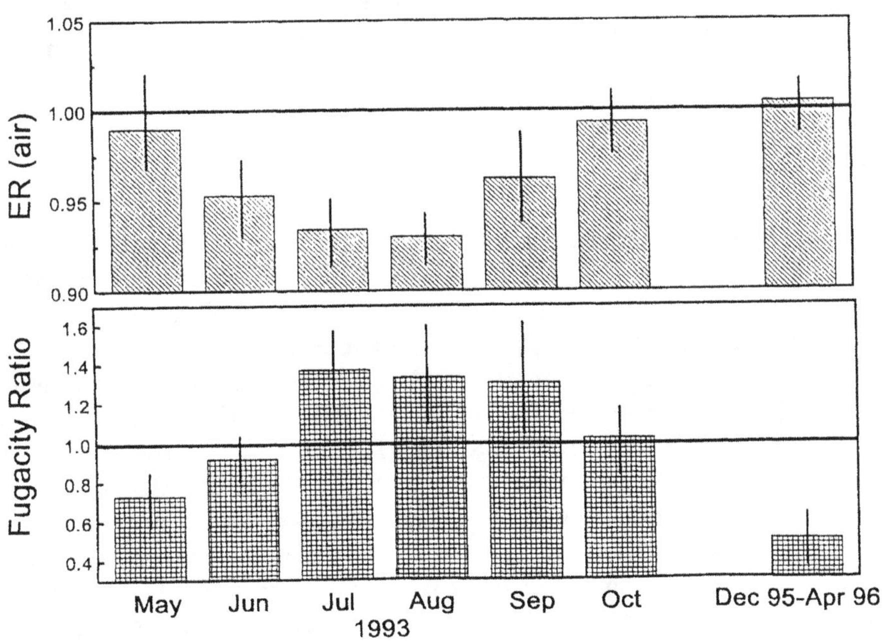

Figure 6. Enantiomer ratios (ER) and water/air fugacity ratios (FR) of α-HCH in Lake Ontario. FR values <1 and >1 indicate net deposition and volatilization, respectively. From Bidleman et al., 1998b and Ridal et al.; 1997.

Green Bay, Wisconsin were different from those at land-based stations a short distance away. Ridal et al. (1997) used the enantiomers of α-HCH to follow air-water exchange in Lake Ontario. When manufactured, the two mirror-image (+) and (-) enantiomers of α-HCH are equally present with an ER = 1.00 (racemic). Selective microbial breakdown of (+)α-HCH in lake water reduced the ER to 0.85. Volatilization in July was evident from the appearance of non-racemic α-HCH in the air above the lake, but in October the α-HCH in air was racemic. When followed over a seasonal cycle, the α-HCH in boundary-layer was racemic during spring, fall and winter (ER = 1) and depleted in the (+) enantiomer (ER < 1) in summer. These ER trends followed the cycle of net deposition and volatilization, as predicted from water/air fugacity ratios (Figure 6). A mass balance based on ER values in water and boundary-layer air indicated that up to 40% of the α-HCH in air over the lake in summertime was due to revolatilization.

5.2. LAKE BAIKAL

Two investigations of gas exchange for OC pesticides and PCBs were done in Lake Baikal, Siberia during June 1991 (McConnell et al., 1996) and May 1992 (Iwata et al., 1995). The net flux directions found in these studies were: deposition of HCHs, DDTs and toxaphene, volatilization of light PCBs, heavy PCBs close to steady state, volatilization or steady state for chlordanes, and either volatilization or deposition of HCB, depending on the study. Although these were only one-month “snapshots” in each case, they pointed out the significance of the gas exchange pathway for this large lake.

5.3. CHESAPEAKE BAY

Chesapeake Bay is the largest estuary in the United States and has recently been the site of several gas exchange studies for PAHs (Gustafson and Dickhut, 1997a; Nelson et al., 1998), PCBs (Nelson et al., 1998) and chlorpyrifos (McConnell et al., 1997). The situation in the bay is complex. Large spatial gradients of temperature, salinity and POPs concentrations occur in the water column, and on- and off-shore winds alternately transport clean and polluted air over the bay. In the McConnell et al. (1997) study, atmospheric loadings of chlorpyrifos were determined from spring to fall, 1993 at five stations located from the north to south of the bay. Gas exchange fluxes exceeded other atmospheric deposition mechanisms at nearly all sites. In March and April, riverine inputs were the most important source of chlorpyrifos to the bay and during this time net volatilization was found at all stations. Between June and September, riverine contributions decreased and higher atmospheric concentrations resulted in net deposition of chlorpyrifos. The study has special significance in that it shows the importance of gas exchange for a modern pesticide.

5.4. TEMPERATE OCEANS

Duce et al. (1991) estimated atmospheric loadings of OC pesticides and PCBs to the world's oceans, using atmospheric concentration data typical of the mid 1970s - early 1980s and considering precipitation scavenging, dry particle deposition and gas exchange. At that time there were few measurements of pesticides and PCBs in ocean surface water, and no realistic estimates of revolatilization could be made. The gas exchange calculations were made assuming that seawater was 90% saturated with respect to atmospheric concentrations; that is, the gas exchange fluxes were 10% of the maximum values. Precipitation and gas exchange accounted for nearly all of the loadings, with dry particle deposition negligible (Figure 7).

Large-scale surveys of OC pesticides and PCBs in ocean air and water were made between 1980-90. A study in 1989-90 covered the Atlantic, Pacific and Indian oceans in the Northern Hemisphere, and the Southern Ocean between Australia and Antarctica (Iwata et al., 1993a). Their distributions of DDT and HCHs in air and surface water showed features which support the "global fractionation" and "cold condensation" hypotheses of Wania and Mackay (1996). Although DDT was found in all locations, concentrations in both air and water were far higher near India and southeast Asia and declined rapidly away from these source regions. HCHs in air were highest in a band extending from India along the east Asian seas between Japan, China and Russia. The heaviest usage of technical HCH during the 1980s was in the People's Republic of China, India and the former Soviet Union (Li et al., 1998a,b). However unlike DDT, the highest concentrations of HCHs in surface water were <u>not</u> found in the seas near sources, but further north in the Bering and Chukchi seas between Alaska and Russia where partitioning from air to water was enhanced by cold surface ocean temperatures.

The Iwata et al. (1993a) study estimated gas exchange fluxes of pesticides and PCBs from paired measurements in air and surface water. Fluxes for two limiting cases were calculated by assuming that a) all of the POPs in surface water were truly dissolved and available to participate in gas exchange and b) all of the POPs were bound to particles or colloidal material and unable to revolatilize. For chlordanes and PCBs, large differences in flux values were estimated for the two situations. Air-to-sea deposition of chlordanes in all ocean regions occurred under case (b), but when the fugacity in surface water was taken into account (case a), chlordanes were near air-water equilibrium or volatilizing in the temperate and tropical oceans and depositing only in the colder regions.

5.5. ARCTIC OCEAN

Several studies of HCH gas exchange in the Arctic Ocean and regional seas have been carried out since 1988. Results and maps of the areas covered are given in Macdonald et al., 1998; Falconer et al., 1995; Harner et al., 1999; Hinckley et al., 1991; Iwata et al., 1993a; and Jantunen and Bidleman, 1995;1996. See also Figure 2. Pacific water entering the Arctic Ocean from the Bering-Chukchi seas contains 2.5 ng/L of α- + γ-HCH. Somewhat lower concentrations, averaging 0.5 - 1 ng/L, occur in Atlantic water which enters via the Barents Sea (Barrie et al., 1997; Harner et al., 1999). Levels in the surface water of the central Arctic Ocean are 2.5 - 3.5 ng/L near the pole and 5 - 6 ng/L in the Beaufort Sea north of Alaska and in the Canadian Archipelago.

Monitoring of HCHs in the arctic atmosphere has been carried out over the last two decades, from stations operated by Norway (since 1980) and Canada (since 1992) and from occasional shipboard measurements (Harner et al., 1999; Hinckley et al., 1991; Iwata et al., 1993a; Jantunen and Bidleman, 1995;1996) (Figure 3). Atmospheric concentrations have shown stepwise decreases over this time, with a three-fold drop since ~1990 (Bidleman et al., 1995a; Jantunen and Bidleman, 1995; Li et al., 1998b). This decline has brought about profound changes in the water/air fugacity ratios (Figure 8). The net exchange direction of α-HCH in the Bering-Chukchi seas and Canada Basin has reversed, from deposition in the 1980s to volatilization in the 1990s, and γ-HCH has gone from net deposition to air-water equilibrium (Falconer et al., 1995; Jantunen and Bidleman, 1995, 1996). In the eastern Arctic Ocean, α-HCH is close to equilibrium and γ-HCH is still undergoing net deposition (Harner et al., 1999). As in the Great Lakes, the enantiomers of α-HCH provided direct evidence of air-water exchange. Depletion of (+)α-HCH in surface water was found over most of the Arctic Ocean except in the Bering-Chukchi seas where (-)α-HCH was preferentially metabolized. The enantiomer signatures in the overlying air from each region showed the same depletions as those in the surface water, indicating sea-to-air transfer (Jantunen and Bidleman, 1996; Harner et al., 1999).

A year-long investigation of air-water gas exchange in the Arctic Ocean was conducted at Resolute Bay, NWT (74°42'N, 94°50'W) in 1993 (Hargrave et al., 1997). Seasonal changes in the concentrations of organochlorine pesticides were found in the upper 50 m of the water column. The lack of concentration gradients between 1 m and 50 m suggested that advection of polar mixed layer water flowing southward from under the polar ice cap was an important factor in controlling seawater concentrations. Net volatilization was found during the ice-free period (June-September) for HCB, HCHs and dieldrin, whereas the net flux direction was depositional for toxaphene and endosulfan. The gas exchange fluxes were influenced by the combined effects of changes in atmospheric concentrations and water mass advection. Potential removal of the pesticides by

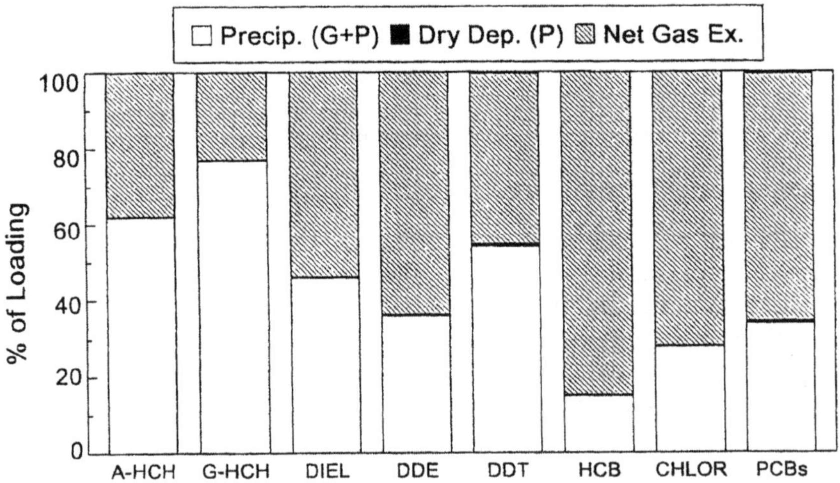

Figure 7. Atmospheric loadings of OCs to the world's oceans by precipitation (gas + particle, G+P), dry particle (P) deposition and net gas exchange, representative of the early 1980s. Based on calculations of Duce et al. (1991).

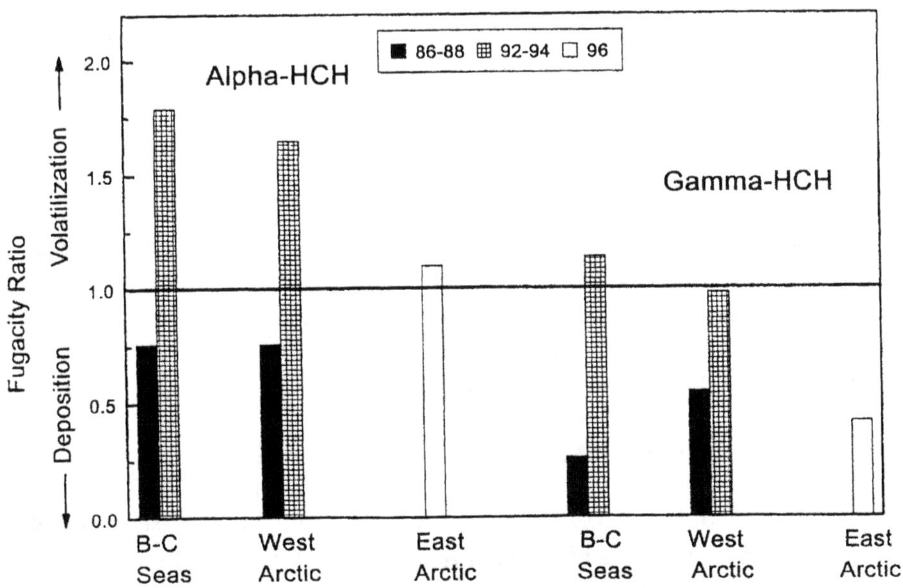

Figure 8. Water/air fugacity ratio and net flux direction of HCHs in arctic waters (B-C = Bering and Chukchi seas). Data from Jantunen and Bidleman, 1995, 1996 and Harner et al., 1999.

sedimenting particles was estimated to be less important than atmospheric processes in modifying concentrations in the surface layer.

6. Factors Influencing the Lifetime of Pesticides in Air

6.1. PARTICLE-GAS PARTITIONING AND ATMOSPHERIC DEPOSITION

In temperate climates, most pesticides are transported largely in the gas phase with a small, but important, fraction sorbed to atmospheric particles. The particle/gas distribution influences atmospheric removal by precipitation, dry particle deposition and gas exchange (Bidleman, 1988; Hoff et al., 1996, Majewski and Capel, 1995; Wania et al., 1998b). The fraction of a semivolatile compound on atmospheric particles (ϕ) is related to the particle/gas partition coefficient ($K_p = C_p/C_g$) and the total suspended particle (TSP, $\mu g/m^3$) load by:

$$\phi = K_p[TSP]/(1 + K_p[TSP]) \qquad (14)$$

C_p and C_g are the concentrations in the particle phase (ng/µg particles) and gas phase (ng/m^3 air), and K_p has units of $m^3/\mu g$.

Methods available for estimating the particulate fraction from physicochemical properties include the Junge-Pankow adsorption, Pankow absorption, and octanol-air partition coefficient models. The adsorption model (Pankow, 1987) is most frequently used:

$$\phi = c\theta/(P_L + c\theta) \qquad (15)$$

where P_L is the saturation liquid-phase vapor pressure of the adsorbing substance (Pa) and θ is the aerosol surface area per unit volume of air (cm^2/cm^3 air). The parameter c is related to the heat of desorption from the particle surface (Q_d, J/mol), the heat of vaporization of the compound (Q_v, J/mol) and the moles of adsorption sites on the aerosol (N_s, mol/cm^2):

$$c = 10^6 RTN_s \exp[(Q_d - Q_v)/RT) \qquad (16)$$

where R = 8.314 J/mol K = 8.314 Pa m^3/mol K. The value of c is usually taken as 17.2 Pa cm. Typical values of θ, taken from Whitby (1971) and compiled by Bidleman (1988), are 1.1×10^{-5} (urban air), 3.5×10^{-6} (rural air), and 4.2×10^{-7} (continental background air). Equipartitioning of a pesticide between the particle and gas phases in rural air is expected to occur at $P_L = 6 \times 10^{-5}$ Pa, the vapor pressure of p,p'-DDT at 7 °C (Hinckley et al., 1990). Combining eq 14 and 15:

$$\log K_p = \log(c\theta/TSP) - \log P_L \qquad (17)$$

A plot of log K_p vs. log P_L has a theoretical slope = -1 and an intercept related to aerosol specific surface area ($A_{tsp} = 10^6\theta/TSP$, $cm^2/\mu g$). Problems with the adsorption model include difficulties in measuring θ or A_{tsp} and uncertainties in the parameter c. Eq. 17 has been used to describe experimental measurements of particle/gas distribution in many studies (reviewed by Bidleman and Harner, 1999; Bidleman et al., 1999; Falconer and Bidleman, 1997; Goss and Schwartzenbach, 1998; Pankow and Bidleman, 1992). The plots are well correlated, but the slopes are often different from the expected value of -1, possibly due to sampling artifacts, non-equilibrium effects, or thermodynamic factors (Pankow and Bidleman, 1992). Goss and Schwartzenbach (1998) and Simcik et al. (1998) argued that slopes differing from -1 do not necessarily mean that the aerosols are out of equilibrium with the gas phase.

The Pankow absorption model (Pankow, 1994a,b) considers partitioning between the gas phase and a liquid-like film on the aerosol:

$$K_p = 10^{-6}RTf_{om}/M_{om}\gamma_{om}\,P_L \qquad (18)$$

where f_{om} is the fraction of the aerosol that consists of absorbing organic matter having molecular weight M_{om}, and γ_{om} is the activity coefficient of the sorbing compound in the film on a Raoult's law basis ($\gamma_{om} \rightarrow 1$ as the mole fraction in the film $\rightarrow$ 1). A portion of this organic matter may be "secondary organic aerosol", which is formed by oxidation of hydrocarbons and is therefore moderately polar (Yu et al., 1998; Jang et al., 1997; Pankow, 1994b). The main uncertainty in applying eq 18 is that γ_{om} is often not known nor easily measured and is likely to vary substantially among classes of compounds. Liang et al. (1996) estimated that γ_{om} was 1.8 for n-alkanes and 12.7 for PAHs partitioning into environmental tobacco smoke (ETS) aerosols. In a subsequent paper (Liang and Pankow, 1997), the authors found that K_p values of PAHs were similar for ETS and secondary aerosol generated from gasoline vapor oxidation. Partitioning of n-alkanes was also similar to the two aerosol types, but weaker than for PAHs. Jang et al. (1997) estimated γ_{om} for different combinations of aerosol compositions and absorbing solutes using group contribution methods. Such methods have not been applied to pesticides however.

The octanol-air partition coefficient model (Finizio et al., 1997; Harner and Bidleman, 1998a) is:

$$K_p = 10^{-9}K_{oa}f_{om}(M_{oct}/M_{om})(\gamma_{oct}/\gamma_{om})/\rho_{oct} \qquad (19)$$

Assuming similar molecular weights for octanol and the organic matter on the aerosol ($M_{oct}/M_{om} \sim 1$) and the density of octanol (ρ_{oct}) = 820 kg/m^3:

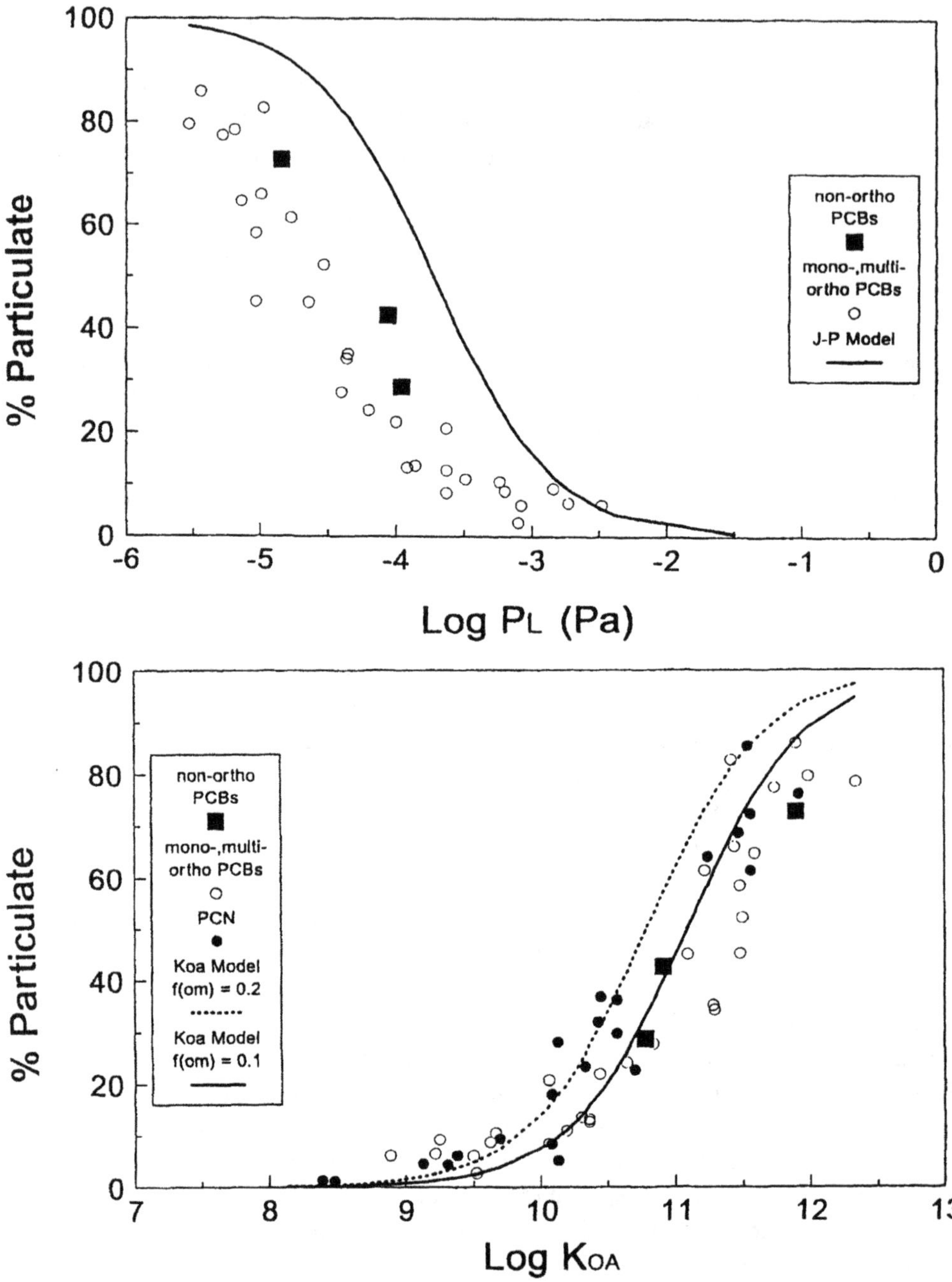

Figure 9. Comparison of the Junge-Pankow (J-P) model (eq 15, $\theta = 3.5 \times 10^{-6}$) and the K_{oa} model (eq 21, $\gamma_{oct}/\gamma_{om} = 1$, $f_{om} = 0.1\text{-}0.2$) for describing the percent of chlorinated aromatic compounds on urban aerosols. Solid and dotted curves are model calculations; points are observed values for PCBs having various numbers of ortho-substituted chlorines, and polychlorinated naphthalenes (PCNs). Modified from Harner and Bidleman (1998a).

$$K_p = 1.2 \times 10^{-12} K_{oa} f_{om}(\gamma_{oct}/\gamma_{om}) \qquad (20)$$

$$\log K_p = \log K_{oa} + \log f_{om}(\gamma_{oct}/\gamma_{om}) - 11.92 \qquad (21)$$

A plot of log K_p vs. log K_{oa} has a theoretical slope of +1 and an intercept related to the absorbing properties of the aerosol. Eq 20 or 21 are an improvement over eq 18 for two reasons. K_{oa} is directly measureable, whereas P_L for solid-phase compounds must be estimated from thermodynamic considerations, or gas chromatography. Uncertainties in doing the latter are discussed in the section on physicochemical properties. Also, the ratio γ_{oct}/γ_{om} is more likely to be constant within a compound class or among different classes than γ_{om} alone (Finizio et al., 1997; Pankow, 1998). Harner and Bidleman (1998a) showed that octanol-air partition coefficient approach (eq 21) was able to explain the observed enhancement of particle/gas partitioning for coplanar (non-ortho substituted) PCBs compared to multi-ortho PCBs that was not explained by vapor pressure (Figure 9). Partitioning of polychloronaphthalenes (PCNs) was also well described by the K_{oa} approach (Figure 9).

Few experimental measurements of particle/gas distribution have been made for pesticides, especially polar compounds. However a large body of information is available on other semivolatile compounds in urban and rural air, particularly polycyclic aromatic hydrocarbons (PAHs) and to a lesser extent PCBs and chlorinated dioxins and furans. Most of these measurements have been made using a high-volume air sampler in which air is drawn through a glass- or quartz-fiber filter to collect particles followed by a vapor trap (typically polyurethane foam or XAD resins). This technique gives the "apparent" phase distribution and is subject to volatilization and adsorption artifacts on the filter. Diffusion denuders have also been used for speciation of particulate and gaseous compounds to avoid these artifact problems. Reviews of particle/gas distributions from field studies have been published by Bidleman and Harner (1999), Bidleman et al. (1999), Falconer and Bidleman (1997), Pankow and Bidleman (1992) and a recent book is devoted to experimental and modelling approaches for particle/gas speciation (Lane, 1999).

The particulate fraction is important for some currently used herbicides (Hawthorne et al., 1996; Waite et al., 1998). Data for non-polar OCs such as DDT, chlordane, HCHs and HCB suggest that they partition to urban air particles similar to PCBs (Bidleman et al., 1986, 1999; Cotham and Bidleman, 1992; Foreman and Bidleman, 1990). It is problematic whether the simple Junge-Pankow or K_{oa} models (eq 15 and 21) will accurately account for the particle/gas distribution of modern agrochemicals, which are more polar than OCs. In view of the quite large differences in sorption that can be expected from various solute - solvent combinations (Jang et al., 1997; Liang et al., 1997), the phase distributions and temperature-dependent physicochemical properties (vapor pressure, K_{oa}, Henry's law constant) should be determined for pesticides that are currently in use.

Very little work has been done on sorption of pesticides to atmospherically suspended soil dust, which is expected to be be prevalent in farming areas. Indeed, some of the earliest work on pesticide transport was done in the early 1960s, when workers measured pesticides on dust that was translocated from Texas to Ohio during a dust storm (Cohen and Pinkerton, 1966). Sorption to mineral aerosols may be important in the transport and deposition of pesticides in continental (Orgill and Sehmel, 1976) and marine (Duce et al., 1991; Seba and Prospero, 1971; Prospero and Seba, 1972) dust regions.

Physical removal of pesticides from the atmosphere by wet and dry deposition depends strongly on their particle/gas distribution (Ballschmiter, 1992; Bidleman, 1988; Ligocki et al., 1985a,b; Majewski and Capel, 1995; Wania et al., 1998b). A relationship first proposed by Ligocki et al. (1985a,b) is widely used to estimate the overall scavenging ratio for wet deposition (W_T):

$$W_T = C_{rain,\,total}/C_{air,\,total} = W_p\phi + W_g(1-\phi) \qquad (22)$$

The particle scavenging ratio by rain or snow, W_p, is usually estimated from field sampling of nonvolatile species such as elemental carbon or trace metals. It is a strong function of the particle size distribution and is on the order of 10^5 - 10^6 (Bidleman, 1988; Cotham and Bidleman, 1991; Franz et al., 1997; Hoff et al., 1996; Poster and Baker, 1997). Equilibration of vapor-phase pesticide with rain and cloud droplets is expressed by the gas scavenging ratio, $W_g = RT/H$. Pesticides with low Henry's law constants (e.g., dieldrin, HCHs) are removed by gas scavenging, while others (e.g., DDT, chlordane) with high Henry's law constants are removed mainly by washout of particles (Bidleman, 1988).

Field measurements of W_T sometimes agree with the predictions of eq 22 (Bidleman, 1988; Dickhut and Gustafson, 1995; Ligocki et al., 1985a,b), but it is not uncommon for the measured values to be higher, sometimes by up to an order of magnitude. Enrichment of pesticides (Glotfelty et al., 1987; Seiber et al., 1993; Schomburg et al., 1991) wood smoke marker compounds (Sagebiel and Seiber, 1993) PCBs and PAHs (Capel et al., 1991) in fog droplets has been observed. Such enrichment may be caused by the presence of organic films on the surface of the droplets (Majewski and Capel, 1995) or to adsorption at the gas-liquid interface (Hoff et al., 1993a, 1995; Pankow, 1997; Thibodeaux et al., 1991). Redistribution of compounds between the particle, gas and rain phases during rain scavenging can bias W_T and phase speciation measurements (Tsai et al., 1991). Another problem is that rain-scavenged compounds may be present on submicrometer aerosols which pass through collection filters. This has been investigated by Poster and Baker (1996a,b, 1997), who proposed a model which includes scavenging of "non-filterable" particles.

The particle dry deposition flux (F, mass/area-time) is usually estimated by multiplying the total, unspeciated atmospheric concentration ($C_{air,\ total}$) by the particulate fraction (ϕ) and the particle dry deposition velocity (V_d).

$$F = \phi V_d C_{air,\ total} \qquad (23)$$

V_d is often taken to be 0.05 - 0.2 cm/s for submicrometer aerosols (Cotham and Bidleman, 1991; Hoff et al., 1996), but can be considerably higher for coarse particles that are found in urban air (Holsen et al., 1991, 1993, 1997).

6.2. ATMOSPHERIC REACTIONS

Reaction with OH radicals is the most important degradation mechanism for many gas-phase organic compounds, and a large body of information is available on experimental and predicted OH radical reaction rates with low- and intermediate weight hydrocarbons, halocarbons and oxygenated compounds (Atkinson, this issue; Atkinson, 1994). Predictions of OH reactivities have also been made for chlorinated dioxins and furans (Atkinson, 1991, Kwok et al., 1995), PCBs (Kwok et al., 1995), and pesticides (Atkinson, in Bidleman et al., 1990; Winer and Atkinson, 1990) (Table III). Second-order reaction rates with OH have been experimentally determined for PCBs (Anderson and Hites, 1996), dioxins and furans (Brubaker and Hites, 1997), HCHs (Brubaker and Hites, 1998) and analogs of organophosphate pesticides (Winer and Atkinson, 1990). Atmospheric lifetimes (1/k', where k' is the pseudo first-order rate constant) based on experimental rate constants for PCBs, dioxins and furans agreed well with predicted values, but were 3-4 times longer for α- and γ-HCH (Table III). Kwok and Atkinson (1995) compared measured and predicted rate constants for 485 organic compounds and found agreement within a factor of two for 90% of the substances. Disagreements most commonly occurred for haloalkanes, haloalkenes, haloethers and some other oxygenated compounds, and the authors discouraged extrapolating their estimation technique to compounds outside classes that have been tested.

Uncertainties in estimating atmospheric lifetimes are also due to variability in OH radical concentrations and temperature. The 24-h averaged OH concentration for the troposphere is 9.7 x 10^5 molecules/cm^3 (Prinn et al., 1995), but this varies diurnally and spatially. The OH concentrations are highest from late morning to mid-afternoon, barely detectable at night, 4-20 times higher in summer than winter, and 2 - 3 times higher at 30°N latitude compared to 60°N (Altschuller, 1989). The rate constant for HCHs is 1.5 - 1.7 times higher at 25° than 0°C (Brubaker and Hites, 1998). Thus, as chemicals travel from temperate to polar regions, their atmospheric lifetimes will increase due to lower temperatures and fewer OH radicals. This may be one reason why substances which have estimated lifetimes of only a few days under temperate conditions are routinely found in arctic air (e.g., dieldrin, chlordane, DDT, Table III).

Little is known about the stability of POPs that are sorbed to atmospheric particles. Most studies have been done for PAHs, and fewer for pesticides. Particulate PAHs display a wide range of reactivities depending on the particle composition and relative humidity (Behymer and Hites, 1988, Kamens et al., 1988, McDow et al., 1994, 1995). Certain oxygenated constituents of atmospheric particulate matter, particularly methoxyphenols, polycyclic aromatic quinones and substituted benzaldehydes and furans, enhance the photodegradation of labile PAHs (Jang and McDow, 1995; McDow, 1994, 1995). Adsorption to fly ash appears to protect chlorinated dioxins and furans from photodegradation (Koester and Hites, 1992). Palm et al. (1997) investigated the reaction between OH radicals and the triazine herbicide terbuthylazine which was adsorbed onto inert silicon dioxide. Comparable measurements in the gas phase were not made for the herbicide, but the authors reported that OH reaction rates were similar in the gas phase and adsorbed to silica for two other compounds, lindane and di-(2-ethylhexyl) phthalate. The authors felt that their rate constant for terbuthylazine was also applicable to the gas phase, and so the atmospheric lifetime based on their value is listed in Table III.

Reaction of pesticides in the atmosphere can lead to products that are themselves of concern because of their toxicity. Photoproducts of many organochlorine pesticides are known, e.g., for heptachlor, dieldrin and mirex. Photoheptachlor has a relatively high toxicity compared to the parent pesticide, and has been found in ringed seal blubber and human plasma in the Arctic (Zhu et al., 1995). Organothiophosphates are oxidized to their oxon analogs (Atkinson, this issue), which are more potent acetylcholinesterase inhibitors. These also have a higher polarity and are thus more likely to be removed by wet deposition and air-water gas exchange (Majewski and Capel, 1995).

6.3. ATMOSPHERIC LIFETIME AND SPATIAL RANGE

The potential for pesticides to be atmospherically transported depends on their ability to be mobilized into air and the removal processes that take place enroute. Thus the spatial range, or transport distance, of a chemical is inversely related to its overall removal rate. Removal of gaseous and particulate species from the atmosphere takes place by wet and dry deposition and by chemical reaction. These processes are strongly influenced by partitioning between the gas and aerosol phases (Ballschmiter, 1992; Bidleman, 1988; Dickhut and Gustafson, 1995; Gustafson and Dickhut, 1997a,b; Hillery et al., 1998; Hoff et al., 1996; Wania et al., 1998b), which was discussed in section 6.1.

TABLE III
Atmospheric lifetimes due to reaction with OH radicals.

Compound	τ, days[a]	Method[b]	Reference[c]
Pesticides			
α-HCH	83	E	1
	20	C	2
γ-HCH	61	E	1
	20	C	2
	3	C	3
HCB	429	E	1
	579	C	2
	1240	C	3
Dieldrin	1	C	4
	1	C	3
DDE	2	C	4
DDT	4	C	4
	6	C	4
Chlordane	8	C	4
	2	C	3
Endosulfan	0.8	C	3
Mirex	6×10^4	C	3
Malathion	0.18	C	5
Parathion	0.12	C	5
Methamidophos	0.44	C	5
Terbuthylazine	~1	E	6
Chlorinated Dioxins and Dibenzofurans			
1,2,3,4-TCDD	14	E	7
	13	C	8
Monochlorodioxins	3.0	E	9
Dichlorodioxins	2.0-2.4	C	9
Trichlorodioxins	2.5-3.3	C	9
Tetrachlorodioxins	2.8-7.2	C	9
Pentachlorodioxins	4.0-8.5	C	9

TABLE III (cont.)
Atmospheric lifetimes due to reaction with OH radicals.

Compound	τ, days[a]	Method[b]	Reference[c]
Monochlorodibenzofurans	2.9	C	9
Dichlorodibenzofurans	4.0-5.5	C	9
Trichlorodibenzofurans	5.5-9.5	C	9
Tetrachlorodibenzofurans	7.7-18	C	9
Pentachlorodibenzofurans	15-29	C	9
Polychlorinated Biphenyls			
monochlorobiphenyls	2-4	E	10
	2.5-3.6	C	9
dichlorobiphenyls	5-6	E	10
	3.7-8.3	C	9
trichlorobiphenyls	9-12	E	10
	6.9-15	C	9
tetrachlorobiphenyls	12-14	E	10
	8.5-40	C	9
pentachlorobiphenyls	19-34	E	10
	16-48	C	9
hexachlorobiphenyls	29-40	C	9

a) $\tau = 1/k'$ at 290-298 K and [OH] = $0.8\text{-}1.0 \times 10^6$ molecule/cm^3.
b) E = experimental, C = calculated
c) 1 = Brubaker and Hites, 1998.; 2 = calculated from structure-activity relationship of Kwok and Atkinson (1995) by Brubaker and Hites (1998). 3 = predicted by Howard and quoted by Scheringer, 1997; 4 = predicted by Atkinson and quoted by Bidleman et al., 1990; 5 = Winer and Atkinson, 1990; 6 = Palm et al., 1997; 7 = Brubaker and Hites, 1997; 8 = predicted by Atkinson and quoted by Brubaker and Hites, 1998.; 9 = Kwok et al., 1995; 10 =Anderson and Hites, 1996.

The relationship between atmospheric removal rate and distance of transport has been considered by several authors. Cohen (1997) divided chemicals into classes based on their atmospheric transport potential over the following ranges 1. $\leq$100 km, 2. >100 - 1000 km, 3. >1000 - 10000 km, and 4. global. The estimates were made by considering physical removal rates (wet, dry deposition) and chemical destruction mechanisms, the latter based on experimental OH radical rate constants or those estimated from structure-reactivity relationships (e.g., Kwok and Atkinson, 1995). Most organochlorine pesticides fell into categories 2 and 3.

Brubaker and Hites (1997) modelled atmospheric removal of PCDDs and PCDFs by gas-phase reaction and wet/dry deposition of gases and particles. Aerosol-bound dioxins and furans were assumed to be unreactive. The homolog profiles of PCDDs and PCDFs which were found in ambient air samples were explained by gas-phase reactions of the lower homologs and physical removal by wet and dry deposition of the higher ones.

Recently the atmospheric transport potential of a chemical has been related to its overall environmental persistence. Overall persistence is determined by the partitioning among different compartments (water, sediment, soil, air), degradation rates in each compartment and mode of emission (into water, soil, air) (Bennett et al., 1999; Müller-Herold, 1996; Müller-Herold et al., 1997; Scheringer, 1996,1997; Wania, 1998; Webster et al., 1998). When calculating overall persistence for a region, advective transport must also be considered. The compartmental distribution and overall persistence can be calculated using fugacity-based multimedia environmental models such as EQC (Mackay et al., 1996a,b,c; Wania, 1998; Webster et al., 1998). The major degradation pathways are metabolism, and hydrolysis in water and soils (photolysis in surface layers), and photolysis and reaction with OH radicals in the atmosphere. The characteristic time, τ, is defined as $1/k$, where k is the pseudo-first order removal constant (Bennett et al., 1999).

Scheringer (1996, 1997) set up a one-dimensional model of global circulation to estimate the effect of atmospheric persistence on spatial range. The model divides a 40000 km circular path (i.e., perimeter of the earth) into boxes containing air, water and soil phases. Partitioning of the chemical within the boxes and transfer between boxes was described similar to the model of Wania and Mackay (1995). Transport time and spatial range (the range containing 95% of the mass distribution) were functions of deposition rates. The atmospheric degradation rate constants were taken as literature values for gaseous species and assumed to be zero for compounds sorbed to aerosols. Thus for particle-bound compounds, the atmospheric lifetime was the same as for the particles themselves and controlled by wet and dry removal processes. The interplay between deposition and degradation was illustrated by comparing two classes of POPs, having long and short degradation times with respect to OH radical attack, allowing the fraction sorbed to aerosols to vary from 0 - 1. For the first class, the spatial range decreased as greater percentages were sorbed to aerosols and the rate of deposition increased. For the second class of compounds that react fairly quickly with OH radicals in the gas phase, the overall persistence and spatial range were increased by sorption to aerosols.

Van Pul et al (1998) modelled the atmospheric residence time of POPs, taking into account chemical reactions, wet and dry deposition and reversible gas exchange with soil surfaces. Transport distances were calculated as the distance over which 50% of the chemical is removed. These ranged from 30- 2000 km for pesticides, 500-800 km for particulate PAHs and 10^5 km for HCB.

Bennett et al. (1998) formulated a fugacity-based multimedia model to calculate the characteristic transport distance (CTD) for semivolatile organic pollutants. The CTD is defined as the distance it takes for the concentration in a moving airstream to fall to e^{-1}, or 37% of its initial value due to degradation in air and net transfer to stationary compartments (water, soil, vegetation). The CTD was estimated from a mocing air parcel, and a non-moving compartment (soil or vegetation). The model included partitioning between air and the stationary phases and chemical transformations. A case study of tetrachlorodibenzo-p-dioxin transport indicated that the CTD was similar to or greater than the distance between urban centers, implying that TCDD should be regulated on a regional or global rather than a local scale. Vegetation was an important compartment for sequestering and degradation of TCDD.

7. Conclusions

This review has emphasized the role of physicochemical properties of pesticides and other POPs in transport and air-surface exchange. A relatively new property, the octanol-air partition coefficient (K_{oa}) is recommended for describing partitioning of non-polar chemicals between air and soil, vegetation and aerosols, and K_{oa} values for more polar pesticides are needed.

Environmental measurements and modelling are being carried out in climates ranging from the tropics to the poles, and therefore determining these physicochemical properties as a function of temperature rates a high priority. Fugacity ratios for POPs in some aquatic and vegetation systems are often close to equilibrium. It is therefore necessary to have precise and accurate values of vapor pressure, K_{oa} and Henry's law constant for making exchange estimates.

Even after international controls on the "dirty dozen" POPs are in place, atmospheric transport will continue from old sources and continued use of existing stocks. Emission inventories for in-place pesticides can be done through modelling and direct measurement of soil-to-air fluxes. The latter are very labor-intensive however, and modelling appears to be the only feasible approach on a regional scale.

Emission estimates for historically used pesticides are limited mainly by the meagre database for residues in soils. Efforts should be made to inventory the reservoir of OC pesticides in agricultural and non-agricultural soils and to further develop "chemical marker" approaches to distinguish these emissions from atmospheric transport of fresh chemical.

A good database for OC pesticides and other POPs in the arctic atmosphere has been built up over the last five years. These programs should be continued for several reasons. Monitoring is needed to check the efficacy of international controls on POPs. Changes in release of POPs to the environment, either by stopping new usage or remediation of old sources, will show up first in the atmosphere. Atmospheric monitoring also serves as a belwether for transport of new chemicals into remote environments. A good database on temporal and spatial trends of POPs in polar atmospheres is needed to estimate loadings by precipitation, dry deposition and air-sea gas exchange.

Evidence is now quite clear that gas exchange is a major, and often the dominant, contributor to atmospheric loadings in large lakes, estuaries and the oceans. Developing quantitative estimates for gas exchange of POPs requires a different strategy from that currently used to determine loadings from precipitation. Atmospheric monitoring programs for POPs usually locate sampling stations near the shore of a lake or estuary, but there is some question of whether these measurements represent conditions over water. In addition, gas exchange has a well-recognized seasonal variability which must be considered in sampling programs. A critical need is to obtain measurements of POPs in surface water. These data are required in all water bodies where air-water gas exchange fluxes are to be estimated, and the lack is especially serious for the open ocean and regional seas.

Speciation of POPs between the particle and gas phases governs atmospheric removal by physical processes (wet/dry deposition and gas exchange) and chemical reaction. Most studies of particle/gas distribution have been carried out for PCBs, PAHs, PCDD/Fs, less frequently for OC pesticides, and hardly at all for modern insecticides and herbicides. To predict the atmospheric fate of pesticides, measurements of the phase distribution are required in both urban and rural environments, and to carbon-rich and mineral aerosols.

Gas-phase reaction of POPs with OH radicals is perhaps the single most important chemical loss mechanism in the atmosphere, yet most second-order OH radical rate constants for pesticides have only been estimated from structure - reactivity relationships. Predictions of atmospheric lifetimes and spatial range may be grossly in error without accurate values for these constants as a function of temperature. There is a need to develop simple experimental methods to measure reaction rate constants for semivolatile compounds and to build up a database for pesticides of different chemical classes. Reactions of pesticides which are sorbed to ambient aerosols should also be investigated.

References

Aigner, E.J., Leone, A.D and. Falconer, R.L. 1998. *Environ. Sci. Technol.* **32**, 1162-1168.

Alegria, H., Bidleman, T.F. and Shaw, T. 1999. *Environ. Sci. Technol.*, submitted.

Altschuller, A.P. 1989. Ambient air hydroxyl radical concentrations: measurements and model predictions. *J. Air Pollut. Control Assoc.* **39**, 704-708.

Anderson, P.N. and Hites, R.A. 1996. *Environ. Sci. Technol.* **30**, 1756-1763.

Aston, L.S. and Seiber, J.N. 1996. *J. Environ. Qual.* **26**, 1483-1492.

Atkinson, R. 1991. *Sci. Total Environ.* **104**, 17-33.

Atkinson, R. 1994. *GasPhase Tropospheric Chemistry of Organic Compounds. J. Phys. Chem. Ref. Data*, Monograph 2, American Chemical Society, American Institute of Physics for the National Institute of Standards and Technology, 216 pages.

Bacci, E., Calamari, D., Gaggi, C. and Vighi, M. 1990. *Environ. Sci. Technol.* **24**, 885-889.

Ballschmiter, K. 1992. *Angew. Chem. Int. Ed. Engl.* **31**, 487-515.

Barrie, L.A., Macdonald, R., Bidleman, T., Diamond, M., Gregor, D., Semkin, R., Strachan, W., Alaee, M., Backus, S., Bewers, M., Gobeil, C., Halsall,C., Hoff, J., Li, Y.-F., Lockhart, L., Mackay, D., Muir, D., Pudykiewicz, J., Reimer, K., Smith, J., Stern, G., Schroeder, W., Wagemann, R., Wania, F. and Yunker, M. 1997. In: Jensen, J., Adair, K. and Shearer, R. (eds.). *Canadian Arctic Contaminants Assessment Report.* Ministry of Public Works and Government Services, Ottawa, QS-8525-000-EE-A1, Catalog No. R72-260/1997E, ISBN 0-662-25704-9, pages 25-182.

Behymer, T.D. and Hites, R.A.. 1988. *Environ. Sci. Technol.* **22**, 1311-1319.

Bennett, D.H., McKone, T.E., Matthies, M. and Kastenberg, W.E. 1998. *Environ. Sci. Technol.* **32**, 4023-4030.

Bennett, D.H., McKone, T.E., Matthies, M. and Kastenberg, W.E. 1999. *Environ. Sci. Technol.* **33**, 503-509.

Bidleman, T.F. 1988. *Environ. Sci. Technol.* **22**, 361-367.

Bidleman, T.F., Jantunen, L.M., Barrie, L.A. and Fellin, P. 1995a. *Geophys. Res. Lett.* **22**, 219-222.

Bidleman, T.F. and Falconer, R.L. 1999. *Environ. Sci. Technol.* **33**, 206A-209A.

Bidleman, T.F., Falconer, R.L. and Harner, T. 1999. In: Lane, D.A. (ed.). *Gas and Particle Measurements of Atmospheric Organic Compounds.* Gordon and Breach Publishers, Newark, NJ, in press.

Bidleman, T.F. and Harner, T. 1999. In: Mackay, D. and Boethling, R.S. (eds.) *Estimating Chemical Properties for the Environmental and Health Sciences, a Handbook of Methods.* Ann Arbor Press, Chelsea, MI, in press.

Bidleman, T.F., Falconer, R.L. and Walla, M.D. 1995b. *Sci. Total Environ.* **160/161**, 55-63.

Bidleman, T.F., Billings, W.N. and Foreman, W.T. 1986. *Environ. Sci. Technol.* **20**, 1038-1043.

Bidleman, T.F., Alegria, H., Ngabe, B. and Green, C. 1998a. *Atmos. Environ.* **32**, 1849-1856.

Bidleman, T.F., Atlas, E.L., Atkinson, R., Bonsang, B., Burns, K., Keene, W.C., Knap, A.H., Miller, J., Rudolph, J. and Tanabe, S. 1990. In: Knap, A.H. and Kaiser, M.-S. (eds.), *The Long-Range Atmospheric Transport of Natural and Contaminant Substances.* NATO ASI Series C, Vol. 297, Kluwer Academic Publishers, Boston, pages 259-301.

Bidleman, T.F., Jantunen, L.M.M., Harner, T., Wiberg, K., Wideman, J.L., Brice, K.A., Su, K., Falconer, R.L., Aigner, E.J., Leone, A.D., Ridal, J.J., Kerman, B., Finizio, A., Alegria, H., Parkhurst, W.J. and Szeto, S.Y. 1998b. *Environ. Pollut.* **102**, 43-49..

Bidleman, T.F., Jantunen, L.M.M., Wiberg, K., Harner, T., Brice, K.A., Su, K., Falconer, R.L., Leone, A.D., Aigner, E.J. and Parkhurst, W.J. 1998c. *Environ. Sci. Technol.* **32**, 1546-1548.

Bidleman, T.F. and McConnell, L.L. 1995. *Sci. Total Environ.* **159**, 101-117.

Bidleman, T.F., Schmidt, S., Roura, R. and Carr, E. 1993. *Mar. Pollut. Bull.* **26**, 258-262.

Blais, J.M., Schindler, D.W., Muir, D.C.G., Kimpe, L.W., Donald, D.B. and Rosenberg, B. 1998. *Nature* **395**, 585-588.

Boul, H.L., Garnham, M.L., Hucker, D., Baird, D. and Aislabie, J. 1994. *Environ. Sci. Technol.* **28**, 1397-1402.

Bowes, G.W. and Jonkel, C.J. 1975. *J. Fish. Res. Board Can.* **32**, 2111-2123.

Brubaker, Jr., W.W. and Hites, R.A. 1997. *Environ. Sci. Technol.* **31**, 1805-1810.

Brubaker, Jr., W.W. and Hites, R.A. 1998. *Environ. Sci. Technol.* **32**, 766-769.

Calamari, D., Bacci, E., Focardi, S., Gaggi, C., Morosini, M. and Vighi, M. 1991. *Environ. Sci. Technol.* **25**, 1489-1495.

Calamari, D., Tremolada, P., DiGuardo, A. and Vighi, M. 1994. *Environ. Sci. Technol.* **28**, 429-434.

Capel, P.D., Leuenberger, C. and Giger, W. 1991. *Atmos. Environ.* **25A**, 1335-1346.

Chernyak, S.M., Rice, C.P. and McConnell, L.L. 1996. *Mar. Pollut. Bull.* **32**, 410-419.

Cohen, J. and Pinkerton, C. 1966 In: Gould, R.F. (ed.), *Organic Pesticides in the Environment*, ACS Adv. in Chem. Ser., Vol. 60, American Chemical Society, Washington, DC, pages 163-176.

Cohen, M.D. 1997. *Review of: I. The Capability of Specific Persistent Toxic Substances to be Subjected to Longe Range Atmospheric Transport, and II. The Status and Capabilities of Associated Emissions Inventories.* Report prepared for the International Joint Commission, International Air Quality Advisory Board, IJC, Windsor, Ontario.

Cortes, D.R., Basu, I., Sweet, C.W., Brice, K.A., Hoff, R.M. and Hites, R.A. 1998. *Environ. Sci. Technol.* **32**, 1920-1927.

Cotham, W.E. and Bidleman, T.F. 1992. *Environ. Sci. Technol.* **26**, 469-478.

Cotham, W.E. and Bidleman, T.F. 1991. *Chemosphere* **22**, 165-188.

Coupe, R.H., Manning, M.A., Foreman, W.T., Goolsby, D.A. and Majewski, M.S. 1998.. In: Daniel, B.J. (ed.), *Proc. 28th Mississippi Water Resources Conference*, Raymond, MS, April, Water Resources Research Institute, Mississippi State University, Mississippi State, MS.

Cousins, I.T. and Jones, K.C. 1998. *Environ. Pollut.* **102**, 105-119.

Cowen-Ellsberry, C., Boethling, R. and Graham, D. 1999. In: *Proceedings of the SETAC Pellston Workshop on Criteria for Persistence and Long-Range Transport of Chemicals in the Environment*, 14-19 July, 1998, Fairmont Hot Springs, BC, Canada. Society for Environmental Toxicology and Chemistry, Pensacola, FL.

Dannenfelser, R.M., Surendran, N. and Yalkowski, S.H. 1993. *SAR and QSAR in Environ. Res.* **1**, 273-292.

Datta, S., Hansen, L., McConnell, L., Baker, J., LeNoir, J. and Seiber, J.N. 1998. *Bull. Environ. Contam. Toxicol.* **60**, 829-836.

DeWulf, J.P., van Langehove, H.R. and van der Auwera, L.F. 1998. *Environ. Sci. Technol.* **32**, 903-911.

Dickhut, R.M. and Gustafson, K.E. 1995. *Environ. Sci. Technol.* **29**, 1518-1525.

Dimond, J.B. and Owen, R.B. 1996. *Environ. Pollut.* **92**, 227-230.

Donnelly, J.R., Drewes, L.A., Johnson, R.L., Munslow, W.D., Knapp, K.K. and Sovocool, G.W. 1990. *Thermochim. Acta* **167**, 155-187.

Dowdy, D.L. and McKone, T.E. 1998. *Environ. Toxicol. Chem.* **16**, 2448-2456.

Duce, R.A., Liss, P.S., Merrill, J.T., Atlas, E.L., Buat-Menard, P., Hicks, B.B., Miller, J.M., Prospero, J.M., Arimoto, R., Church, T.M., Ellis, W., Galloway, J.N., Hansen, L., Jickells, T.D., Knap, A.H., Reinhardt, K.H., Schneider, B., Soudine, A., Tokos, J.J., Tsunogai, S., Wollast, R. and Zhou, M. 1991. *Global Biogeochem.Cycles* **5**, 193-259.

Eisenreich, S.J., Hornbuckle, K.C. and Achman, D.R. 1997. In: Baker, J.E. (ed.), *Atmospheric Deposition of Contaminants to the Great Lakes and Coastal Water*, SETAC Press, Pensacola, Florida, pp. 109-135.

Eriksson, G., Jensen, S., Kylin, H. and Strachan, W. 1989. Nature **341**, 42-44.

Falconer, R.L. and Bidleman, T.F. 1997. In: Baker, J.E. (ed.) *Atmospheric Deposition of Contaminants to the Great Lakes and Coastal Waters.* Society for Environmental Toxicology and Chemistry, Pensacola, Florida, pp. 151-169.

Falconer, R.L., Bidleman, T.F. and Gregor, D.J. 1995. *Sci. Total Environ.* **160/161**, 65-74.

Falconer, R.L., Bidleman, T.F. and Szeto, S.Y. 1997. *J. Agric. Food Chem.* **45**, 1946-1951.

Falconer, R., Leone, A., Bodner, C., Wiberg, K., Bidleman, T., Jantunen, L., Harner, T., Parkhurst, W., Alegria, H., Brice, K. and Su, K. 1998. *Organohalogen Compounds* **35**, 331-334.

Fellin, P., Barrie, L.A., Dougherty, D., Toom, D., Muir, D., Grift, N., Lockhart, L. and Billeck, B. 1996. *Environ. Toxicol. Chem.* **15**, 253-261.

Finizio, A., Bidleman, T.F. and Szeto, S.Y. 1998. *Chemosphere* **36**, 345-355.

Finizio, A., Mackay, D., Bidleman, T.F. and Harner, T.. 1997. *Atmos. Environ.* **31**, 2289-2296.

Foreman, W.T. and Bidleman, T.F. 1990. *Atmos. Environ.* **24A**, 2405-2416.

France, R.L., Muir, D.C.G. and Segstro, M.D. 1997. *Environ. Sci. Technol.* **31**, 1879-1882.

Frank, R., Braun, H.E., Chapman, N. and Burchat, C. 1991. *Bull. Environ. Contam. Toxicol.* **47**, 374-380.

Franz, T.P., Gregor, D.J. and Eisenreich, S.J. 1997. In: Baker, J.E. (ed.) *Atmospheric Deposition of Contaminants to the Great Lakes and Coastal Waters.* Society for Environmental Toxicology and Chemistry, Pensacola, FL, pp. 73-107.

Glotfelty, D.E., Seiber, J.N. and Liljedahl, L.A.. 1987. *Nature* **325**, 602-605.

Goldberg, E.D. 1975. *Proc. Royal Soc. London, Ser. B.* **189**, 277-289.

Goolsby, D.A., Thurman, E.M., Pomes, M.L., Meyer, M.T. and Batteglin, W.A. 1997. *Environ. Sci. Technol.* **31**, 1325-1333.

Goss, K.-U. and Eisenreich, S.J. 1997. *Atmos. Environ.* **31**, 2827-2834.

Goss, K.-U. and Schwartzenbach, R.P. 1998. *Environ. Sci. Technol.* **32**, 2025-2032.

Gregor, D.J. 1991. In: Sturges, W.T. (ed.) *Pollution of the Arctic Atmosphere.* Elsevier Applied Science, New York, pp. 217-254.

Gustafson, K.E. and Dickhut, R.M. 1997a. *Environ. Sci. Technol.* **31**, 1623-1629.

Gustafson, K.E. and Dickhut, R.M. 1997b. *Environ. Sci. Technol.* **31**, 140-147.

Halsall, C.J., Lee, R.G.M., Coleman, P.J., Burnett, V., Harding-Jones, P. and Jones, K.C. 1995. *Environ. Sci. Technol.* **29**, 2368-2376.

Halsall, C.J., Bailey, R., Stern, G.A., Barrie, L.A., Fellin, P., Muir, D.C.G., Rovinsky, F. Ya., Kononov, E. Ya. and Pastukhov, B. 1998. *Environ. Pollut.* **102**, 51-62.

Halsall, C.J., Barrie, L.A., Fellin, P., Muir, D.C.G., Billeck, B., Lockhart, L., Rovinsky, F. Ya., Kononov, E. Ya. and Pastukhov, B. 1997. *Environ. Sci. Technol.* **31**, 3593-3599.

Hargrave, B.T., Barrie, L.A., Bidleman, T.F. and Welch, H.E. 1997. *Environ. Sci. Technol.* **31**, 3258-3266.

Harner, T. and Bidleman, T.F. 1996. *J. Chem. Eng. Data* **41**, 895-899.

Harner, T. and Bidleman, T.F. 1998a. *Environ. Sci. Technol.* **32**, 1494-1502.

Harner, T. and Bidleman, T.F. 1998b. *J. Chem. Eng. Data* **43**, 40-46.

Harner, T., Bidleman, T.F., Wideman, J.L. and Parkhurst, W.J.. 1998b. *Environ. Pollut.*, submitted.

Harner, T., Jantunen, L., Bidleman, T., Falconer, R., Parkhurst, W. and Mackay, D.. 1998a. *Organohalogen Compounds* **36**, 397-400.

Harner, T., Kylin, H., Bidleman, T.F., Halsall, C.J., Strachan, W.M.J., Barrie, L.A. and Fellin, P. 1998c. *Environ. Sci. Technol.* **32**, 3257-3265.

Harner, T., Kylin, H., Bidleman, T.F. and Strachan, W.M.J. 1999. *Environ. Sci. Technol.* **33**, 1157-1164.

Harner, T. and Mackay, D. 1995. *Environ. Sci. Technol.* **29**, 1599-1606.

Harner, T., Mackay, D. and Jones, K.C. 1995. *Environ. Sci. Technol.* **29**, 1200-1209.

Haugen, J.-E., Wania, F., Ritter, N. and Schlabach, M. 1998. *Environ. Sci. Technol.* **32**, 217-224.

Hawthorne, S.B., Miller, D.J., Louie, P.K.K., Butler, R.D. and Mayer, G.G. 1996. *J. Environ. Qual.* **25**, 594-600.

Hillery, B.R., Basu, I., Sweet, C.W. and Hites, R.A. 1997. *Environ. Sci. Technol.* **31**, 1811-1816.

Hillery, B.R., Simcik, M.F., Basu, I., Hoff, R.M., Strachan, W.M.J., Burniston, D., Chan, C.H., Brice, K.A., Sweet, C. and Hites, R.A. 1998. *Environ. Sci. Technol.* **32**, 2216-2221.

Hinckley, D.A., Bidleman, T.F., Foreman, W.T. and Tuschall, J. 1990.. *J. Chem. Eng. Data* **35**, 232-237.

Hinckley, D.A., Bidleman, T.F. and Rice, C.P. 1991. *J. Geophys. Res.* **96**, 7201-7213.

Hippelein, M. and McLachlan, M.S. 1998. *Environ. Sci. Technol.* **32**, 310-316.

Hitch, R.K. and Day, H.R. 1992. *Bull. Environ. Contam. Toxicol.* **48**, 259-264.

Hoff, J.T., Mackay, D., Gillham, R. and Shiu, W.-Y. 1993a. *Environ. Sci. Technol.* **27**, 2174-2180.

Hoff, J.T., Wania, F., Mackay, D. and Gillham, R. 1995. *Environ. Sci. Technol.* **29**, 1982-1989.

Hoff, J.T., Gregor, D.J., Mackay, D., Wania, F. and Jia, C.Q. 1998a. *Environ. Sci. Technol.* **32**, 58-62.

Hoff, R.M., Brice, K.A. and Halsall, C.J. 1998b. *Environ. Sci. Technol.* **32**, 1793-1798.

Hoff, R.M. and Chan, K.-W.. 1986. *Chemosphere* **15**, 449-452.

Hoff, R.M., Muir, D.C.G., Grift, N.P. and Brice, K.A. 1992a. *Environ. Sci. Technol.* **27**, 266-275.

Hoff, R.M., Muir, D.C.G., Grift, N.P. and Brice, K.A. 1992b. *Environ. Sci. Technol.* **27**, 276-283.

Hoff, R.M., Muir, D.C.G., Grift, N.P. and Brice, K.A.. 1993b. *Chemosphere* **27**, 2057-2062.

Hoff, R.M., Strachan, W.M.J., Sweet, C.W., Chan, C.H., Shackleton, M., Bidleman, T.F., Brice, K.A., Burniston, D.A., Cussion, S., Gatz, D.F., Harlin, K. and Schroeder, W.H. 1996. *Atmos. Environ.* **30**, 3505-3527.

Holsen, T.M., Noll, K.E., Fang, G., Lee, W. and Lin, J.J. 1993. *Environ. Sci. Technol.* **27**, 1327-1333.

Holsen, T.M., Noll, K.E., Liu, S. and Lee, W. 1991. *Environ. Sci. Technol.* **25**, 1075-1081.

Holsen, T.M., Zhu, X., Khalili, N.R., Lin, J.J., Lestari, P., Lu, C.-S. and Noll, K.E. 1997. In: Baker, J.E. (ed.) *Atmospheric Deposition of Contaminants to the Great Lakes and Coastal Waters.* Society for Environmental Toxicology and Chemistry, Pensacola, FL, pp. 35-50.

Honrath, R.E., Sweet, C.W. and Plouff, C.J. 1997. *Environ. Sci. Technol.* **31**, 842-852.

Hornbuckle, K.C., Achman, D.A. and Eisenreich, S.J. 1993. *Environ. Sci. Technol.* **27**, 87-98.

Hornbuckle, K.C. and Eisenreich, S.J. 1996. *Atmos. Environ.* **30**, 3935-3945.

Hornbuckle, K.C. and Eisenreich, S.J. 1995. *Environ. Sci. Technol.* **29**, 848.

Hornbuckle, K.C., Jeremiason, J.D. and Eisenreich, S.J. 1994. *Environ. Sci. Technol.* **28**, 1491-1501.

Hornbuckle, K.C., Sweet, C.W., Pearson, R.F., Swackhamer, D.L. and Eisenreich, S.J. 1995. *Environ. Sci. Technol.* **29**, 869-877.

Hung, H. and Mackay. D. 1998. *Chemosphere* **35**, 959-977.

Iwata, H., Tanabe, S., Norio, S. and Tatsukawa, R. 1993a. *Environ. Sci. Technol.* **27**, 1080-1098.

Iwata, H., Tanabe, S., Norio, S. and Tatsukawa, R. 1993b. *Mar. Pollut. Bull.* **26**, 302-305.

Iwata, H., Tanabe, S., Ueda, K. and Tatsukawa, R. 1995. *Environ. Sci. Technol.* **29**, 792-801.

Jang, M., Kamens, R.M., Leach, K.B. and Strommen, M.R. 1997. *Environ. Sci. Technol.* **31**, 2805-2811.

Jang, M. and McDow, S.R. 1995. *Environ. Sci. Technol.* **29**, 2654-2660.

Jantunen, L.M. and Bidleman, T.F. 1995. *Environ. Sci. Technol.* **29**, 1081-1089.

Jantunen, L.M. and Bidleman, T.F. 1996. *J. Geophys. Res.* **101**, 28837-28846; corrections *Ibid.* 1997, **102**, 19279-19282.

Jantunen, L.M.M. and Bidleman, T.F. 1998. *Arch. Environ. Contam. Toxicol.* **35**, 218-228.

Jantunen, L.M.M., Bidleman, T.F. and Wideman, J. 1998. Presented at the 41st conference of the International Association for Great Lakes Research, Hamilton, Ontario, May.

Jensen, S., Eriksson, G. and Kylin, H. 1992. *Chemosphere* **24**:,229-245.

Jeremiason, J.D., Hornbuckle, K.C. and Eisenreich, S.J. 1994. *Environ. Sci. Technol.* **28**, 903-914.

Jury, W.A., Spencer, W.F. and Farmer, W.J. 1983. *J. Environ. Qual.* **12**, 558-564.

Jury, W.A., Spencer, W.F. and Farmer, W.J. 1984. *J. Environ. Qual.* **13**, 567-572.

Jury, W.A., Spencer, W.F. and Farmer, W.J. 1984. *J. Environ. Qual.* **13**, 573-579.

Jury, W.A., Spencer, W.F. and Farmer, W.J. 1984. *J. Environ. Qual.* **13**, 580-586.

Kallenborn, R., Oehme, M., Wynn-Williams, D.D., Schlabach, M. and Harris, J. 1998. *Sci. Total Environ.* **220**, 167-180.

Kamens, R.M., Guo, Z., Fulcher, J.N. and Bell, D.A. 1988. *Environ. Sci. Technol.* **22**, 103-108.

Karickhoff, S.W. 1981. *Chemosphere* **10**, 833-846.

Kawano, M., Ramesh, A., Thao, V.D., Tatsukawa, R. and Subramanian, A.N. 1992. *Internat. J. Environ. Anal. Chem.* **48**, 163-174.

Kim, Y.-H., Woodrow, J.E. and Seiber, J.N. 1984. *J. Chromatogr.* **314**, 37-53.

Koester, C.J. and Hites, R.A. 1992. *Environ. Sci. Technol.* **26**, 502-507.

Kömp, P. and McLachlan, M. 1997a. *Environ. Sci. Technol.* **31**, 2944-2948.

Kömp, P. and McLachlan, M. 1997b. *Environ. Sci. Technol.* **31**, 886-890.

Kömp, P. and McLachlan, M. 1997c. *Environ. Toxicol. Chem.* **16**, 2433-2437.

Kwok, E.S.C. and Atkinson, R. 1995. *Atmos. Environ.* **29**, 1685-1695.

Kwok, E.S.C., Atkinson, R. and Arey, J. 1995. *Environ. Sci. Technol.* **29**, 1591-1598.

Kylin, H. and Sjödin, A. 1998. *Organohalogen Compounds.* **39**, 133-135.

Lane, D.A. 1999. *Gas and Particle Measurements of Atmospheric Organic Compounds.* Gordon and Breach Publishers, Newark, NJ, in press.

Larsson, P., Järnmark, C. and Södergren, A. 1992. *Mar. Pollut. Bull.* **25**, 281-287.

Lartiges, S.B. and Garrigues, P.P. 1995. *Environ. Sci. Technol.* **29**, 1246-1254.

Lee, R.G.M., Hung, H., Mackay, D. and Jones, K.C. 1998. *Environ. Sci. Technol.* **32**, 2172-2179.

Leone, A.D. 1998. *Enantiomeric Composition of Chiral Pesticides in Soil and Air from the U.S. Cornbelt Region.* M.S. thesis, Dept. of Chemistry, Youngstown State University, Youngstown, Ohio, U.S.A.

Li, Y.-F., Barrie, L.A., Bidleman, T.F. and McConnell, L.L. 1998b. Global HCH use trends and their impact on the arctic atmospheric environment. *Geophys. Res. Lett.* **25**, 39-41.

Li, Y.-F., Cai, D.J. and Singh, A. 1998a. *Arch. Environ. Contam. Toxicol.* **35**, 688-697.

Li, Y.-F., Cai, D.J. and Singh, A. 1999. *Adv. Environ. Res.* **2**, 497-506.

Liang, C. and Pankow, J.F.1996. *Environ. Sci. Technol.* **30**, 2800-2805.

Liang, C., Pankow, J.F., Odum, J.R. and Seinfeld, J.H. 1997. *Environ. Sci. Technol.* **31**, 3086-3092.

Ligocki, M.P., Leuenberger, C. and Pankow, J.F. 1985a. *Atmos. Environ.* **19**, 1609-1617.

Ligocki, M.P., Leuenberger, C. and Pankow, J.F. 1985b. *Atmos. Environ.* **19**, 1619-1626.

Livingstone, D.M. and Imboden, D.M. 1993. *Tellus* **42B**, 364-377.

Lorber, M., Cleverly, D., Schaum, J., Phillips, L., Schweer, G. and Leighton, T. 1994. *Sci. Total Environ.* **156**, 39-65.

Macdonald, R.W., McLaughlin, F.A. and Carmack, E.C. 1998. In: Jensen, J. (ed.), *Synopsis of Research Conducted under the 1995-97 Northern Contaminants Program*, Minister of Public Works and Government Affairs Canada, Ottawa, QS-8573-000-EF-A1, catalog number R71-19/74-1997E, pages 89-95.

Mackay, D. 1982. *Environ. Sci. Technol.* **16**, 274-278.

Mackay, D. 1991. *Multimedia Environmental Models, the Fugacity Approach.* Lewis Publishers, Chelsea, Michigan, 257 pp.

Mackay, D. and Bentzen, E. 1997. *Atmos. Environ.* **31**, 4045-4047.

Mackay, D., DiGuardo, A., Paterson, S., Kicsi, G, and Cowen, C.E. 1996a. *Environ. Toxicol. Chem.* **15**, 1618-1626.

Mackay, D., DiGuardo, A., Paterson, S., Kicsi, G, and Cowen, C.E. 1996b. *Environ. Toxicol. Chem.* **15**, 1627-1637.

Mackay, D., DiGuardo, A., Paterson, S., Kicsi, G., Cowen, C.E. and Kane, D.M. 1996c. *Environ. Toxicol. Chem.* **15**, 1638-48.

Mackay, D., Shiu, W.-Y. and Ma, K.-C. 1997. *Illustrated Handbook of Physical-Chemical Properties and Environmental Fate for Organic Chemicals.* Vol. 5, Pesticide Chemicals. Lewis Publishers, Boca Raton, Florida, 812 pages.

Majewski, M.S. and Capel, P.D. 1995. *Pesticides in the Atmosphere, Distribution, Trends and Governing Factors.* Ann Arbor Press, Chelsea, MI, 214 pages.

Majewski, M.S., Desjardins, R.L., Rochette, P., Pattey, E., Seiber, J.N. and Glotfelty, D.E. 1993. *Environ. Sci. Technol.* **27**, 121-128.

Majewski, M.S., Foreman, W.T., Goolsby, D.A. and Nakagaki, N. 1998. *Environ. Sci. Technol.* **32**, 3689-3698.

Martijn, A., Bakker, H. and Schreuder, R.H. 1993. *Bull. Environ. Contam. Toxicol.* **51**, 178-184.

McConnell, L.L., Bidleman, T.F. and Walla, M.D. 1998a. *Environ. Pollut.* **101**, 391-399.

McConnell, L.L., Kucklick, J.R., Bidleman, T.F., Ivanov, G.P. and Chernyak, S.M. 1996. *Environ. Sci. Technol.* **30**, 2975-2983.

McConnell, L.L., LeNoir, J.S., Datta, S. and Seiber, J.N. 1998b. *Environ. Toxicol. Chem.* **17**, 1908-1916.

McConnell, L.L., Nelson, E., Rice, C.P., Baker, J.E., Johnson, W.E., Harman, J.A. and Bialek, K. 1997. *Environ. Sci. Technol.* **31**, 1390-1398.

McDow, S.R., Sun, Q., Vartiainen, M., Hong, Y., Yao, Y., Fister, T., Yao, R. and Kamens, R.M. 1994. *Environ. Sci. Technol.* **28**:,2147-2153.

McDow, S.R., Vartiainen, M., Sun, Q., Kong, Y. and Kamens, R.M. 1995. *Atmos. Environ.* **29**, 791-797.

McLachlan, M.S. 1996. *Environ. Sci. Technol.* **30**, 252-259.

McLachlan, M.S. and Horstmann, M. 1998 *Environ. Sci. Technol.* **32**, 413-420.

Mischke, T., Brunetti, K., Acosta, V., Weaver, D. and Brown, M. 1985. *Agricultural Sources of DDT Residues in California's Environment.* California Dept. of Food and Agriculture, Sacramento, California, 42 pages.

Muir, D.C.G. and Grift, N.P. 1995. In: Ragsdale, N., Kearney, P.C. and Plimmer, J.R. (eds.) *Eighth International Congress of Pesticide Chemistry: Options 2000.* American Chemical Society. 141-156.

Müller-Herold, U. 1996. *Environ. Sci. Technol.* **30**, 586-591.

Müller-Herold, U., Caderal, D. and Funck, P. 1997. *Environ. Sci. Technol.* **31**, 3511-3515.

Murphy, T.J. 1995. *Environ. Sci. Technol.* **29**, 846-847.

Nelson, E., McConnell, L.L. and Baker, J.E. 1998. *Environ. Sci. Technol.* **32**, 912-919.

Ngabe, B., Bidleman, T.F. and Falconer, R.L. 1993. *Environ. Sci. Technol.* **27**, 1930-1933.

Ockenden, W.A., Steinnes, E., Parker, C. and Jones, K.C. 1998a. *Environ. Sci. Technol.* **32**,2721-2726.

Ockenden, W.A., Sweetman, A.J., Prest, H.F., Steinnes, E. and Jones, K.C. 1998b. *Environ. Sci. Technol.* **32**, 2795-2803.

Oehme, M. 1991. *Ambio* **20**, 293-297.

Oehme, M., Haugen, J.-E. and Schlabach, M. 1996. *Environ. Sci. Technol.* **30**, 2294-2304.

Orgill, M.M. and Sehmel, G.A. 1976. *Atmos. Environ.* **10**:,813-825.

Orgill, M.M., Sehmel, G.A. and Petersen, M.R. 1976. *Atmos. Environ.* **10**, 827-834.

Pacyna, J.M., Broman, D. and Lipiatou, E. 1998. *Sea-Air Exchange: Processes and Modelling. Scientific report of the workshop on sea-air exchange,* Kjeller, Norway, June, 1997. European Commission, EUR 17660, ISBN 92-828-2577-9, Office for Official Publications of the European Commun., Luxembourg, 313 pp.

Palm, W.-U., Elend, M., Krueger, H-U. and Zetzsch, C. 1997. *Environ. Sci. Technol.* **31**, 3389-3396.

Pankow, J.F. 1987. *Atmos. Environ.* **21**, 2275-2283.

Pankow, J.F. 1993. *Atmos. Environ.* **27A**, 1139-1152.

Pankow, J.F. 1994a. *Atmos. Environ.* **28**, 185-188.

Pankow, J.F. 1994b. *Atmos. Environ.* **28**, 189-193.

Pankow, J.F. 1997. *Atmos. Environ.* **31**, 927-929.

Pankow, J.F. 1998. *Atmos. Environ.* **32**, 1493-1497.

Pankow, J.F. and Bidleman, T.F. 1992. *Atmos. Environ.* **25A**, 2241-2249.

Paterson, S., Mackay, D., Bacci, E. and Calamari, D. 1991. *Environ. Sci. Technol.* **25**, 866-871.

Patton, G.W., Hinckley, D.A., Walla, M.D., Bidleman, T.F., Hargrave, B.T. 1989. *Tellus* **41B**, 243-255.

Patton, G.W., Walla, M.D., Bidleman, T.F. and Barrie, L.A. 1991. *J. Geophys. Res.* **96**, 10867-10877.

Peterle, T.J. 1969. *Nature* **224**, 620

Poissant, L. and Koprivnjak, J.-F. 1996. *Environ. Sci. Technol.* **30**, 845-851.

Poster, D.L. and Baker, J.T. 1996a. *Environ. Sci. Technol.* **30**, 341-348.

Poster, D.L. and Baker, J.T. 1996b. *Environ. Sci. Technol.* **30**, 349-354.

Poster, D.L. and Baker, J.T.. 1997. In: Baker, J.E. (ed.) *Atmospheric Deposition of Contaminants to the Great Lakes and Coastal Waters.* Society for Environmental Toxicology and Chemistry, Pensacola, FL, pp. 51-72.

Prinn, R.G., Weiss, R.F., Miller, B.R., Huang, J., Alyea, F.N., Cunnold, D.M., Fraser, P.J., Hartley, D.E. and Simmonds, P.G. 1995. *Science* **269**, 187-192.

Prospero, J.M. and Seba, D.B. 1972. *Atmos. Environ.* **6**, 363-364.

Ramesh, A., Tanabe, S., Murase, H., Subramanian, A.N. and Tatsukawa, R. 1991. *Environ. Pollut.* **74**, 293-307.

Rapaport, R.A., Urban, N.R., Capel, P.D., Baker, J.E., Looney, B.B., Eisenreich, S.J. and Gorham, E. 1985. *Chemosphere* **14**, 1167-1173.

Rawn, D.F.K., Halldorson, T.H.J. and Muir, D.C.G. 1998. In: Ballantine, L., McFarland, J. and Hackett, D. (eds.) *Triazine Herbicides: Risk Assessment.* American Chemical Society. Washington D.C., in press.

Rice, C.P. and Chernyak, S.M. 1997 *Chemosphere* **35**, 867-878.

Rice, C.P., Samson, P.J. and Noguchi, G.E. 1986. *Environ. Sci. Technol.* **20**, 1109-1116.

Ridal, J.J., Bidleman, T.F., Kerman, B., Fox, M.E. and Strachan, W.M.J. 1997. *Environ. Sci. Technol.* **31**, 1940-1945.

Ridal, J.J., Kerman, B., Durham, L. and Fox, M.E. 1996. *Environ. Sci. Technol.* **30**, 852-858.

Risebrough, R.W. 1990. In: Kurtz, D.A. (ed.), *Long-Range Transport of Pesticides,* Lewis Publishers, Chelsea, MI, pp. 417-426.

Risebrough, R.W., Huggett, R.J., Griffin, J.J. and Goldberg, E.D. 1968. *Science* **159**, 1233-1236.

Sagebiel, J.C. and Seiber, J.N. 1993. *Environ. Toxicol. Chem.* **12**, 813-822.

Scheringer, M. 1996. *Environ. Sci. Technol.* **30**, 1652-1659.

Scheringer, M. 1997. *Environ. Sci. Technol.* **31**, 2891-2897.

Scholtz, M.T., McMillan, A.C., Slama, C., Li, Y.-F., Ting, N. and Davidson, K. 1997. *Pesticides Emissions Modelling. Development of a North American Pesticides Emissions Inventory.* Canadian Global Emissions Interpretation Centre, ORTEC Corporation, Mississauga, Ontario, 242 pages.

Schomburg, C.J., Glotfelty, D.E. and Seiber, J.N.. 1991. *Environ. Sci. Technol.* **25**, 155-160.

Schreitmüller, J. and Ballschmiter, K. 1995. *Environ. Sci. Technol.* **29**, 205-215.

Seba, D.B., and Prospero, J.M.. 1971. *Atmos. Environ.* **5**, 1043-1050.

Seiber, J.N., Wilson, B.W. and McChesney, M.M. 1993. *Environ. Sci. Technol.* **27**, 2236-2243.

Shoeib, M., Brice, K.A. and Hoff, R.M. 1999. *Chemosphere,* in press.

Simcik, M.F., Zhang, H., Eisenreich, S.J. and Franz, T.P. 1997. *Environ. Sci. Technol.* **31**, 2141-2147.

Simcik, M.F., Zhang, H., Eisenreich, S.J. and Franz, T.P. 1998. *Environ. Sci. Technol.* **32**, 251-257.

Simonich, S.L. and Hites, R.A. 1995. *Science* **269**, 1851-1854.

Sladen, W.L.J., Menzie, C.M. and Reichel, W.L. 1966. *Nature* **210**, 670-673.

Spencer, W.F., Singh, G., Taylor, C.D., LeMert, R.A., Cliath, M.M. and Farmer, W.J. 1996. *J. Environ. Qual.* **25**, 815-821.

Stern, G.A., Halsall, C.J., Barrie, L.A., Muir, D.C.G., Fellin, P., Rosenberg, B., Rovinski, F. Ya., Kononov, E. Ya. and Pastuhov, B. 1997. *Environ. Sci. Technol.* **31**, 3619-3628.

Swackhamer, D.L., Pearson, R.F. and Schlotter, S.P. 1998. *Chemosphere* **37**, 2547-2561.

Sweet, C.W., Harlin, K.S., Gatz, D.F. and Bauer, S. 1996. *Great Lakes Integrated Atmospheric Deposition (IADN) Data Report, 1993-94.* Illinois State Water Survey, Champaign, Illinois.

Szeto, S.Y. and Price, P.M. 1991. *J. Agric. Food Chem.* **39**, 1679-1684.

Tatton, J. O'G. and Ruzicka, J.H.A. 1967. *Nature* **215**, 346-348.

Thao, V.D., Kawano, M. and Tatsukawa, R. 1993a. *Environ. Pollut.* **81**, 61-71.

Thao, V.D., Kawano, M., Matsuda, M., Wakimoto, T., Tatsukawa, R., Cau, H.D. and Quyatt, H.T. 1993b. *Internat. J. Environ. Anal. Chem.* **50**, 147-159.

Thibodeaux, L.J., Nadler, K.C., Valsaraj, K.T. and Reible, D.D. 1991. *Atmos. Environ.* **25A**, 1649-1656.

Thomas, G.O., Sweetman, A.J., Lohmann, R. and Jones, K.C. 1998a. *Environ. Sci. Technol.* **32**, 3522-3528.

Thomas, G.O., Sweetman, A.J., Ockenden, W.A., Mackay, D. and Jones, K.C. 1998b. *Environ. Sci. Technol.* **32**, 936-942.

Thomas, G.O., Smith, K.E.C., Sweetman, A.J., and Jones, K.C. 1998c. *Environ. Pollut.* **102**, 119-128.

Tremolada, P., Burnett, V., Calamari, D. and Jones, K.C.. 1996.. *Environ. Sci. Technol.* **30**, 3570-3577.

Tsai, W., Cohen, Y., Sakugawa, H. and Kaplan, I.R. 1991. *Environ. Sci. Technol.* **25**, 2012-2023.

Ulrich, E.M. and Hites, R.A. 1998. *Environ. Sci. Technol.* **32**, 1870-1874.

Van Jaarsveld, J.A. 1996. In: Gryning, S.E. and Schiermeier, F. (eds.), *Air Pollution Modeling and Its Application XI*, Plenum Press, New York, pages 333-343.

Van Pul, W.A.J., de Leeuw, F.A.A.M., van Jaarsveld, J.A., van der Gaag, M.A. and Sliggers, C.J. 1998. *Chemosphere* **37**, 113-141.

Vetter, W. and Schurig, V. 1997. *J. Chromatogr.* **774**, 143-175.

Voldner, E.C. and Schroeder, W.H. 1989. *Atmos. Environ.* **23**, 1949-1961.

Wagrowski, D.M. and Hites, R.A. 1998. *Environ. Sci. Technol.* **32**, 2389-2393.

Waite, D.T., Cessna, A.J., Gurprasad, N.P. and Banner, J. 1998. *Atmos. Environ.* **33**, 1513-1523.

Wallace, J.C. and Hites, R.A. 1996. *Environ. Sci. Technol.* **31**, 2389-2393.

Wania, 1997. *Chemosphere* **35**, 2345-2363.

Wania, F. 1998. *An Integrated Criterion for the Persistence of Organic Chemicals Based on Model Calculations.* WECC report 1/98, WECC Ltd., Toronto, Ontario, Canada.

Wania, F., Axelman, J. and Broman, D. 1998b. *Environ. Pollut.* **102**, 3-23.

Wania, F., Haugen, J.E., Lei, Y.D. and Mackay, D. 1998a. *Environ. Sci. Technol.* **32**, 1013-1021.

Wania, F., Hoff,J.T., Jia, C.Q. and Mackay, D. 1998c. *Environ. Pollut.* **102**, 25-41.

Wania, F. and Mackay, D. 1995. *Sci. Total Environ.* **160/161**, 211-232.

Wania, F. and Mackay, D. 1996. *Environ. Sci. Technol.* **30**, 390A-396A.

Wania, F., Mackay, D. and Hoff, J.T. 1999. *Environ. Sci. Technol.* **33**, 195-197.

Webber, M.D. and Wang, C. 1995. *Canad. J. Soil Sci.* **75**, 513-524.

Webster, E., Mackay, D. and Wania, F. 1998. *Environ. Toxicol. Chem.* **17**, 2148-2158.

Welsch-Paulsch, K. and McLachlan, M.S. 1998. *Environ. Pollut.* **102**, 129-137.

Whang, J.M., Schomburg, C.J., Glotfelty, D.E. and Taylor, A.W. 1993. *J. Environ. Qual.* **22**, 173-180.

Whitby, K.T. 1978. *Atmos. Environ.* **12**, 135-159.

Wiberg, K., Jantunen, L.M., Harner, T., Wideman, J.L., Bidleman, T.F., Brice, K., Su, K., Falconer, R.L., Leone, A.D., Parkhurst, W. and Alegria, H. 1997. *Organohalogen Compounds* **33**, 209-213.

Willis, G.H., McDowell, L.L., Harper, L.A., Southwick, L.M. and Smith, S. 1983. *J. Environ. Qual.* **12**, 80-85.

Winer, A.M. and Atkinson, R. 1990. In: Kurtz, D.A. (ed.), *Long Range Transport of Pesticides*, Lewis Publishers, Chelsea, Michigan, pages 115-126.

Woodrow, J.E., Seiber, J.N. and Baker, L.W. 1997. *Environ. Sci. Technol.* **31**, 523-529.

Yu, J., Flagan, R.C. and Seinfeld, J.H. 1998. *Environ. Sci. Technol.* **32**, 2357-2370.

Zabik, J.M. and Seiber, J.N. 1993. *J. Environ. Qual.* **22**, 80-90.

Zhu, J., Norstrom, R.J., Muir, D.C.G., Ferron, L.A., Weber, J.-P. and Dewailly, E. 1995. *Environ. Sci. Technol.* **29**, 267-271.

MODELLING OF ATMOSPHERIC TRANSPORT AND DEPOSITION OF PESTICIDES

J. (HANS) A. VAN JAARSVELD AND W. ADDO J. VAN PUL*
Air Research Laboratory, RIVM, PO Box 1, 3720 BA, Bilthoven, The Netherlands
*(*corresponding author, e-mail: addo.van.pul@rivm.nl)*

Abstract. Modelling of atmospheric transport and deposition of pesticides is presented and discussed. Modelling on regional scale builds on the existing knowledge gained in other air pollution fields. An overview of current modelling studies on transport and deposition on a regional scale (typically 30-3000 km) is given. From these studies it is concluded that the models are capable in simulating the spatial distribution of the concentrations and depositions. However, large uncertainties are present in this type of modelling and are for the greater part induced by the uncertainty in the emissions and subsequently in the exchange process parameterisations and the physicochemical properties needed in the parameterisations. Many more measurement data are needed to validate the models.

Key words: pesticides, atmospheric transport, modelling, deposition

1. Introduction

Pesticides are emitted into the atmosphere in a direct way during application by drift and after application by volatilisation from treated plant and soil surfaces. Subsequently, they are dispersed in the atmosphere and deposited to the earth's surface where they enter the water, soil and vegetation (eco-)systems from where they may re-volatilise.

Modelling of the atmospheric transport and deposition cycle of pesticides is in the first place an attempt to combine knowledge on relevant processes in a way that a quantitative link between the use of pesticides and environmental levels can be established. Modelling on a regional scale builds on the knowledge gained in the modelling of other air pollutants such as acidifying substances and ozone. That is, the framework in which the dispersion and transport of these compounds is described, is used also for pesticides. For pesticide modelling subsequently additional information is needed on measurements of environmental levels, the physicochemical properties and emission amounts and application characteristics. From comparisons between predicted and observed levels and their variability in space and time one may find indications on missing or inadequate knowledge. In this stage of model development the availability of high quality field and monitoring data is crucial. In such a way present or historical situations can be diagnosed in terms of sources and source contributions and also the consequences of future developments in emissions may be predicted. Moreover, these modelling tools

Water, Air, and Soil Pollution **115:** 167–182, 1999.

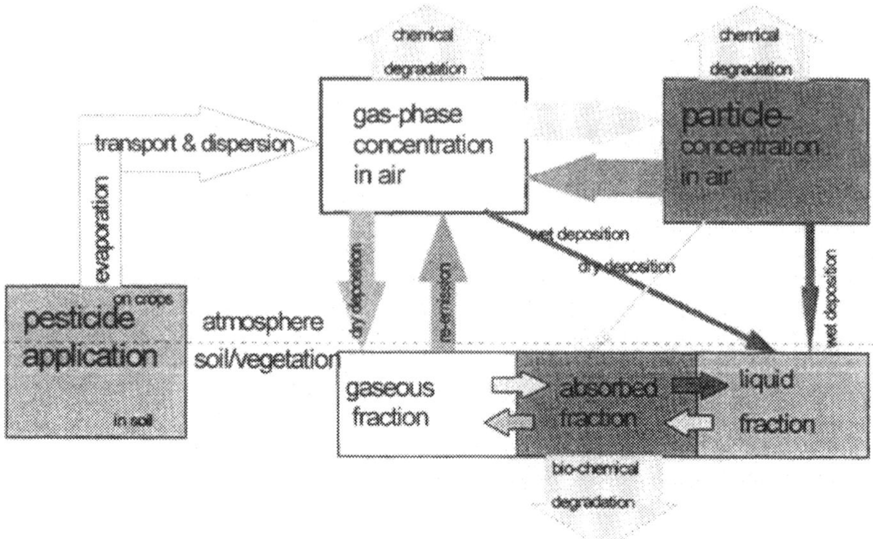

Figure 1. The main atmospheric processes involved in modelling the pesticide levels in the environment.

may then be used to obtain a better spatial and temporal coverage of the pesticide levels as found by measurements alone. Some aspects on the use of atmospheric dispersion models in the risk assessment of pesticides are discussed by De Leeuw *et al.* (1999).

The aim of this paper is to give an overview of the modelling studies on the transport and deposition of pesticides and the performance of the models. First we will briefly address the processes which are parameterised in describing the transport and deposition in models. We will focus on the regional scale models (typical spatial scale 30-3000 km) and present studies on this scale so far. Overviews concerning the emission and the transport and exchange of pesticides are presented by Majewski and by Bidleman (this issue).

2. Atmospheric processes

An overview of the processes generally addressed in modelling the environmental pesticide levels is presented in Figure 1.

The pesticide is applied at the soil and crop surface. During and after the application part of the pesticide evaporates. The pesticide is subsequently dispersed and transported by turbulence and mean wind flow to greater altitudes and distances.

In the atmosphere the pesticide may adsorb to particles and so partitions over the gas and particle phases. The gaseous and particle attached pesticides are removed from the atmosphere by dry and wet deposition and by (photo)chemical degradation. Once deposited at the soil or vegetation, the pesticide distributes over the gas, water and organic matter or lipid phases in the soil and vegetation. In water, the pesticide may also adsorb to suspended material and deposit to the sediment. Dependent on the gas concentrations of the various compartments the pesticide may re-volatilise again and be transported further via the atmosphere.

2.1 EMISSIONS

The first requirement for a good emission estimate of pesticides is to have reliable information on the application volumes and the circumstances under which they are used. The amount of pesticide which becomes airborne depends not only on the application method, but is also a function of, among others, the physicochemical properties of the pesticide, environmental conditions and the interaction of the pesticide with soil and vegetation. The total emission of the pesticide to air in a certain period is in fact the sum of the evaporation of the pesticide during application and afterwards from vegetation and soil. These total emission estimates are very uncertain (e.g. Majewski, this issue) and often the main source of uncertainty in the model calculations (see Section on uncertainties). A mismatch may occur between the environmental conditions used for the emission estimates and the conditions used in the transport and deposition calculations. An option therefore is to incorporate the emission estimates as a part of the modelling system.

2.2 TRANSPORT AND DISPERSION

Both dispersion and transport processes have been studied since the sixties (e.g. Turner, 1969, Pasquill, 1974). Recently, activities in EU framework have been started to harmonise the input and calculation schemes of the dispersion in local scale dispersion models (EU-Cost 710 program, Fisher *et al.*, 1998). We will give a brief overview of the relevant processes in and requirements for dispersion modelling. This is split into a local and a regional scale part.

2.2.1 Local scale (0.1-30 km)

The most important parameters influencing the dispersion at a local scale are:

emission rate as a function of atmospheric stability and wind speed
lateral and vertical dispersion
atmospheric boundary layer height
wind speed (as a function of height).

The time scale at which the dispersion takes place is typically a few minutes to an hour. At these time scales substances released in the atmospheric boundary layer show large gradients in space and time which are a strong function of the thermal stability of the atmosphere.

Particularly the lateral and vertical dispersion, vertical wind speed profile and the boundary layer height are strong functions of the atmospheric stability. Parameterisations of these variables are relatively well-known.

Note that at the local scale the removal processes do not play an important role in describing the transport and dispersion of a substance since the time scales of these removal processes generally are much larger than the transport and dispersion time scales.

2.2.2 Regional scale (30-3000 km)

At a regional scale the following parameters are important:

emission characteristics (in time such as seasonal distribution)
vertical transport to higher layers
removal and exchange processes
land-sea differences.

The time scale associated with the regional scale is typically from hours to several days. At these time scales the substance is mixed rather homogeneously over the atmospheric boundary layer and so the local dispersion conditions are of less importance for describing the concentrations at longer distances. During daytime the atmospheric boundary layer is typically 1 kilometre and is well-mixed in the vertical. During night time the atmospheric boundary layer is typically 100 metres and steep vertical gradients of the substance concentrations are found due to a strongly reduced mixing of the substance. Due to this diurnal cycle of the atmospheric boundary layer and synoptic weather systems the substance is transported out of the boundary layer to higher layers, i.e. the free troposphere.

2.3 GAS-PARTICLE PARTITIONING

The partitioning between the gas and particle phases is a phenomenon that is typical for organic compounds (Bidleman, 1988). For organics the partitioning as a function of vapour pressure and aerosol surface is often used (e.g. Junge, 1977). The partitioning is important in terms of the removal of the pesticides from the atmosphere (see next sections and Table I).

TABLE I.
Removal and exchange characteristics of pesticides in gas and particle phase.

Process	gas phase	particle phase
dry deposition	depends on surface saturation, both deposition and re-emission possible	depends on particle size (V_d=0.01-1cm/s), pesticides distribute over particles according to particle surface area, therefore characteristic deposition velocity in the order of 0.1 cm/s
wet deposition	depending on Henry's law constant	depending on particle size but very efficient
chemical reaction in atmosphere	usually estimated on the basis of chemical structure	lower reaction rates for particles ?

2.4 DEPOSITION AND RE-EMISSION

The most important process parameters determining the removal of the pesticide by dry and wet deposition are presented in Table I. In the dry deposition process of gaseous pesticides the interaction between the pesticide and the surface, *i.e.* soil, vegetation, water, is very important. This process is not very well known for pesticides, however, some information on this process for polar organics is available (see also Bidleman, this issue). This surface interaction is typical for organic compounds and is different from most of the air pollutant compounds studied so far. For most other air pollutant compounds, generally, the surface concentration is assumed to be zero. For pesticides, however, the dry deposition process is strongly affected by the surface saturation of the pesticides. In this way the dry deposition is reduced and can even be reversed, that is, re-emission of the pesticide takes place. For sea water re-emission of lindane is observed in Arctic regions because of the recent drop in atmospheric concentrations (Bidleman *et al.*, 1995). In a number of models, an effective deposition velocity is used in which the surface saturation of a pesticide is taken into account (Van Jaarsveld *et al.*, 1997, Baart *et al.*, 1995, Persson and Ullerstig, 1996). These deposition velocities were calculated off-line using a detailed air-soil and air-sea exchange model (DEPASS; Van Jaarsveld, 1995). Recently, Jacobs and Van Pul (1996) and Pekar *et al.*, (1998) used a dynamic model to estimate the exchange at the soil and sea water interface, using soil and sea water compartments coupled to the atmospheric transport models. Modelling of the soil surface concentration and profiles with depth was carried out following the equilibrium partitioning formulations by Jury *et al.* (1983). The surface water concentrations were calculated using the two-film theory by Liss and Slater, (1974). In modelling the surface concentrations physicochemical properties of the pesticides, such as Henry's law constant, the octanol-water partitioning coefficient, degradation rates in the soil and water, are required.

For non-reactive gaseous compounds the wet scavenging process is relatively well known. It is assumed that the equilibrium between the gas and rain water phase is established rather quickly according to the Henry's law constant. For a number of organics this was demonstrated by Ligocki *et al.*, (1985).

The dry and wet deposition of pesticides bound to particles are determined by the deposition properties of the particles themselves. Both the dry and wet deposition are strongly dependent on the particle size. The pesticides distribute over the particles according to their surface area. This means that pesticides are more attached at the finer aerosol particles (typically $<1\ \mu m$) that have a dry deposition velocity in the order of 0.1 cm s^{-1}. Wet deposition of particles is also particle size dependent but is a very efficient process. Wet scavenging ratios are typically 10^5-10^6 (Van Jaarsveld *et al.*, 1997).

2.5 CHEMICAL TRANSFORMATION

Estimates of the transformation (degradation) rate of pesticides in the atmosphere are obtained usually on the basis of chemical structure. The most important reaction is the reaction with the OH-radical (Atkinson *et al.*, this issue). Parameterisations of the removal rates are given as first order reaction constants assuming a certain OH-radical concentration.

Little is known about the transformation of organics at particle surfaces. In most model studies only the gaseous phase transformation is taken into account.

3. Modelling studies on the atmospheric transport and deposition of pesticides

A few model approaches exist with which the atmospheric transport and deposition of pesticides are described. Roughly the models can be split into fate and regional scale models. In these models the processes are parameterised and to a certain level simplified according to the purpose of the modelling exercise. It should be noted that for both modelling types the parameterisations of, specifically, the removal processes are similar and so is their relation with the physicochemical properties. The main difference between the two model approaches is primarily the spatial scale at which the concentrations are calculated. We will shortly discuss the general features of both approaches but we will focus on modelling of pesticides on a regional scale.

3.1 FATE MODELS

The general purpose of fate models is to show where a chemical ends up in the environment. The environment therefore is split into a number of compartments such as air, water, water sediment, fish, soil, which have uniform environmental characteristics and chemical concentrations. The exchange fluxes between the compartments are calculated from the fugacities of the chemical in the compartments and exchange coefficients. Examples are the Mackay environmental fate models levels 1-4 (Mackay, 1991, Mackay *et al.* 1992) such as Simplebox as is used in the European Union System for Evaluating Substances (EC, 1996). Applications of this model approach have been made on a global scale for zonal distributions of POPs such as lindane and DDT (Wania and Mackay, 1995); for the Baltic Sea region for POPs such as lindane, DDT, DDE, chlordane, toxaphene and HCB (Pacyna *et al.*, 1996); and for local to regional distribution of PCBs (Harner et al, 1995).

A somewhat different approach was presented by Strand and Hov (1996) for the global distribution of HCHs by using only three compartments; atmosphere, ocean and soil but with more detail in the transport processes in the atmosphere and ocean. It was concluded that the model could reproduce the environmental levels reasonably well and can be used to increase our understanding of the environmental fate of HCHs. For simulating concentration levels in specific areas the model was found to be too coarse and the simplifications too extensive.

3.2 REGIONAL SCALE MODELLING

The purpose of the regional scale models is to obtain a spatial and temporal distribution of the concentration in air and the deposition at the earth's surface. With these models the spatial distribution of sources and environmental conditions (*e.g.*, surface characteristics, meteorological conditions) over the model domain can be accounted for. The models for pesticides are mainly extensions from existing modelling tools for air pollutants. The models are also used in describing the transport and deposition of heavy metals and POPs as in the UN-ECE/CLRTAP.

Usually the models have only one compartment, i.e. the atmosphere, in which the transport is described in detail. The earth's surface is considered as the lower boundary of the model domain at which the deposition flux is defined. Roughly the framework wherein the transport is described can be split into trajectory or Lagrangian models and 3D grid or Eulerian models. In the trajectory model all processes are calculated along the trajectory or travelling path of a column of air. The annual concentration and deposition at a certain point is then calculated from the sum of all the trajectories arriving at that location every 6h for instance in a year (ASTRAP: Voldner and Schroeder, 1989) or from a statistical distribution of trajectories classified according to meteorological conditions (EUTREND model, Van Jaarsveld *et al.*, 1997). In Eulerian models the transport from a grid cell to another grid cell is calculated for the entire model domain simultaneously and at every grid cell and time step (typically 1 h) the removal processes are calculated. The Eulerian models offer the best framework in describing the transport and deposition of semi-volatile substances. Recently, the Eulerian EUROS model has been extended with water and soil compartments to describe the exchange with those compartments dynamically (Jacobs and Van Pul, 1996). An example of the net-deposition of lindane over Europe calculated with the EUROS model is depicted in Figure 2.

For modelling of the more persistent pesticides it is necessary to take into account the "residual" concentrations in all compartments. Since hardly any measurement data are available, this often is solved for the soil and water concentrations by running the model over the period the pesticide was used or the model is run until steady state of the concentrations in soil and water is achieved (Jacobs and Van Pul, 1996, Pekar *et al.*, 1998).

An overview and key findings of the modelling studies on pesticides on the European and North American region are shown in Table II. General results from the studies are that atmospheric deposition is an important and mostly dominant input pathway of pesticides for large water bodies and drainage areas of rivers. Deposition rates of gaseous pesticides to surface waters are much higher than to soil and vegetation. With the Eulerian models (EUROS, ASIMD) re-emission fluxes of the pesticide lindane at a large scale were calculated.

3.2.1 Validation of the regional models

Very few data are available for validation of the model calculations. However, a comparison of modelled annual mean concentrations of lindane in rain water with measurements from the OSPAR network could be made (Figure 3, calculations with TREND in the ESQUAD study; Van Jaarsveld *et al.* (1997)).

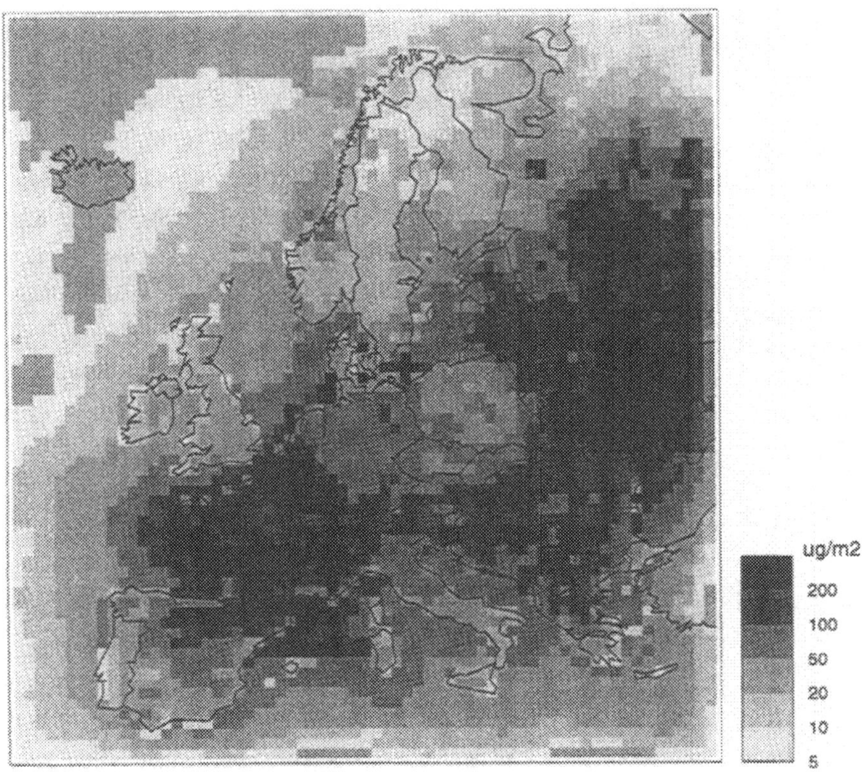

Figure 2. Annual net-deposition of lindane (sum of deposition and re-emission) over Europe (Jacobs and Van Pul, 1996, study #6, Table II).

TABLE II.
Summary of model studies on pesticides carried out for the European and North American scale.
In the column on Model applied; 1 indicates Lagrangian type models, 2 Eulerian type models.

Nr.	Compounds	Receptor area	Model applied	Emissions	Key results	Reference
1	Toxaphene	Great Lakes	ASTRAP model[1]	based on information on use patterns, methods and time of application, predictions form spray models, experimental studies on volatilisation	* conformation that atmosphere is a major transport route of toxaphene to great lakes * major source regions in Southern US. * wet deposition contributes 70-80% of deposition	Voldner and Schroeder (1989)
2	atrazine, endosulfan, fentin compounds, lindane, parathion	North Sea	OPS[1]	reported data supple-mented by estimates based on emission fac-tors and production/usequantities	First estimates of deposition to North Sea * Contribution of North Sea countries * Indicates the atmospheric pathway as the main pollution route	Warmen-hoven *et al.* (1989)
3	same as 2. Plus: azinphos-methyl, pentachlorophenol, bentazone, simazine, dichlorvos, trifluralin	River Rhine drainage area	OPS[1]	as above	as above, for Rhine drainage area	Baart & Diederen (1991)
4	Lindane	Europe	EU-TREND[1]	as above.	* Deposition over Europe and marine areas * Importance of soil saturation on deposition rates * Predicted levels in rain water are a factor 3 lower than observations indicating underestimated emissions	Van Jaarsveld *et al.* (1997) ESQUAD -project

TABLE II. Continued.

5	2,4-D, atrazine, azinphos-methyl, bentazon, dichlorvos, diuron, endosulfan, fentin-hydroxide, lindane, mecoprop (MCPP), mevinphos, parathion, pentachlorophenol, simazine, trifluralin	Netherlands, North Sea, Wadden Sea, Dutch coastal waters, some Dutch inland waters	EU-TREND[1]	based on activity data in 1990, point source characteristics, emission factors	* atmospheric deposition to North Sea important source of contamination * 90% of the deposition originates from emissions in North Sea littoral states * 90% of deposition on the Netherlands results from emissions in Benelux, France, Germany and the UK * Predicted levels of lindane in rain water are a factor 2 higher than observations	Baart *et al.* (1995) This study is partly a continuation of Warmenhoven *et al.* (1989)
6	Lindane	Europe	EUROS[2]	as above	* re-volatilisation of lindane over large parts of Europe * wet deposition is the major input to land surfaces but up to 50% can be re-emitted	Van Pul *et al.* (1996), Jacobs and Van Pul (1996)
7	Lindane	Europe	MATCH[2]	as above	*tentative conclusions: emissions estimates too high in Finland and Sweden	Persson and Ullerstig (1996)
8	Lindane	Europe	ASIMD[2]	TNO/UBA (Berdowski *et al.* 1997)	*40% of lindane transported out model domain *15% of the deposited lindane at soil was re-emitted	Pekar and Van Pul, (1998), Pekar *et al.* (1998)

Here an underestimation is found but with a high correlation which means that the spatial distribution is rather well described. Tentatively it was concluded that the underestimation is caused by an underestimation of the emissions. Second estimates of the lindane emissions were much higher resulting in much higher calculated concentrations in rain water (Baart *et al.*, 1995, Jacobs and Van Pul, 1996).

A preliminary conclusion from a comparison between calculated and measured atrazine concentrations in rain water in the Netherlands was that no major discrepancies were found (Baart *et al.*, 1995). Voldner and Schroeder (1989) concluded for toxaphene that their predicted concentrations in air and rain

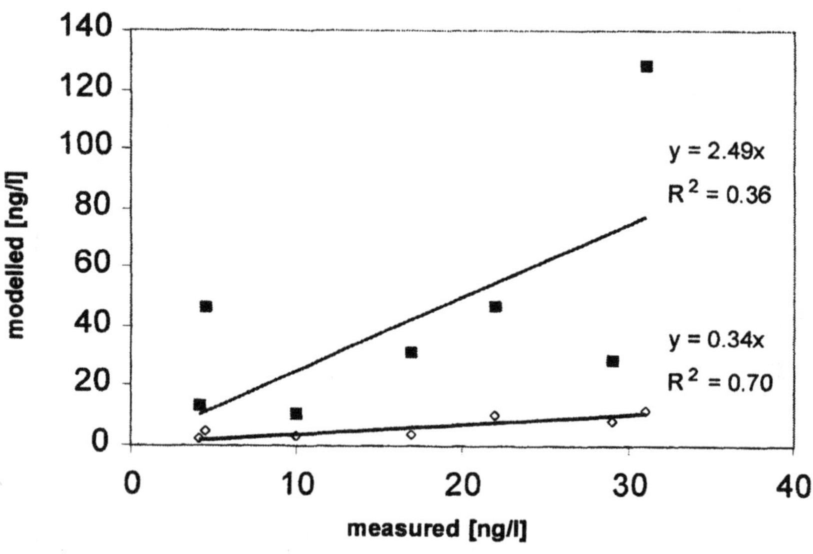

Figure 3. Measured versus modelled lindane concentrations in rain water. Measurements taken from the OSPAR/CAMP monitoring network. Diamonds are data from ESQUAD study (study #4, Table II) and squares those from Baart *et al.* (1995; study #5, Table II). Also indicated are he regression lines through the origin (y=a x) and the square of the correlation coefficient (R^2).

water were in reasonable agreement. For PCBs it was concluded that measured and modelled concentrations for a number of locations in the UK, Sweden and Netherlands agreed reasonable well, *i.e.* within typically a factor of 2 (Baart *et al.*, 1995).

3.2.2 Uncertainties in modelling the concentration and deposition of pesticides
Uncertainties are present in all modelling steps. In the validation of the models, uncertainties in the measurements play also an important role. Not for all steps it is possible to give a good estimate of the uncertainty. We will discuss the main sources of uncertainty below.

- emissions: the uncertainty in the emissions is quite large and estimated at a factor of 2-5 (Berdowski *et al.*, 1997); this lies in the estimate of the volume as well as in the emission factors of the pesticide.
- transport and dispersion: the uncertainty in the modelling of the transport and dispersion is rather well known from other air pollution studies. The uncertainty in the annual deposition calculations, evaluated from a comparison with measurements, is in the order of 30% (Van Jaarsveld *et al.*, 1997). An intercomparison of transport and deposition models for lead, which is particle bound, showed that the models gave results on the concentrations in air and rain water and deposition typically within a factor of 2 of each other (Sofiev *et al.*, 1996). Note that here the uncertainty in the deposition parameterisations for particles is included.
- deposition and exchange process: the transfer of substances to the earth's surface is also known relatively well from other air pollution studies. However, the uncertainty in modelling of the accumulation at the surface or the concentration in soil, vegetation or water is rather poorly known and strongly dependent on the physicochemical properties of the pesticide. Sensitivity studies show that by varying the deposition velocity and wet scavenging coefficient independently by a factor of 2 results in a variation of less than 30% in the total deposition of lindane on the North Sea (Van Jaarsveld *et al.*, 1997). Voldner and Schroeder (1989) found similar results in their sensitivity analysis and concluded that the change in deposition to the Great Lakes and basins was less than 10%. The reason for this non-linear effect is the feed back between the deposition processes and the substance mass in the atmospheric compartment. If for instance the dry deposition velocity is increased the dry deposition will increase but will reduce the mass in the atmosphere and consequently leads also to a reduction in the wet deposition. In this way the initial increase in dry deposition is compensated to a certain extent by the reduction in the wet deposition.
- physicochemical properties: these can be identified as a large source of uncertainty in the modelling of atmospheric transport and deposition of chemicals as a whole and typical for organics such as pesticides in particular. Differences in estimates of the same property originating from different databases can amount up to an order of magnitude. In particular the transformation in air (degradation) is quite uncertain. Moreover the

uncertainty increases considerably from much studied pesticides to relatively new pesticides. A conservative estimate of the uncertainty in the physicochemical properties for a widely-used pesticide such as lindane amounts to a factor of 2 (see Mackay *et al.*, 1997). The uncertainty in the physicochemical properties is translated into the estimate of the deposition. The impact of this uncertainty on the deposition is strongly dependent on the importance of the physicochemical property in the parameterisations of the removal processes and soil and water concentrations and the importance of these processes in describing the deposition of the pesticide. This means that not all physicochemcal properties have to be known with the same accuracy for all pesticides. A discussion on the impact of the physicochemical properties on the calculated atmospheric residence time of a number of POP and pesticides is given in Van Pul *et al.* (1998). In general the vapour pressure, Henry's law constant and the transformation rate in air are the most important properties.

- measurements: uncertainty sources in concentration data on air and precipitation can be manifold and originate, among others, from the sampling and analysis techniques. Moreover, measurements are representative for a certain area and can be locally influenced. From the IADN study it was estimated that the uncertainty in the annual deposition fluxes lies in the range 30-100% based on measurements only (Hoff *et al.*, 1996).

4. Conclusions

The model studies presented in Table II should be viewed as first attempts to link measured concentrations in the environment to emissions estimates of pesticides. In general the models are capable of simulating the spatial distribution of concentrations and deposition. However, the uncertainties in the modelled concentrations and depositions are rather large. These uncertainties originate for the greater part from the emissions and to a lesser degree in the exchange process description and the physicochemical properties.

The models are based on simple parameterisations of the atmospheric processes and on strong simplifications of the water and soil compartments. The uncertainty in the individual processes can be large. However, due to the feed back between the removal processes which are parameterised as a function of the concentration in air, the uncertainty in the total deposition is significantly smaller than in the individual processes. This is an advantage of estimating the deposition by modelling.

Examples of shortcomings in models are : a) the neglect of the uptake of pesticides by vegetation, b) scavenging by snow, c) the neglect of currents in the water compartment.

References

Baart, A.C. and Diederen, H.S.M.A.: 1991. *Calculation of the atmospheric deposition of 29 contaminants to the Rhine catchment area.*, TNO 91/219 Delft, the Netherlands.

Baart, A.C., Berdowski, J.J.M. and Van Jaarsveld, J.A.: 1995. *Calculation of atmospheric deposition of contaminants on the North Sea.* TNO 95/138, Delft, the Netherlands.

Berdowski, J.J.M., Baas, J., Bloos, J-P.J., Visschedijk, A.J.H. and Zandveld, P.Y.J.: 1997. *The European emission inventory of heavy metals and persistent organic pollutants.* TNO/UBA 104 02 672/03.

Bidleman, T.F.: 1988. *Environ. Sci. Technol.* **22**: 361-367.

Bidleman, T.F., Jantunen, L.M., Falconer, R.L. and Barrie, L.A.: 1995. *Geophys. Res. Lett.* **22**, 219-222.

De Leeuw, F.A.A.M., Van Pul, W.A.J., Van Den Berg, F. and Gilbert, A.J.: 1999. *Environ. Monitor. Assess.* In press.

EC: 1996. *EUSES, the European Union System for the Evaluation of Substances.* RIVM. Available from European Chemicals Bureau EC/DGXI, Ispra, Italy.

Fisher, B.E.A., Erbrink, J.J., Finardi, S., Jeannet, P., Joffre, S., Morselli, M.G., Pechinger, U., Seibert, P. and Thomson, D.J.: 1998, *Harmonisation of the pre-processing of meteorological data for atmospheric dispersion models.* COST 710, Final report, EC.

Harner, T., Mackay, D. and K.C. Jones: 1995. *Environ. Sci Technol.* **29**, 1200-1209.

Hoff, R.M., Strachan, W.M.J., Sweet, C.W., Chan, C.H., Shackleton, M., Bidleman, T.F., Brice, K.A., Burniston, D.A., Cussion, S., Gatz, D.F., Harlin, K. and Schroeder, W.H.: 1996, *Atm. Environ.* **30**, 3505-3527.

Jacobs, C.M.J. and Van Pul, W.A.J.: 1996. *Long-range transport of persistent organic pollutants I: Description of surface-atmosphere exchange modules and implementation in EUROS.* RIVM 722401013 Bilthoven, the Netherlands.

Junge, C.E.: 1977. *In: Fate of pollutants in the air and water environment. Part I.* Advances in *Environ. Sci. Technol.*, Vol. **8**, 7-25.

Jury W.A., Spencer, W.F. and Farmer W.J.: 1983, *J. Environ. Qual.*, **12**, 558-564.

Ligocki, M. Leunenberger, C. and Pankow, J.F.: 1985, *Atm. Environ.* **19**, 1609-1617.

Liss, P.S. and Slater P.G.: 1974, *Nature* **247**, 181.

Mackay, D.: 1991. *Multimedia Environmental Models. The Fugacity Approach.* Lewis Publishers, Inc., Chelsea, Michigan.

Mackay, D., Paterson, S. and Shiu, W.Y.: 1992. *Chemosphere* **24**, 695-717.

Mackay, D., Shiu, W-Y. and Ma, K-C.: 1997. *Illustrated handbook of physical-chemical properties and environmental fate for organic chemicals. Vol. V: Pesticide Chemicals.* Lewis Publ. Chelsea, Michigan

Pacyna, J. M., Brorström-Lundén, Paasivirta, J., Runge, E. and Wania, F.: 1996. *Emissions and behaviour of selected persistent organic pollutants (POP) in the northern environment. Part I.* NILU, Kjeller, Norway.

Pasquill, F.: 1974, *Atmospheric Diffusion*, John Wiley & Sons, Chichester.

Pekar, M. and Van Pul, W.A.J.: 1998. *Modelling of lindane and PCBs transport in the European region.* In: Proceedings EUROTRAC symposium'98 Garmisch-Partenkirchen. In press.

Pekar, M. Pavlova, N., Erdman, L. Ilyin, I. Strukov, B., Gusev, A. and Dutchak, S.: 1998. *Development of transport models for lindane, PCB, B(a)P.* EMEP/MSC-E 2/98, Moscow.

Persson, C. and Ullerstig, A.: 1996. *Model calculations of dispersion of lindane over Europe.* SMHI 68.

Sofiev, M., Maslyaev, A. and Gusev, A.: 1996. *Heavy metal model intercomparison. Methodology and results for Pb in 1990.* EMEP/MSC-E 2/96, Moscow.

Strand, A., and Hov, Ø.: 1995. *Water, Air Soil Pollut.* **86**, 283-316.

Turner, D.B.: 1969, *Workbook of atmospheric dispersion estimates, Public Health Service Publication No.* 999-AP-26 US Department of Health, Education and Welfare, Cincinnati, Ohio.

Van Jaarsveld, J.A., Van Pul, W.A.J. and De Leeuw, F.A.A.M.: 1997. *Atm. Environ.* **31**, 1011-1024.

Van Jaarsveld, J. A., 1997, *In* (ed. Gryning and Schiermeier)*: Air Pollution Modelling and its Application XI.* Plenum Press, New York, 333-334.

Van Pul, W.A.J., Jacobs, C.M.J., and Van Jaarsveld, J.A.: 1996. *In (Ed. Gryning and Schiermeyer): Air Pollution Modelling and its Application. XI.* Plenum Press, New York, 203-212.

Van Pul, W. A. J, De Leeuw, F. A. A. M., Van Jaarsveld, J. A., Van der Gaag, M. A. and Sliggers, C. J.: 1998. *Chemosphere* **37**, 113-141.

Voldner, E., and Schroeder, W.H.: 1989. *Atm. Environ.* **23**, 1949-1961.

Wania, F. and Mackay, D.: 1995. *Sci. Total Environ.* **160/161**, 211-232.

Warmenhoven, J.P., Duiser, J.A., de Leu, L.Th. and Veldt, C.: 1989. *The contribution of the input from the atmosphere to the contamination of the North Sea and the Dutch Wadden Sea.* TNO 89/349A, Delft, the Netherlands.

REGULATORY RISK ASSESSMENT OF PESTICIDE RESIDUES IN AIR

A. J. GILBERT
Pesticides Group, Central Science Laboratory, MAFF, Sand Hutton, York YO41 1LZ, UK.
E-mail: a.gilbert@csl.gov.uk

Note: This paper is a personal contribution directed towards assisting discussion. It does not necessarily reflect the views of the author's parent organisation the Central Science Laboratory, MAFF, UK

Abstract. Background information describing current approaches taken by different countries to risk assessment in regard of pesticide residues in air is presented with a view to stimulating discussion of the subject in order to identify means for improving both the underlying science and administrative procedures concerned. Fundamental regulatory objectives are explored, alternative guidelines for conduct of exposure evaluation, risk assessment and regulatory decision making are examined and the basic components of the technical framework in which this complex work is carried out are discussed. The criteria which are capable of being used for regulatory decision support are considered with a view to questioning their feasibility for practical use and relevance for operation to provide a robust basis for regulatory risk assessment for plant protection products. Brief recommendations are given for aspects of discussion where attention should be focused.

Keywords: air, authorisation, drift, exposure, pesticide, regulation, risk, transport, vapour

1. Introduction

This paper is a written text of a presentation given at the workshop on 'Fate of pesticides in the atmosphere; implications for risk assessment' organised by the Health Council of the Netherlands at Driebergen, the Netherlands on 22-24 April 1998 and focuses on the demands that regulatory risk assessment makes regarding pesticide residues in the air. It has a double purpose. First it seeks generally to provide background information on current approaches to risk assessment in respect of atmospheric transport of pesticides. Secondly it is intended to stimulate discussion of the subject in order to identify necessary improvements in the science underlying risk assessment and the associated mechanisms for conduct of risk assessment itself.

For its context the paper may refer to aspects of discussion dealt with more fully in other workshop session documents concerning the associated processes of emission, transformation, transport and exchange of pesticides, which lead to potential environmental exposure by airborne routes. It explores and compares conceivable approaches to risk assessment by describing criteria currently used, or under development, in pesticide approval schemes in different countries or

Water, Air, and Soil Pollution **115:** 183–194, 1999.

international frameworks. It leaves open to question whether such approaches are feasible, practical, economical or even desirable, as means for assessment or limitation of the risks posed by atmospheric transport of pesticides. It draws together components of a comprehensive cross-sectional view of the subject, ranging from regulatory guidelines and scientific publications to less formal publications, together with other comment and experience. In advancing a personal view on the successfulness or suitability of different approaches it is hoped to stimulate discussion leading to identification of desired improvements to the mechanisms available for attaining worthwhile policy goals.

2. Fundamental regulatory objectives

From the public perception, as revealed by discussion within the mass media, there is a demand for some guarantee of keeping a pesticide residue free environment. From a policy-makers perception that goal may remain an ideal, while there is more justification for the case to sanction approved uses of pesticides that satisfy a risk/benefit analysis made within a rational framework. In turn, regulators, who are tasked with carrying out such risk assessment for pesticides approval, must take a perspective which can balance many alternative views. The public perception and the policy-makers view, as well as expert opinions drawn from among a wide range of professions including commercial, academic, advisory and legal, have all to be reconciled. Moreover, the regulators perspective has to underpin the transparent and objective framework within which risk assessment and regulatory decision making has to take place.

Within the European Union approval of pesticides carried out under the terms of the Authorisations Directive (Council Directive 91/414/EEC, 1991) calls on regulatory authorities of member states to consider the fate and behaviour of pesticides in air. A further Council Directive (Council Directive 95/36/EC, 1995) requires prediction of 'the level of residues in air, to which man, animals and other non-target organisms may be exposed (acute and chronic)'. Such data are evaluated following revised uniform principles of the relevant Directive (Council Directive 97/57/EC, 1997). In the UK, the Pesticides Safety Directorate has developed guidance (PSD, undated) for those seeking approval of pesticides. This outlines a variety of alternative approaches available for provision of data pertinent to the fate and behaviour of pesticides in air, with which environmental risk assessment is to be carried out. Alternative approaches to provision of relevant data have been published as guidelines by the Organisation for Economic Co-operation and Development (OECD, 1981), the Society of Environmental Toxicology and Chemistry (SETAC, 1995) and the European and Mediterranean Plant Protection Organisation/Council of Europe (EPPO/CoE, 1993). The broad objective of such data provision is to enable a quantitative prediction of potential environmental exposure. This is compared with concentration values for which likely levels of risk to non-target species are known. Thus a margin of safety may be maintained between the predicted environmental concentration (PEC) and the predicted no-effect concentration (PNEC), where, at values <1, the lower the ratio of PEC/PNEC the greater is the margin of safety. In general it is both difficult and expensive to obtain great accuracy in PEC values. It is a fundamental characteristic of regulatory risk assessment schemes to be coupled with decision making mechanisms that enable less accurate data to be used, while ensuring that error lies on the side of caution. Thus valid approval decisions can expediently be made in cases where potential risk is demonstrably very low, although the exact degree of risk based on more accurate values for PEC and PNEC has not been necessary to work out.

3. Setting the scenario

To provide context for the discussion of regulatory risk assessment and decision making criteria, it is helpful to set out some fundamental terms of reference to describe the alternative physical transport pathways and the respective routes for potential environmental exposure. The EPPO/CoE decision making sub-scheme for Air (Chapter 12, EPPO/CoE, 1993) has adopted a convention for defining terms which describe alternative pathways for environmental exposure by airborne routes (Figure 1).

Distinction among such key terms is important in order to avoid confusion between emissions to air by particle drift and by volatilisation losses. These losses occur by separate (although not unrelated) processes which tend to arise predominantly during or after application, respectively, and it is advantageous to make an assessment of each in its own right. It is also helpful to distinguish between the short- and long-ranges for both spatial and temporal scales because particle drift tends to lead to exposure in the short range while gaseous transport may extend further. Particle and vapour losses that are separable at the emission stage may become indistinguishable as airborne residues in the form of an aerosol in the longer term. This diagram may help to identify and define the subject under discussion at this workshop, which intends to focus upon the longer term, longer range effects.

The complex transport and transformation processes involved make quantification of exposure by airborne routes a most challenging aspect of pesticide risk assessment. It is notable that the units suggested for expression of exposure effects in figure 1 do not include specific reference to underlying time span. Deposition of spray drift by fallout of airborne particles in the short range should happen over a time span short enough to be considered instantaneous.

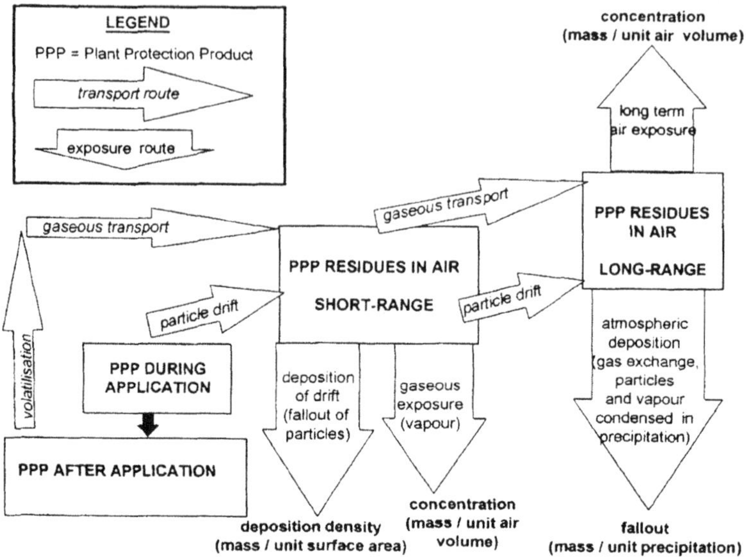

Figure 1. Definition of terms for alternative pathways for environmental exposure by airborne routes.

However, the time span of gaseous exposure processes is likely to be longer and the product of airborne concentration (which may vary) and exposure time could play a critical role in determining ultimate risk to non-target species. So a framework to link the space-time continuum and identify reasonable and pertinent incremental limits for relative range scales would be a useful common reference tool for regulatory risk assessment, although there are no absolute ways of gauging where such relative ranges should begin or end (Table I).

This workshop concerns 'long range' effects, corresponding to the horizontal space scale in the range of 1 - 1000 km (or more), a vertical space scale in the range of >30m above ground level and a corresponding time scale of an hour or more following application. It is interesting to note this could be described by terms of 'near-medium' to 'global' range and might best be defined by exclusion of the 'short' range and below.

4. Quantified risk assessment

In the EPPO/CoE decision-making scheme there are several chapters which take potential exposure data from the air sub-scheme. Each chapter has different requirements in respect of routes of exposure and various bases for the

TABLE I
A framework for relating temporal and spatial scales

Temporal Scale	Horizontal Spatial Scale	Vertical Spatial scale	Relative range term	Example usage event
Second	0 - 100 m	0 - 10m (surface)	micro range (adjacent)	treatment of single swath
Minute	100 - 3000 m	10 - 100 m	short range (local)	treatment of single field
Hour	3 - 100 km	100 - 3000 m (boundary layer)	near-medium range	treat multiple fields
Day	100 - 1000 km	3000 - 6000 m	far-medium range (regional)	treat all fields
Month	hemisphere	15 km (troposphere)	long range (continental)	treat total regional area
Year	global	50 km (stratosphere)	full range (global)	net annual use

units in which exposure values should preferably be expressed (Table II). The relationships listed in this table are under review along with the whole EPPO/CoE scheme and may be amended to reflect the demands of routine regulatory decision-making, as identified within the technical remit. Reference to human (e.g. bystander) exposure is not within that remit (being evaluated separately in the context of human risk assessment). It is notable that no exposure time scale is discussed.

Exposure, usually as predicted environmental concentration (PEC) level based upon evaluation of these data, must be weighed against eco-toxicological data (not discussed here in detail) in order to establish that critical environmental loads are not likely to be exceeded. It is quite likely that for most categories of exposure evaluation a time-weighted average concentration (TWAC) approach generally provides a sufficient basis for toxicity/exposure comparison. However, this approach might mask transient peak concentrations which may be critical in some cases where more acute toxic effects should be addressed. Such mechanisms of toxicity may demand actual concentration vs. time profiles to be quantified to provide a more accurate assessment of dose-response for certain products.

TABLE II
Chapters of the EPPO / CoE scheme requiring air exposure data.

Chapter	Particle drift exposure	Vapour exposure	Atmospheric deposition
Surface Water	YES (short term) fallout deposition, mass/area	NO	YES (long term) precipitation, mm/day & concentration, mass/volume precipitation
Terrestrial Vertebrates	YES (short term) fallout deposition, mass/area plus inhalation of spray drift, mass/volume air	YES (short term) absorption & inhalation, mass/volume air	NO
Non-target Higher Plants	YES (short term) fallout deposition, mass/area plus interception, mass/plant	YES (short term) mass/volume air	NO
Non-target Arthropods	YES (short term) fallout deposition, mass/organism	YES (short term) absorption & inhalation, mass/volume air	NO

The current level of scientific knowledge makes the operation of predictive environmental exposure models, involving quantification of emission, transformation, transport and exchange of airborne pesticide residues, very difficult at present. The report of an international workshop on application of simple models for environmental exposure assessment (OECD, 1993) identified, with respect to air, major processes as being dispersion and dilution, while optional processes were deposition and degradation. Sound modelling chiefly requires good input data for quantity and rate of emission to the air, the emission source height and pattern (point source or larger area) associated with the application method, along with very reliable environmental data, particularly for meteorological variables. Such models have not, to date, been applied to agriculture as intensively as for other industrial processes, and calculation models for long range dissipation of pesticides in air are only scarcely validated.

Preliminary findings of an ongoing desk study commissioned by the UK Pesticides Safety Directorate (Dubus et al, 1997) indicate that very little field data appears to be available on long distance aerial transport and subsequent atmospheric deposition of pesticides in the UK. Not only have few reliable data been collected from field studies, but where pesticides have been found it was not always possible to rule out short range transport processes as plausible routes of exposure. Where pesticide residues were found in rainfall, consistent with long range transport mechanisms, the field data were usually qualitative (identifying pesticide) but often lacked quantification, so limiting the interpretation of results and their potential use for wider predictive modelling.

Predictive exposure evaluation for regulatory risk assessment is typically associated with use of data for physical and chemical properties related to emission. Much also needs to be known about exposure processes such as horizontal transport, downward vertical flux and deposition. Local micro-meteorological effects (e.g. atmospheric temperature inversion) may be of great importance in determining the level of exposure (hence risk) likely to result from airborne routes. Moreover, there may be a need to study the potential for non-target species to accumulate contaminants, either physically or biologically. For example, it is possible that some plants might act as linear integrators of atmospheric residues (especially in the vapour phase) which they collect from the air moving past them. It is likely that the current level of scientific knowledge leaves a high degree of uncertainty in the conclusions possible to draw from results of field studies. This, in turn will adversely affect the reliability of associated modelling and can limit the usefulness of simple criteria for decision making with regard to such highly complex natural processes.

5. Decision-making criteria

The factors described above illustrate the sources of difficulty in accurately calculating PEC values. This opens a discussion of alternative approaches to regulatory decision making, such as using simple criteria to screen cases where the risk assessment can be accomplished with least difficulty. Several types of relatively simple data are possible to evaluate in order to conduct a conservative assessment of risk and, using relevant threshold criteria, support decision-making for approval of pesticides. However, an international survey (OECD, 1994) which summarised data requirements for pesticide registration in OECD member countries at that time demonstrated that there was no uniformity among regulators in different countries regarding the detailed procedures concerned. The case by case approach (e.g. formerly taken by the United Kingdom) suggests quantification of risk in any case according to its own circumstances, whereas a more formalised approach to assessment (taken e.g. by Germany and Canada) reflects use of an hierarchical decision making system which can consistently screen out cases needing less assessment if potential risk is demonstrably low. Although there is a commonality in taking a tiered approach and in the inclusion of a need to identify major metabolites and breakdown products (characterising these where appropriate), there is a variation in the frequency with which data pertinent to airborne routes of exposure may be called upon by different countries; both for physical and chemical properties from laboratory studies, and relating to fate and behaviour measured in the environment (Table III).

Table III
Pattern of requirement for data for physical-chemical properties and relating to fate and behaviour in the environment with emphasis on air routes of exposure.

Country / Data type	European Union	Germany	UK	CANADA
Hydrolysis rate	Always required	Frequently required (outdoor)	Always required	Always required
Photodegradation in water	Always required	Frequently required (outdoor)	Always required	Always required
Photodegradation on soil	Not required	Not required	Frequently required	Always required
Volatility (laboratory)	Always required	Frequently required	Occasionally required	Always required
Rate and route of photochemical degradation in air	Always required	Occasionally required	Occasionally required	Occasionally required

Most countries, including Germany, UK and Canada consider hydrolysis rate and photodegradation rate in water of the active substance, although not of formulated products. Volatility (vapour pressure) is required with varying frequency and only one OECD member country cites Henry's constant, as does tier 1 of the EPPO/CoE air scheme. Generally, the call for more complex data is more varied between countries, reflecting alternative emphasis in their approach to decision making.

To illustrate the formalised screening approach it is useful to describe the mechanism adopted by the BBA in Germany (Gottschild et al, 1990). Using a routine series of steps, the basic properties of candidate pesticides are compared with key criteria that determine their propensity to accumulate in the environment to levels where risk may be caused to non-target organisms (Figure 2). First the rate of hydrolysis and direct photolysis in water is compared with the threshold value of a half life of four days. Authorisation is possible for those active ingredients which are degraded more quickly than this. In other cases, where degradation is slower, the rate of volatilisation from plant and soil surfaces is examined. Those with a volatilisation rate less than 20% within 24 hrs may be accepted for authorisation because they evaporate at a slow enough rate that accumulation in air is unlikely. For other cases their photochemical- oxidative degradation in air is examined and compared with a threshold value, such that those having a half life of less than four days may be authorised because accumulation of residues emitted to air is not likely.

Finally, for those cases where none of the above criteria can be met, a specific examination of the particular proposed circumstances of use would be carried out. It is possible to find factors that may limit the scope for emission to air, such as products which are used only in closed environments. In the final analysis it would be possible to conduct field studies to assess distribution and dissipation in air in use.

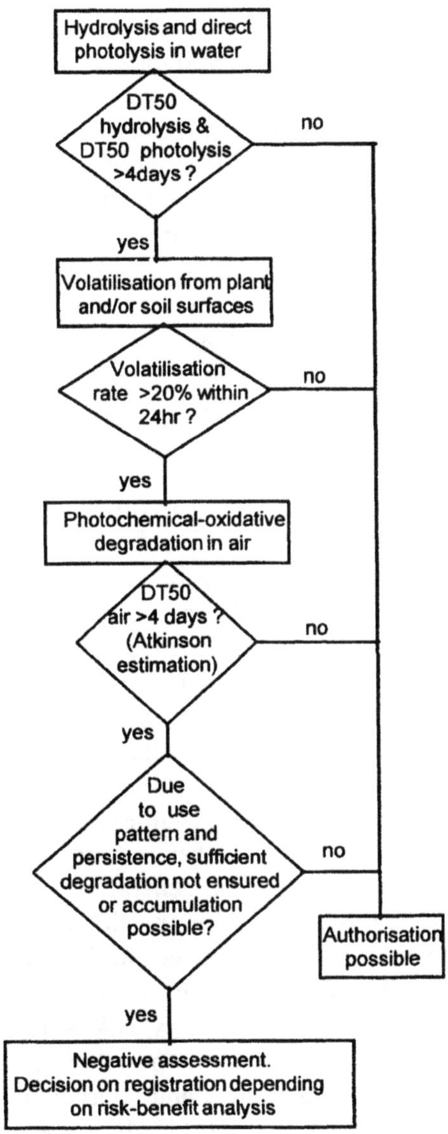

Figure 2. Behaviour of Plant Protection Product active ingredients in the Air - Examination according to BBA Guideline Part IV, 6-1

6. Tiered risk assessment

It is a general feature of quantitative risk assessment / decision-making schemes to operate at alternative tiers, whereby assessments are carried out in a successively more rigorous manner at higher tiers if no satisfactory outcome has been reached at lower tiers of assessment. Lower tiers of assessment generally rely on modelled predictions of exposure, which may be based upon surrogate values and/or use of conservative assumptions, whereas higher tiers of assessment rely more upon measured data to supplement assumptions used in models, and ultimately call for specific field studies. The tiered approach to risk assessment calls for careful control over setting of trigger values used to carry an assessment up to a higher tier. All cases need to be addressed at an appropriate level of stringency, balancing the need for a sound assessment with sensible economy in obtaining necessary data. Selection of which data to evaluate depends on the potential risk(s) needing to be assessed, which in turn depends on both the inherent hazards of candidate pesticides and their potential environmental impact. The broad aim of this is to protect all non-target species, in a way consistent with maximising the selectivity in action that is a requirement of useful pesticidal agents.

7. Future directions

Policy targets relating to safeguarding of people and the environment can be technically achieved by successful procedures for exposure evaluation, risk assessment and decision-making. Objectives of 91/414/EEC seek to protect the organisms living in, or depending upon such environmental compartments as the soil, surface water and air. That protection is afforded by ensuring that environmental concentrations of pesticides do not exceed threshold levels beyond which untoward toxic effects could occur.

Policy targets aiming to provide genuine public assurance that pesticide residues can exist at safe levels in their environment cannot necessarily be achieved so simply, however. The general public probably remain to be convinced that 'tolerable' levels of pesticides can be set, where the environment is generally still perceived to be either 'untainted', or not. Therefore it is probably the case that to completely satisfy the different objectives of all parties in the regulatory risk assessment process is not an achievable political aim. However, it is possible, if challenging, to achieve a pragmatic outcome to risk assessment and consequently to define a place and role for pesticides in the modern world. It is the task of all concerned to consider all the detail of this complex discussion in order to bring the pragmatism of everyday practice more toward the ideal at this time, and in the future.

It is difficult to draw clear conclusions within this paper, because the aim is to stimulate questions and discussion. However the following list of points can be made, which all need to be addressed to some degree in order to bring about desirable improvement. These are not presented in any order of priority, but simply reflect the order of points of view taken in the original list of likely regulatory objectives:

1. A more objective public appreciation of risk and benefit may be helpful in bringing about an acceptance of the place and role of pesticides, even with their attendant risks, within a practical world which is never truly risk free.

2. Consistency in the fundamental basis for discussion of technical issues would aid the systematic analysis of complex processes.

3. The regulatory decision making process is likely to be best served by a tiered approach, which would integrate the best features of formalised screening out of least risk cases, using simple criteria, and procedures for predictive quantitative exposure assessment to deal case by case with other products where risk cannot be ruled out using conservative assumptions.

4. Better use of 'simple' and relatively inexpensive data using appropriate threshold criteria to operate screening mechanisms would assist in prioritising cases needing less complicated assessment at lower tiers.

5. Better developed models for environmental exposure prediction would aid risk assessment at higher tiers. Some detailed points about air models include:

5.1 It would be helpful to include a summary sheet for each model proposed to be used for predictive environmental exposure calculation, giving its range of validity (with justification), likely levels of accuracy of prediction (with justification for conclusion), and criteria for applicability to ensure that the model suits the scenario setting(s) it is used for.

5.2 It is useful to have detailed information on which environmental concentration(s) vs. time (with a basis for TWAC aspects) can be derived from which models for which scenarios, also to have detailed information of the variability of these values when calculated using different models from a standard dataset.

5.3 It is best to have a standard operating procedure for using any model, to remove "need for judgement" of the modeller / regulatory user as far as possible. Deviations from a SOP should become less frequent and should always be fully justified if done.

5.4 Modellers / Regulators / Other users need to liaise more, through a dedicated forum to maintain and develop a consistent approach to the subject.

6. Better means for gathering relevant field data should be found, capable not only of addressing individual cases, as required, but ideally able to reconcile the more indirect and theoretically based approaches in 4 and 5.

References

Council Directive: 1991, *Council Directive 91/414/EEC of 15 July 1991 concerning the placing of plant protection products on the market,* O.J. No L230, 19.8.91

Council Directive: 1995, *Council Directive 95/36/EC of 14 July 1995 amending Council Directive 91/414/EEC of 15 July 1991 concerning the placing of plant protection products on the market,* O.J. No L172 22.7.95.

Council Directive: 1997, *Council Directive 97/57/EC of 22 September 1997 establishing Annex VI to Directive 91/414/EEC concerning the placing of plant protection products on the market,* O.J. No L265 27.9.97.

Dubus I, Hollis J M & Brown C D,: 1997, *Long distance aerial transport and subsequent atmospheric deposition of pesticides in the UK: Desk study results.* SSLRC Report to the Pesticides Safety Directorate of the Ministry of Agriculture, Fisheries and Food of project PL0528, Nov 1997.

EPPO /CoE: 1993, *European and Mediterranean Plant Protection Organisation/Council of Europe (EPPO/CoE) decision making schemes for the environmental risk assessment of plant protection products,* EPPO Paris 1993 (with additional Air sub-scheme)

Gottschild D, Siebers J and Nolting H -G.: 1990, *Richtlinien für die Prüfung von Pflanzenschutzmitteln im Zulassungsverfahren, Teil IV, 6-1, Prüfung des Verflüchtigungsverhaltens und des Verbleibs von Pflanzenschutzmitteln in der Luft,* Biologische Bundesanstalt für Land- und Forstwirtschaft, (BBA) Bundesrepublik Deutschland, Juli 1990, Saphir Verlag, 38551 Ribbesbüttel.

OECD: 1981, *Organisation for Economic Co-operation and Development (OECD) guidelines for the testing of chemicals,* OECD Paris, 1981 and revisions

OECD: 1993, *OECD Environment Monographs No 69 - Report on the OECD workshop on the application of simple models or environmental exposure assessment.* OECD Paris 1993

OECD: 1994, *OECD Environment Monographs No 77 - Data Requirements for pesticide registration in OECD member countries: Survey Results* OECD Paris 1994.

Pesticides Safety Directorate (PSD): undated,. *The Registration Handbook: Pesticides, Biocides, Plant Protection Products. A guide to the policies, procedures and data requirements relating to their control in the United Kingdom.* Available from PSD, Mallard House, Kings Pool, 3 Peasholme green, York YO1 2PX, UK.

SETAC: 1995, *Society of Environmental Toxicology and Chemistry (SETAC) procedures for assessing the environment fate and ecotoxicology of pesticides,* SETAC-Europe Brussels, 1995

EMISSION OF PESTICIDES INTO THE AIR[1]

F. VAN DEN BERG[1,*], R. KUBIAK[2], W. G. BENJEY[3,**], M.S. MAJEWSKI[4], S.R. YATES[5], G. L. REEVES[6], J.H. SMELT[1], A.M.A VAN DER LINDEN[7]

[1] DLO Winand Staring Centre for Integrated Land, Soil and Water Research (SC-DLO), PO Box 125, 6700 AC, Wageningen, The Netherlands, [2] Staatliche Lehr und Forschungs Anstalt für Landwirtschaft Weinbau und Gartenbau (SLFA), Breitenweg 71, D 67435, Neustadt/W., Germany, [3] Atmospheric Sciences Modeling Division, Air Resources Laboratory, National Oceanic and Atmospheric Administration, Research Triangle Park, NC, 27711, USA, NERL, [4] US Geological Survey, Water Research Department, Placer Hall, 6000 J Street, Sacramento, CA 95819, USA, [5] US Salinity Laboratory, Agricultural Research Service, USDA, 450 West Big Springs Road, Riverside, CA 92507, USA, [6] Dow AgroSciences, Letcombe Regis, Wantage, Oxon., OX12 9JT, UK, [1 7] National Institute of Public Health and the Environment (RIVM), PO Box 1, 3720 BA, Bilthoven, The Netherlands

([*] author for correspondence, e-mail: f.vandenberg@sc.dlo.nl; [**] on assignment to the National Exposure Research Laboratory, U.S. Environmental Protection Agency)

Abstract. During and after the application of a pesticide in agriculture, a substantial fraction of the dosage may enter the atmosphere and be transported over varying distances downwind of the target. The rate and extent of the emission during application, predominantly as spray particle drift, depends primarily on the application method (equipment and technique), the formulation and environmental conditions, whereas the emission after application depends primarily on the properties of the pesticide, soils, crops and environmental conditions. The fraction of the dosage that misses the target area may be high in some cases and more experimental data on this loss term are needed for various application types and weather conditions. Such data are necessary to test spray drift models, and for further model development and verification as well. Following application, the emission of soil fumigants and soil incorporated pesticides into the air can be measured and computed with reasonable accuracy, but further model development is needed to improve the reliability of the model predictions. For soil surface applied pesticides reliable measurement methods are available, but there is not yet a reliable model. Further model development is required which must be verified by field experiments. Few data are available on pesticide volatilization from plants and more field experiments are also needed to study the fate processes on the plants. Once this information is available, a model needs to be developed to predict the volatilization of pesticides from plants, which, again, should be verified with field measurements. For regional emission estimates, a link between data on the temporal and spatial pesticide use and a geographical information system for crops and soils with their characteristics is needed.

Keywords: air quality, application technique, emission factor, emission model, glasshouse, measurement method, pesticides, regional emissions, spray drift, volatilization

---

[1] This paper has been subjected to review by the U.S. Environmental Protection Agency and approved for publication. Mention of trade names or commercial products does not constitute endorsement or recommendation for use.

Water, Air, and Soil Pollution **115:** 195–218, 1999.

1. Introduction

In agriculture, pesticides are applied to the soil or to a crop. The application can be made using different techniques, which depends on the formulation type, the pest to be controlled and the timing of the application. The pesticide can be injected into the soil for use as a fumigant or sprayed onto the soil surface, possibly followed by it's incorporation into the soil top layer. In addition, seeds are sometimes treated with pesticides prior to planting. Crops can be sprayed, for example, with boom sprayers, tunnel sprayers or by aerial application or treated with systemic pesticides.

During the application, a fraction of the dosage is lost to the atmosphere. The application loss is defined as the total fraction of the dosage applied that does not reach the target area. The portion of this loss, in the form of droplets moving off-target (crossing the field border) through the air is referred to, here, as spray drift. Spraying pesticides through spray nozzles produces a spectrum of droplet diameters. Those droplets of smallest size within the spectrum are prone to become lost as spray drift. The larger droplets are carried away by the wind and may be deposited either just outside the target area or at some downwind distance. During aerial transport, the diameter of the droplets can decrease through evaporation of the carrier formulation, and/or the pesticide, and when the diameter is sufficiently small, the droplets or particles can remain airborne with the potential for long-range transport. The fine droplets (diameter less than 100 μm) may evaporate very rapidly. The evaporated component can also travel long distances. Loss during application through spray drift is not usually related to the physico-chemical properties of the pesticide itself, but is largely dependent on the application method, properties of the formulation and carrier (diluent) used and the environmental conditions.

In some cases it is not necessary to spray the whole cropped area (row application, spot spraying). Herbicides can be applied by spot spraying, which means that the sprayer (one or more nozzles) is only turned on when a weeded area is detected. This application method can result in a much lower total loss by both spray drift and volatilization than if the whole field were treated. The extent of reduction depends on the fraction of the field surface sprayed. The use of other formulation types, e.g. solid granules, will eliminate spray drift altogether.

After the pesticide is deposited on the target area, a fraction of this mass volatilizes in the course of time. Some of this volatilization loss may occur during the application period, because no application is instantaneous (e.g. about 15 min per hectare for soil surface spraying; fumigant application is much slower). Consequently, it is difficult to accurately assess the fraction of the dosage not reaching the target area through flux measurements, as the volatilization process can occur as soon as the surface has been sprayed and some time elapses before measuring equipment can be put in place. In particular for the more volatile compounds, the fraction of the dosage lost to the atmosphere by volatilization from the target surface can be substantial during the period of application. Majewski and Capel (1995) present an overview of the sources of emission, measurements and processes determining the emission of pesticides into the atmosphere.

The pesticide applied may be transformed into degradation products which are more toxic than the parent compound (e.g. the transformation of some organophosphorus insecticides to their oxygen analogs). The environmental effect of relevant transformation products should be included in any risk assessment. Moreover, the transformation products may have a much higher vapor pressure than the parent compound (e.g. the formation of the active compound methyl isothiocyanate from metam-sodium). For example, when assessing the health risks from the use of metam-sodium, the volatilization of methyl isothiocyanate and its subsequent dispersion in air should also be assessed. However, there is little information on the rate and the extent of the formation of transformation products for many of the pesticides applied to plants and soils. The transformation products may depend on soil and environmental conditions, which make it very difficult to predict under general conditions. In addition, much more specific chemical, soil and environmental information would be required. Under Directive 91/414/EEC of the European Union (EU, 1991), industry now has to consider relevant metabolites (transformation products) in the risk assessment process. Such information could be obtained using laboratory studies with radiolabelled compounds. Moreover, guidance is needed on the selection of relevant transformation products to be included in the risk assessment.

In this paper a review is presented of the research on the emission of pesticides into the air resulting from applications in agriculture. In addition, the information needed to improve the ability to evaluate the emission potential of pesticide is defined.

2. Emission during Application

To date, many measurements of pesticide emissions during application have been done on the spray drift of droplets or particles. In the United States, a consortium of agrochemical companies has established a Spray Drift Task Force, to collect drift data for current application techniques. These data have been used to develop an assessment method for aerial spray drift (Bird *et al.*, 1995a). In Germany, data on spray drift to adjacent watercourses for various techniques and crops have been collected by Ganzelmeier *et al.* (1995). In the United Kingdom (UK) a database on spray drift has been compiled by the Central Science Laboratory, which is used to underpin regulatory risk assessment, including the recent Local Environmental Risk Assessment for Pesticides (LERAP) arrangements (A Gilbert, personal communication 1999). The UK database has also supported development of the British Crop Protection Council (BCPC) Spray / Nozzle classification scheme (Doble *et al.*, 1985). In particular, the ability to measure a drift potential factor for different spray qualities under controlled conditions in the wind tunnel (Miller *et al.*, 1993) and to reconcile these results against field measured drift levels has enabled a revision of the BCPC scheme to include a drift potential factor (Southcombe *et al.*, 1997). As well as the scheme for classification of size spectra and drift potential of sprays, a more comprehensive basis for classification of application equipment and techniques by hazard has been made (Parkin *et al.*, 1994). Many measurements of spray drift to ditches just downwind of the treated field have been made in the Netherlands (Huijsmans *et al.*, 1997). A draft protocol for the measurement of drift is being discussed by the International Organization for Standardization (ISO, 1999), which would enable a (better) comparison of drift data from various sources that were obtained with different methods.

In most cases drift measurements have been limited to the determination of the mass of pesticide deposited on the surface adjacent to the treated field and on measurements of droplets in the air close to the ground leaving the target area using passive drift collectors. Little is known about the total fraction of the dosage which does not reach the target area. However, data from field experiments indicate that the emission during application can typically range from a few percent (e.g. Maybank *et al.*, 1974; Nordby and Skuterud, 1975; Grover *et al.*, 1988; Glotfelty *et al.*, 1990) to 20-30% (e.g. Nordby and Skuterud, 1977; De Heer *et al.*, 1985; Bird *et al.*, 1995b). In some cases it may be as high as 50% or even more (Warren, 1972; Armstrong, 1973; Wood and Stewart, 1976; Symons, 1977). The amount of pesticide not reaching the target area depends strongly on the application technique, the formulation and the environmental conditions (wind speed, temperature, humidity, atmospheric stability). Although the overall range for this loss is wide, it can be substantially narrowed for a given combination of application technique and environmental conditions.

A portion of the pesticide that does not reach the target area consists of gas-phase pesticide and small droplets or particles (aerosols) which are or have become so small that they cannot be captured effectively by drift collectors. For this fraction, measurement may be possible, but would require some form of iso-kinetic sampling (the flow rate of air into the orifice of the sampling unit is approximately the same as the flow rate of air around it).

3. Emission from Soil

After being deposited on the target area (soil surface and/or plant surface) or after injection or incorporation into the soil, volatilization is one of the processes which can affect the fate of the pesticide. The dominant factors that affect volatilization are the physico-chemical properties of a pesticide (e.g. vapor pressure, water solubility), its persistence in the soil, and environmental conditions (soil and air temperature, soil water content and soil organic matter). Some pesticides can occur both in a neutral and an ionic form in the soil solution. Depending on the pK_a of the pesticide, the volatilization from soil can also be substantially affected by the soil pH (Müller *et al.*, 1998). To illustrate the range in values for physicochemical properties of pesticides, data on basic properties for some common pesticides are listed in Table 1. Depending on the properties of the pesticide, soil and environmental conditions, cumulative volatilization losses range from a few percent to 50 % of the dosage (e.g. Glotfelty *et al.*, 1989; Majewski *et al.*, 1993; Bor *et al.*, 1995; Stork *et al.*, 1998a), or even more in some cases (Glotfelty *et al.*, 1984). The emission of fumigants from soil into the air is strongly affected by the application technique and volatilization losses can be as high as 90% of the dosage (e.g. Gan *et al.*, 1997; Gan *et al.*, 1998).

Soil fumigants form a class of pesticides that are unique since their vapor pressures are many orders of magnitude higher than those of other pesticides. Due to their high volatility, most of the dosage is lost to the atmosphere if no appropriate measures are taken, e.g. deeper injection, covering soil surface with plastic sheeting, soil compaction of the top layer or forming a water seal at the soil surface (Jury *et al*, 1997; Yates *et al.*, 1997b). For example, Yates *et al.* (1996) measured a loss of 64% of the dosage of methyl bromide after injection at a depth of 0.25 m and covering the soil surface with a polyethene film. Deeper injection of the fumigant into the soil results in a higher residence time in the soil. Consequently, a greater part of the dosage may be transformed into non-volatile compounds and volatilization loss is reduced (Yates *et al.*, 1996; Yates *et al.*, 1997a; Yates *et al.*, 1997b; Gan *et al.*, 1997; Gan *et al.*, 1998). Soil compaction (e.g. by rolling) and the supply of water to the soil reduce the air-filled porosity, thereby decreasing the diffusion of the pesticide through the gas phase of the soil system (Gan *et al.*, 1996, Gan *et al.*, 1998). A more drastic measure is wetting the surface soil (common practice in California), which results in a water cap on top of the soil profile.

TABLE 1

Physico-chemical properties of some common pesticides. Values for 25 °C if not stated otherwise. K_{om} = coefficient for the adsorption on organic matter.

Compound	Vapor pressure (Pa)	Water solubility (mg L^{-1})	K_{om} (L kg^{-1})
Atrazine	0.000039[†]	33[¶]	70[§]
Alachlor	0.0021[†]	242[†]	117[§]
Chlorpyrifos	0.0027[†]	1.4[†]	293[§]
Cyanazine	0.0000002[¶]	171[†]	55[§]
EPTC	4.53[‡]	375[†]	61[§]
Methyl bromide	190000[¶]	17500[¶]	2.4[§]
Methyl isothiocyanate	1700[‡‡]	8900[††]	3[§]
Metolachlor	0.0042[†]	488[†]	103[§]
Parathion	0.00089[¶]	11[¶]	1746[§]
Tri-allate	0.016[†]	4[†]	1164[§]
Trifluralin	0.0061[†]	0.18-0.22[†,¶¶]	3775[§]
2,4-D	0.000011[¶]	311[†,§§]	26-230[§,¶¶]

[†]Tomlin, 1997; [¶]at 20 °C, except for atrazine: 22°C, Tomlin, 1997; [§]Linders et al., 1994; [‡]Baker et al., 1996; [‡‡]at 20 °C, Siebering and Leistra, 1979; [††] at 20 °C, Smelt and Leistra (1974); [§§] at pH 1; [¶¶]value depends on pH.

Apart from the fumigants, there are other pesticides whose volatility requires their incorporation into the soil (e.g. tri-allate, vapor pressure of 16 mPa at 25 °C). However, most of the soil-applied pesticides are less volatile than tri-allate, and they can be sprayed onto the soil surface. Soil-incorporation may also be required to eliminate effectively harmful insects in the topsoil.

Another emission pathway for pesticides into the atmosphere occurs when pesticides are sorbed to soil particles and entrained into the atmosphere on wind blown particles (Glotfelty *et al.*, 1989). There are few data on the significance of this transport pathway and on the quantitative effects of soil and environmental factors that influence this process.

3.1 MEASUREMENT METHODS

Several methods have been used to determine the volatilization of fumigant after injection into the soil in the field, such as the aerodynamic-gradient method (Majewski, 1995; Yates *et al.*, 1996; Yates *et al.*, 1997b), a theoretical profile method (Yates *et al.*, 1997b), the integrated horizontal flux method (Yates *et al.*, 1997b) and the box method (e.g. Smelt and Smidt, 1994). The temporal uncertainty in the volatilization rate can be a factor of two. There is less uncertainty in the total loss of the fumigant into the air; this uncertainty has been estimated to be about 10% of the dosage (Yates *et al.*, 1996). The uncertainty in the total loss is a bulk uncertainty which is integrated over time. Therefore, errors are averaged out to some degree.

For pesticides which are sprayed on the soil surface or incorporated, the rate of volatilization can be measured in the field with different micro-meteorological methods, such as the aerodynamic-gradient (Parmele *et al.*, 1972, Glotfelty *et al.*, 1984; Majewski *et al.*, 1993), the Bowen-Ratio (except in moisture limiting situations; Majewski *et al.*, 1990), the theoretical profile (Majewski *et al.*, 1989; Whang *et al.*, 1993), the integrated horizontal flux (Glotfelty *et al.*, 1990, Majewski *et al.*, 1990) and the Eddy accumulation (Majewski *et al.*, 1993) methods. Majewski (1999) presents an overview of these methods and discusses their use and limitations. The difference in the measured total loss by volatilization between these methods is generally less than 20 - 25% (Majewski *et al.*, 1990). At a specific time, the measured volatilization rates resulting from using these methods may differ by a factor 2 - 5 from each other. Even for the same method such differences may occur. As weather conditions change with time, the volatilization loss can be expected to be different for each field experiment. Both temperature and soil moisture conditions at the soil surface have a great effect on the volatilization rate (e.g. Spencer and Cliath, 1974, Taylor and Spencer, 1990). For example, upon re-wetting of the soil surface, the volatilization rate can increase by a factor 2 - 5 (Spencer and Cliath, 1973; Spencer *et al.*, 1982; Spencer and Cliath, 1990). The variability in weather conditions makes it difficult to study the effect of only one factor on the volatilization rate. It should be noted that many of the micrometeorological methods that rely on measurements of horizontal wind speed produce highly uncertain flux estimates when the winds are calm. In fact, a zero wind gradient produces an unrealistic Richardson's number that affects the stability correction and the flux. Most of these methods were developed for turbulent conditions and, probably, should not be used when turbulence is practically absent (nighttime stable/inversion conditions).

In pesticide volatilization field experiments, measurements are mostly done under ideal conditions. The typical experimental field is situated in a flat area with no significant wind obstacles in the vicinity of the field. In practice, however, agricultural fields are also found in hilly areas. Hills and windbreaks may have an effect on pesticide volatilization but these factors have yet to be fully investigated in field experiments.

Field volatility experiments are costly, and volatilization chambers and wind tunnel systems containing a lysimeter (semi-field systems) have been developed to simulate field applications at a lower cost. In these systems, weather variables such as wind speed, air temperature and humidity and solar radiation can be controlled, so field conditions can be approximated by simulating the weather conditions as closely as possible (Stork *et al.*, 1998a; Stork *et al.*, 1998b, Kubiak *et al.*, 1993). The advantage of semi-field systems is that experiments can be reproduced and the factors affecting the volatilization process can be studied. Radio-labeled pesticides can be used to facilitate the analysis of the parent compound as well as any transformation products that are formed. In the volatilization chamber all terms of the mass balance of the pesticide in the soil and air compartments can be quantified, and the experimental time period can be easily lengthened (e.g. hours, days, weeks), therefore, the temporal variation in the volatilization rate as well as the total volatilization for a single application can be quantified.

The volatilization chamber method also has limitations. Substantial differences may occur between the atmospheric conditions in the field and the simulated weather conditions in the volatilization chamber. First, air turbulence in the chamber is likely to be different from that in the field (e.g. no larger scale turbulence). Further, there is always a flow of air in the chamber, so the situation in the field in which there is no wind cannot be simulated. Secondly, photo-degradation of pesticide at the soil surface can be investigated, but as air travel distances in the chamber are very short, photo-degradation in air cannot be effectively studied. Thirdly, the flow of air through the chamber might result in a light vacuum in the volatilization chamber, which might cause an advective transport component and result in measurement error unless similar gas advection occurs elsewhere in the field (i.e. not just in the chamber). A vacuum can be avoided by installing a pressing and a drawing transport blower (Maurer and Kubiak, 1994). The air pressure entering the chamber should be the same as that leaving the chamber. The influence of comparatively small pressure changes with time on the volatilization loss of pesticide is unknown.

Experimental results from volatilization chambers can be equivalent to results obtained from field studies, provided field soil and weather conditions are simulated accurately. Kubiak *et al.* (1995) simulated field applications of isoproturon and methyl parathion in the volatilization chamber and they measured the volatilization in the chamber system during the first 24 hours after application. Volatilization rates were not measured in the field, but there was a good correlation between the pesticide residues on the plants in both systems. However, when simulating field applications in the semi-field system, significant differences between the conditions in this system and those prevailing in the field can occur that may make evaluation of the experimental results more difficult. These differences can be due to differences in the initial penetration of the pesticide, differences in the soil surface temperature (caused by shielding of the soil surface from solar radiation in the semi-field system), and by soil puddling and compaction due to the higher intensity of the sprinkling events in the semi-field system than that of rainfall in the field. For example, due to a combination of such differences, Stork *et al.* (1998a) measured differences between the volatilization rates in the field and those in the wind tunnel system of up to about a factor 10.

The assessment of the volatilization potential of a pesticide using a volatilization chamber has been adopted by the Biologische Bundesanstalt (BBA, Germany) in their registration procedure (BBA, 1990). If the trigger value for the hydrolysis or photolysis half-life for the pesticide is exceeded, then the cumulative volatilization of the pesticide must be determined for the 24-hour period after application. Then further assessment, involving the subsequent stability of the pesticide in air, has to be made if a trigger value of 20% loss is exceeded. For the conditions mentioned in the guideline, the error in the cumulative volatilization measured was estimated to be ± 5%.

To evaluate the various methods to assess the volatilization potential of a pesticide after application, 18 laboratories (in Germany, Switzerland and England) measured the volatilization of three pesticides during the first 24 hours after application under the conditions required by the BBA (Walter *et al.*, 1996). The vapor pressures of these pesticides were $2{\cdot}10^{-6}$, $2{\cdot}10^{-4}$ and $3.5{\cdot}10^{-3}$ Pa (at 20 °C). The water solubilities (at 20 °C) of these compounds were 0.001-0.25 (only range reported), 55 and 4.3 mg L^{-1} respectively. For all the methods used in this inter-laboratory comparison (a ring test in which different laboratories measure the volatilization of a specific pesticide under conditions specified in a guideline), the initial volatilization loss for the compound with the lowest vapor pressure was well below the trigger value of 20%. For the other two compounds, some studies resulted in a value higher than the trigger value and some resulted in a lower value. For those compounds, the method to be used must be evaluated in more detail to assess whether it can give a representative and reliable value for the initial volatilization loss. Further guidance on the use of the method may also be needed.

4. Emission from Crops

The dominant factors that influence the volatilization of pesticides from crops are the physico-chemical properties (see Table 1), the persistence on the plant surface and the environmental conditions (atmospheric stability, wind, temperature and humidity). The persistence on the leaf surface depends on the various dissipation processes, such as photo-degradation (e.g. Liang and Lichtenstein, 1976, Devlin *et al.*, 1987), wash-off from the leaves by rainfall or irrigation (McDowell *et al.*, 1987; Willis *et al.*, 1992), and uptake of the pesticide by the plant leaves. Case studies on photodegradation have been briefly discussed by Leistra (1998). The fate processes on the leaf surface that affect the pesticide are not well understood. Some important factors include the nature of the plant, the age of the plant, the stage of development of the plant (e.g. seedling vs. fully mature), the characteristics of the leaf surface (e.g. waxy cuticle type), and the density and the height of the canopy (Taylor and Glotfelty, 1988). Furthermore, little is known on the effect of the formulation type on the waxy layer. The waxy layer may be partially destroyed when using a specific formulation.

A substantial fraction of the mass of pesticide may be sorbed onto the waxy layer of the plant leaves or, if no waxy layer is present, the sorption may be on other organic materials at the leaf surface layer. Volatilization can be affected by the initial distribution of the pesticide on the leaf surface. Depending on the formulation used for spraying, the pesticide may be uniformly distributed over the leaf surface or it may be concentrated in a number of areas that represent a fraction of the total leaf surface. Therefore, more research is needed on the factors that influence the volatilization of pesticides from plant surfaces. Volatilization experiments have been done for vegetative surfaces (e.g. Turner *et al.*, 1977; Taylor *et al.*, 1977; Grover *et al.*, 1985, Breeze *et al.*, 1992; Kubiak *et al.*, 1995; Van den Berg *et al.*, 1995; Smelt *et al.*, 1997, Stork *et al.*, 1998a) with cumulative volatilization losses ranging from less than a few percent to 60% or more in some cases. However, more experiments are needed to collect data on the volatilization of pesticides in the field for a range of environmental conditions and crop types.

The evaluation of pesticide emissions may be made more complex by the potential of several plant species to produce halogenated pesticides. For example, Gan *et al.* (1998) measured the production of methyl bromide by live *Brassica* plants in significant quantities (daily rates ranging from 20 to 40 ng methyl bromide per g dry plant material) as a result of uptake of Br^- by the plant roots from the soil.

4.1 MEASUREMENT METHODS

The rate of volatilization of pesticide from the plant leaves of most crop types can be measured in the field using the same micro-meteorological methods as those for the measurement of the volatilization rate from soil, e.g. the aerodynamic-gradient methods, the Bowen-Ratio method, and the theoretical profile shape methods. However, these methods will simultaneously measure any volatilization from the soil beneath the plants as well. The measurements on volatilization from a sprayed orchard are more complex because of the structure and height of the trees.

For assessment of the potential volatilization of pesticides from low crops the volatilization chamber and the wind tunnel-lysimeter methods as described in Section 3 are measurement methods that can be used at comparatively low costs. Volatilization can be studied under standard conditions or worst-case conditions. Since conditions can be controlled when chambers are used, experiments can be reproduced so the effect of the various factors governing the volatilization from plant surfaces can be studied. An advantage of the chamber method is that the sprayed surface can be limited to the plant surfaces only, since chambers sample small areas, but, small areal samples also introduce high uncertainty. Before spraying, surfaces other than the plant surfaces can be covered with paper that is removed after spraying (Kubiak *et al.*, 1993). The error in the cumulative volatilization measured with this system has been estimated to be less than 10%. The limitations of the chamber method have been mentioned in Section 3. To date, several volatilization experiments using a chamber or wind tunnel-lysimeter system have been reported (e.g. Stork *et al.*, 1998a; Müller *et al.*, 1998). However, more simultaneous field and chamber experiments are needed to verify the data obtained with the volatilization chamber.

5. Emission from Glasshouses

The use of pesticides in glasshouses can result in substantial emissions to the outdoor environment (Baas and Huygen, 1992). The rate of this emission is dependent on the ventilation rate of air in the glasshouse and the concentration of pesticide in the glasshouse air. The initial concentrations in the glasshouse air depend to a large extent on the application technique used. For example, much higher concentrations will occur when using a low-volume mister than when using a high volume technique (i.e. spray bar with 6 nozzles, pressure 1200 kPa). After application, a fraction of the mass deposited on the surfaces within the glasshouse volatilizes. Although the key processes that affect the volatilization are largely the same as those in the field, the environmental conditions in a glasshouse differ greatly, as they are much more controlled. Firstly, the short wavelength UV light (290-310 nm) responsible for many outdoor photolytic reactions may be filtered out by the glass, so photo-degradation occurring outdoors may not happen in the glasshouse. In some glasshouses, artificial light sources are used to promote the growth of the crop. Whether the light from these sources causes photo-degradation of the pesticide will depend on the UV spectrum that the lights emit. Secondly, there is little flow of air except when windows are opened. Furthermore, there is a tendency towards higher air temperature and humidity in the glasshouse compared with conditions in the field. In addition, the glasshouse may contain plastic materials, which may adsorb substantial amounts of pesticide in the glasshouse air following application (Bor *et al.*, 1994).

Because of the limited air movement in the glasshouse, limited ventilation with outside air, and relatively long residence time inside the glasshouse, the pesticide concentration in the glasshouse air can be expected to be high. The built-up of the concentration inside the glasshouse may result in a somewhat lower rate of volatilization from the plant leaves than under similar outdoor conditions which may be offset, somewhat, by higher temperature conditions inside the glasshouse.

The rate of emission of the pesticide from the glasshouse into the atmosphere can be determined by measurement of the fractional rate of ventilation of the glasshouse air (by leakage in the glasshouse structure, open windows and/or open doors) and the concentration of the pesticide in the glasshouse air. In the Netherlands, such measurements have been done to estimate the emission of pesticides into the air in a region with many glasshouses (e.g. Baas and Bakker, 1996). Models to predict the ventilation rate have been discussed by Fernandez and Bailey (1992).

The emission of pesticides from a glasshouse can be reduced in several ways. Firstly, the rate of ventilation in the glasshouse can be lowered by improvement of the structure. Secondly, comparatively high concentrations in air may be avoided by selecting an application technique with coarser spray droplets in combination with a less volatile pesticide (lower vapor pressure). The opening of windows and doors should be avoided when there are high concentrations of the pesticide in the glasshouse air, e.g. during application and the first few hours thereafter. Ultimately, emissions can even be avoided by technical means; this would necessitate the installation of a device to clean all exhaust air from the glasshouse but this would prove to be expensive. A feasibility study by Van Os *et al.* (1993) indicated that the reduction in emission out of the glasshouse by cleaning the air by carbon filters may be limited and not very practical. However, when assessing the health risk of glasshouse workers, the effect of measures reducing the emission of pesticide out of the glasshouse into the atmosphere should be taken into consideration, because a lower emission out of the glasshouse may result in greater concentrations in air prevailing in the glasshouse.

In most countries only about 1% of the agricultural area or less is under glass, so the contribution of the emission from glasshouses to air compared to the total emission of agricultural pesticides may be of limited importance. For the risk evaluation of the exposure of people living in the neighborhood of glasshouses as well as that of nearby ecosystems, the emission of pesticide from the glasshouse into the atmosphere can be an important issue.

6. Estimation and Modeling

6.1 Emission during Application

To date several spray drift models have been developed, such as AgDrift (Bird *et al.*, 1997), PEDRIMO (PEsticide DRift MOdel, Kaul *et al.*, 1996), IDEFICS (IMAG program for Drift Evaluation from FIeld sprayers by Computer Simulation, Holterman *et al.*, 1994, Holterman *et al.*, 1997) and others (Thompson and Ley, 1983; Walklate, 1992; Hashem and Parkin, 1991). The AgDrift 1.0 model has been tested and documented (Bird *et al.*, 1997). The PEDRIMO model has also been tested against measurements and a summary of these tests is given by Kaul *et al.* (1996). The IDEFICS model is being evaluated and documentation is not yet available. Although these models were not developed for the assessment of the fraction of the particles and droplets that remains air-borne and the vapor phase pesticide formed by evaporation of the droplets, they can be used to estimate this loss. To do this assessment some adjustment of the model may be required. Moreover, standardization of the definition and description of alternative pesticide application methods would promote a better use of available models and datasets (Gilbert, 1999, personal communication).

6.2 Emission after Application

An estimate of the initial volatilization rate after spraying on the soil surface can be made using the physico-chemical properties of the pesticide: vapor pressure, water solubility and the coefficient of the sorption on organic carbon (or organic matter). A good indicator of volatilization is the effective vapor pressure of the pesticide, i.e. the vapor pressure that is in equilibrium with the concentration in the liquid phase and the mass sorbed on the soil surface. A good correlation (n =

12, $r^2 = 0.988$) between the logarithm of the ratio of the vapor pressure divided by the water solubility and the organic carbon sorption coefficient (K_{oc}) and the logarithm of the volatilization rate as measured during the first day after application, was obtained by Woodrow *et al.* (1997), although more data sets are needed to verify this method. Further, some of the data used in the correlation were obtained by residue analysis, which provides an indirect measurement of the total flux. For soil incorporated pesticides, the above mentioned ratio had to be multiplied by the application rate divided by the depth of incorporation ($r^2 = 0.93$). For surface applied pesticides, a similar approach has been made by Smit *et al.* (1997), who estimated the cumulative loss by volatilization during the first 21 days after application from the fraction of the pesticide in the gas phase of the soil system. For normal to moist field conditions r^2 was calculated to be 0.76 ($n = 22$) and, for dry soil conditions, it was 0.89 ($n = 7$). For this correlation only flux measurements were used. Both approaches take the effect of temperature on the physico-chemical properties of the pesticide into account. The uncertainty in the calculated pesticide flux density and that in the cumulative loss depends on the quality of the underlying data used in the respective correlation.

A screening-level estimate of the initial volatilization rate after spraying of the crop can be made using the vapor pressure of the pesticide. Woodrow *et al.* (1997) observed a good correlation between the logarithm of the volatilization rate and the logarithm of the vapor pressure. For the cumulative loss from plant surfaces an estimation method has been developed by Smit *et al.* (1998). Using literature data on volatilization rates from plant surfaces as measured in the field or in volatilization chambers, the best correlation was found between the logarithm of cumulative loss by volatilization (over a period of the first 7 days after application) and the vapor pressure of the pesticide ($n = 19$, $r^2 = 0.78$). The correlation between the cumulative loss and the ratio between the vapor pressure and the sorption coefficient on organic carbon was less clear ($r^2 = 0.55$).

For a more accurate calculation of the volatilization flux a model is needed which describes a pesticide's fate in the soil and the exchange with the lower part of the atmospheric boundary layer. For soil fumigants, a number of models have been used (e.g. Leistra, 1972; Jury *et al.*, 1983; Van den Berg, 1992; Baker *et al.*, 1996; Freijer *et al.*, 1996; Wang *et al.*, 1997; Wang *et al.*, 1998). In these models processes occurring in the soil are described at various levels of detail. Relevant processes are the transformation of fumigant in the soil and the diffusion and convection of fumigant in the gas phase. The transformation of fumigant is mostly described with first-order kinetics (Siebering and Leistra, 1979; Wagenet *et al.*, 1989; Leistra and Crum, 1990; Van den Berg, 1992). For the diffusion in the gas phase of the soil, several models have been used to calculate the diffusion coefficient (Troeh *et al.*, 1982; Millington and Quirk, 1960; Currie, 1965; Freijer,

1994). Convective transport of fumigant can be caused by changes in soil water content, temperature and air pressure at the soil surface. A model for air flow driven by air pressure changes at the soil surface has been described by Chen *et al.* (1995). The description of the process of exchange of soil fumigant across the soil-air interface is mostly simple. At the interface, a zero concentration of pesticide is incorrectly assumed or a thin stagnant air layer is assumed to exist through which the pesticide must diffuse (molecular diffusion), before it can enter the turbulent air. Van den Berg *et al.* (1999) compared volatilization rates of methyl isothiocyanate from soil after application of metam-sodium into the soil computed with a model based on this simple air boundary layer concept with measured data. The computed rates corresponded roughly to those measured in the field. However, the existence of a stagnant boundary layer is questionable because it precludes mass flow of air across the soil-air interface. Therefore, this concept should be considered a simplification of the processes that really occur at the boundary layer. For an adequate description, a direct coupling of the soil profile with the atmosphere is needed, which would require a sophisticated model. In a recent study on the volatilization from bare soil, Baker *et al.* (1996) developed a method to estimate the transport resistance of the boundary layer for general atmospheric conditions. This approach is rather complex (the resistance is expressed as a function of the Schmidt and Reynolds numbers) and it requires many meteorological input data. During the first day after application, the computed volatilization flux of EPTC (S-ethyl dipropylthiocarbamate) corresponded well to the measurements, but thereafter the model overestimated the volatilization flux. Further testing of the model is needed under different weather and soil conditions.

For soil surface applied pesticides, reliable models are not yet available, although a comprehensive model is being tested (Scholtz *et al.*, 1997). The initial distribution of the pesticide in the soil needs to be described adequately. Freijer *et al.* (1996) proposed the concept of a thin pesticide layer on the surface from which all pesticide is leached into the soil at the time of the first rainfall. The model should also take into account other loss processes at the soil surface, such as photodegradation and the occurrence of non-equilibrium conditions in the partitioning of the pesticide over the soil phases. The most important factors that determine the volatilization rate from soil are the soil moisture conditions and the soil surface temperature. These two variables change considerably, not only from day to day but during the course of a 24-hour period, as well. When the top few mm of the soil surface layer dries out, the moisture content can decrease below a few percent. At these low moisture conditions, pesticides are much more strongly adsorbed to soil and so the volatilization flux can be expected to be at a low level (Spencer and Cliath, 1973). Upon rewetting of the soil surface, the volatilization flux increases substantially (Spencer and Cliath, 1973; Spencer *et al.*, 1982;

Spencer and Cliath, 1990). Further research is needed to determine the importance of this process with respect to the vapor pressure of the pesticide. Temperature can have a large effect on the partitioning of the pesticide between the gas and liquid phases (e.g. the Henry coefficient). A higher temperature results in higher concentrations in the gas phase. Diurnal temperature differences can be substantial (more than 10 °C), in particular under clear skies. More research is needed on the description of the soil water content and temperature gradients and their temporal behavior in the top few mm of the soil profile to improve the volatilization model for soil surface applications of pesticide.

A model in which exchange between the air and soil compartment is described with the simple concept of a stagnant air boundary layer (e.g. as in the Behavior Assessment Model (BAM) (Jury *et al.*, 1983, Jury *et al.*, 1984), the PESTicide Leaching and Accumulation model (PESTLA) (Van den Berg and Boesten (1999), Van den Berg *et al.*, (1999)) is useful for screening purposes. However, it should be noted that rough estimates may not be good enough for a further step in the risk assessment procedure (higher tier), so computations would be needed with a model which describes all relevant processes adequately and which has been tested and validated against field measurements.

No models are currently available for estimating the volatilization flux of pesticides from plants. More research is needed on describing the processes and factors that affect the fate of the pesticide on the plant leaf, such as the development stage of the plant, the characteristics of the plant leaf surface, photo-degradation, uptake by the plant and wash-off of the pesticide by rainfall and irrigation. There is no model yet available that handles photo-degradation on plant surfaces. A plant growth model may be a good starting point for modeling processes occurring on the plant leaf surface.

After spraying a pesticide on the crop, a fraction of the dosage is deposited on the soil surface. Furthermore, a part of the mass deposited on the plant surface may be washed off and be deposited on the soil surface. When measuring the volatilization rate, losses of pesticide from both the soil and the plant surfaces contribute to this flux. In the field it is not possible to distinguish between the two volatilization sources, which would make model testing more difficult. However, the volatilization of pesticide from solely the plant surface can be measured in a volatilization chamber.

7. Regional Emissions

Information is needed on the location and timing of the applications of a pesticide in an agricultural area to assess the total risk of transport of the pesticide via the air to great distances from this area. This involves information on the type of application of the pesticide (crop, soil), the mode of application (equipment and technique), the dosage and the temporal and spatial distribution of the application of the pesticide. The degree of detail in the data requirements for this type of assessment depends on the scale for which the assessment is made. An emission estimate has been made for 9 pesticides used in North America (Scholtz and Voldner, 1992; Scholtz *et al.*, 1997). Sales and use information were collected or disaggregated to a county level by crop type (county diameter ranging from 30 to > 140 km) and this information was aggregated on the surface area of the relevant crop for the area studied, using 127 km grid cells on a polar stereographic projection. For each grid cell, annual and seasonal emission factors were calculated for each pesticide -mode of application combination using submodels for volatilization from soil and crops. In general, the distribution of pesticide emissions corresponded to the usage pattern. As expected, the largest emissions occurred in spring and summer. Because simplifications were made in the procedure for the calculation of the emission factors, the results should be treated with caution. Further model development and testing is needed, in particular on the volatilization from crops.

The availability of detailed information on pesticide use differs from country to country. In the State of California, US, pesticide use permit data are compiled in each township. Using this information, the mass of pesticide used can be calculated on a weekly or monthly basis. For the entire US, county level pesticide use data were compiled by the US Geological Survey based on the 1992 Agricultural Census. The data are available in both map and digitized geographical information system–compatible data bases. In the Netherlands, the use of pesticides is estimated from relevant factors, such as the crop type, the total area of the crop, the dosage, the time of application and the frequency of use (ISBEST: Information System for Pesticides, Lentjes and Denneboom, 1996). This information is linked with a geographical information system in which data are available on the total crop area for each county. For several groups of pesticides, the total estimated use has been compared with information on pesticide sales. For soil fumigants the correlation was good (105%), but for the fungicides (72%), insecticides (51%) and herbicides (62%), there was a substantial difference between estimated use and pesticide sales (Smidt *et al.*, 1999). At present, such data on regional pesticide use are not available in Germany.

For emission of pesticides from bare soils, the organic matter content of the topsoil should be taken into account. The geographical information system could provide data on the occurrence of a soil type in the area studied, together with the organic matter content in the topsoil. In the US, a national data base, the State Soil Geographic Database (STATSGO), containing organic matter estimates at a map scale of 1:250,000, with some assumptions of spatial homogeneity and intended for regional application has been developed (United States Department of Agriculture, 1994). A more detailed data set named the Soil Survey Geographic (SSURGO), which is as detailed as map scale 1:12000, is under development. For a screening assessment, the emission of pesticide from bare soil can be estimated with a model such as the Behavior Assessment Model (Jury *et al.*, 1983). Using this model, output on the emission of pesticide into the air can be provided on a daily basis.

Using information from a geographical database and linking these data to information on the volatilization flux of pesticide from each soil type under a given set of environmental conditions, estimates on the total emission into the atmosphere from applications in the region can be obtained for each pesticide-soil combination. A similar approach can be followed to obtain estimates of the total emission for each pesticide-crop combination.

8. Conclusions and Recommendations

Little information is available on the fraction of the dosage of the pesticide that misses the target surface and is lost, mostly via spray drift, during application. Because 30% or more (greater than 50% in some cases) of the dosage can be lost during application, depending upon application technique, formulation and environmental conditions, more data are needed for a range of application types and weather conditions. In addition, these data are also necessary to verify the computed results of drift models and in further model development and verification. Standardization of the definition and description of alternative application techniques would promote a better use of both models and data. Accurate data on the loss during application are necessary for an adequate risk evaluation and they are also needed to identify measures which could reduce pesticide losses during application.

The volatilization of soil fumigants and soil incorporated pesticides following application can be reasonably well estimated with models. With the help of a model for fumigant behavior in soil, the effect of possible measures to reduce the emission into the air can be quantified. However, as the differences observed between the computed and measured rates were sometimes substantial, further development and testing is needed to improve the reliability of both models and measurement methods.

The rate and extent of volatilization of many soil surface-applied pesticides is uncertain. Further model development is needed as well as accompanying laboratory and field studies investigating those factors that can affect the volatilization process, in particular the soil water content and the temperature gradients within the top few mm of the soil. This should result in better descriptions relating the pesticide volatilization rate with the physico-chemical properties (vapor pressure and water solubility) and the sorption to soil particles under variable temperature and soil water conditions.

The stagnant air-boundary layer concept is commonly used for screening pesticides on their volatilization potential. For a more advanced risk assessment, a volatilization model is needed which describes the processes at the soil surface adequately, in particular for soil surface-applied pesticides.

More data are needed on the volatilization of pesticides from plant surfaces. More data are also needed about the effect of the various processes on the fate of the pesticide on the plants, such as photo-degradation, uptake by plant leaves, sorption and wash-off. These data are required for the development of a model describing the fate of pesticides on the crop.

Depending on pesticide properties, application technique, soil, crop and environmental conditions, volatilization losses of pesticides after application range from less than a few percent to 50-60% of the dosage or even more in some cases. It should be noted that incorporation of the pesticide into the soil directly after spraying greatly reduces the volatilization of the pesticide compared to soil surface applications.

Chamber-lysimeter experiments are useful tools to determine the volatilization potential of pesticides from plants and soils. The data obtained in such experiments can also be used for further model development. The chambers are also useful tools to asses the fate of the transformation products. The duration of the experiments in such systems can be prolonged to at least several weeks, so a better insight into the processes that affect pesticide volatilization can be obtained with varying time periods and controlled environmental conditions. Many important weather conditions in the field (e.g. wind speed, air temperature, solar radiation) can be reasonably well simulated in volatilization chambers, but the simulation of some processes in the chamber, such as rainfall and turbulence needs to be improved. The degree and type of air turbulence in the field cannot be representatively simulated in the volatilization chamber.

Because in most countries only about 1% of the agricultural area or less is under glass, the emission from glasshouses may be of limited importance. However, for the risk evaluation of exposure of man and ecosystems to pesticides in air in the neighborhood of a glasshouse area, the emission of pesticide from the glasshouse into the atmosphere should be considered.

For regional emissions detailed and accurate input data are needed which involves information on the spatial and temporal pesticide use, soil and/or crop type, the dosage applied and the mode of application. This information should be linked with a geographical information system that provides information on the area of a specific soil or crop with its characteristics within a predefined area.

Acknowlegdements

The authors would like to thank Dr A.J. Gilbert for useful suggestions on sections 2 and 6.1.

References

Armstrong, J.: 1973, Meteorological aspects of drift, paper presented at the Pesticide Accountancy Workshop, Ottawa, Canada, April, 1973.

Baas, J. and C. Huygen: 1992, *Emission of crop protection products from glasshouses into the air (in Dutch)*, TNO-IMW Report IMW-R 92-304, Netherlands Organization for Applied Scientific Research, Delft, The Netherlands.

Baas, J. and D.J. Bakker: 1996, *Exposure to pesticides: Concentrations in air, soil, water and vegetation by emission from glasshouses (in Dutch)*, TNO Report R 96-313, Netherlands Organization for Applied Scientific Research, Apeldoorn, The Netherlands.

Baker, J.M., W.C. Koskinen and R.H. Dowdy: 1996, *J. Environ. Qual.* **25**, 169-177.

BBA: 1990, *Richtlinien für die Prüfung von Pflanzenschutzmitteln im Zulassungsverfahren*, Teil IV, 6-1, Prüfung des Verflüchtigungsverhaltens und des Verbleibs von Pflanzenschutzmitteln in der Luft, Biologische Bundesanstalt für Land- und Forstwirtschaft, Bundesrepublik Deutschland, Juli 1990, Saphir Verlag, Ribbesbüttel.

Bird, S.L., S.L. Ray, M.E. Teske, D.M. Esterley, S.G. Perry and D.I. Gustafson: 1995a, *A proposed screening level assessment method for aerial spray drift of pesticides*, National Exposure Research Laboratory, US EPA, Athens, GA, USA.

Bird, S.L., D.M. Esterly, and S.G. Perry: 1995b, *J. Environ. Qual.* **25**, 1095-1104.

Bird, L., S.G. Perry, S.L. Ray, M.E. Teske and P.N. Scherer: 1997, *An evaluation of AgDrift 1.0 model for use in aerial applications*, National Exposure Research Laboratory, US EPA, Athens, GA, USA.

Bor, G., F. Van den Berg, J.H. Smelt, A.E. Van de Peppel-Groen, M. Leistra and R.A. Smidt: 1994, *Deposition patterns of dichlorvos and parathion in a glasshouse and discharge of parathion with condensation water*, DLO-Winand Staring Report 84, Wageningen, The Netherlands.

Bor, G., F. van den Berg, J.H. Smelt, R.A. Smidt, A.E. Van de Peppel-Groen and M. Leistra: 1995, *Volatilization of tri-allate, ethoprophos and parathion measured with four methods after spraying on a sandy soil*, DLO-Winand Staring Report 104, Wageningen, The Netherlands.

Breeze, V.G., Simmons, J.C. and M.O. Roberts: 1992, *Pest. Sci.* **36**, 101-107.

Chen, C., R.E. Green, D.M. Thomas and J.A. Knuteson: 1995, *Environ. Sci. Technol.* **29**, 1816-

1821.

Currie, J.A.: 1965, *J. Soil Sci.* **16**, 279-289.

De Heer, H., C.J. Schut, H.A.J. Porskamp and L.M. Lumkes: 1985, *Gewasbescherming (in Dutch)*, **16**, 185-197.

Devlin, D.L., D.R. Gealy and L.A. Morrow: 1987, *Weed Sci.* **35**, 775-779.

Doble, S.J., G. A. Matthews, I.. Rutherford and E. S. E. Southcombe: 1985, A system for classifying hydraulic nozzles and other atomisers into categories of spray quality. *Proceedings British Crop Protection Conference - Weeds* pp. 1125 - 1134.

EU, 1991: Council directive 91/414/EEC of 15 July 1991 concerning the placing of plant protection products on the market, *Official Journal of the European Communities*, L230, 1-32.

Fernandez, J.E. and B.J. Bailey: 1992, *Agric. Forest Meteorol.* **58**, 229-245.

Freijer, J.I.: 1994, *Soil Sci. Soc. Am. J.* **58**, 1067-1076.

Freijer, J.I., A. Tiktak, S.M. Hassanizadeh and A.M.A. van der Linden: 1996. *PESTRAS v3.1: A one-dimensional model for the assessment of the fate of pesticides in soil*, RIVM Report 715501007, Bilthoven, The Netherlands.

Gan, J., S.R. Yates, D. Wang and W.F. Spencer: 1996, *Environ. Sci. Technol.* **30**, 1629-1636.

Gan, J., S.R. Yates, W.F. Spencer, M.V. Yates and W.A. Jury: 1997, *J. Environ. Qual.* **26**, 310-317.

Gan, J., S.R. Yates, H.D. Ohr and J.J. Sims: 1998, *Geophys. Res. Letters* **25**, 3595-3598.

Gan, J., S.R. Yates, D. Wang and F.F. Ernst: 1998, *J. Environ. Qual.* **27**, 432-438.

Ganzelmeier H., D. Rautmann, R. Spangenberg, M. Sreloke, M. Herrmann, H-J. Wenzelburger and H-F. Walter: 1995, *Studies on the spray drift of plant protection products*, Deutsche Bundesanstalt für Land- und Forstwirtschaft, Berlin-Dahlem, Heft 305, Blackwell Wissenschaftverlag GmbH Berlin.

Gilbert, A. J.: 1999, Personal communication. Pesticides Group, Central Science Laboratory, Sand Hutton,York YO41 1LZ, UK

Glotfelty, D.E., A.W. Taylor, B.C. Turner and W.H. Zoller: 1984, *J. Agric. Food Chem.* **32**: 638-643.

Glotfelty, D.E., M.M. Leech, J. Jersey and A.W. Taylor: 1989, *J. Agric. Food Chem.* **37**, 546-555.

Glotfelty, D.E., C.J.Schomburg, M.M. McChesney, J.C. Sagebiel and J.N. Seiber: 1990, *Chemosphere* **21**, 1303-1314.

Grover, R., S.R. Shewchuk , A.J. Cessna, A.E. Smith and J.H. Hunter: 1985, *J. Environ. Qual.* **14**, 203-210.

Grover, R., A.E. Smith, S.R. Shewchuk, A.J. Cessna, and J.H. Hunter: 1988, *J. Environ. Qual.* **17**, 543-550.

Hashem, A. and C.S. Parkin: 1991, *Atmos. Environm.* **25A**, 1609-1614.

Holterman, H.J., H.A.J. Porskamp and J.F.M. Huijsmans: 1994, *Modelling spray drift from boom sprayers*, Report N. 94-D-148 Ag. Eng. 1994 Milan.

Holterman, H.J., J.C. Van de Zande, H.A.J. Porskamp and J.F.M. Huijsmans: 1997, *Computers and Electronics in Agric.* **19**, 1-22.

Huijsmans, J.F.M., H.A.J. Porskamp and J.C. van de Zande: 1997. *Spray drift (reduction) in crop protection application technology. Evaluation of spray drift in orchards, field crops and nursery tree crops spraying.* IMAG-DLO Report 97-04, 41 p. (in Dutch with summary in English), DLO Institute of Agricultural and Environmental Engineering, Wageningen, the Netherlands.

ISO: 1999, ISO/CD 12057 *Equipment for crop protection - Methods for the field measurement of spray drift.* Document ISO/TC23/SC6 N299.

Jury, W.A., W.F. Spencer and W.J. Farmer: 1983, *J. Environ. Qual.* **12**, 558-564.

Jury, W.A., W.J. Farmer and W.F. Spencer: 1984, *J. Environ. Qual.* **13**, 567-572.

Jury, W.A., Y. Jin, J. Gan and T. Gimmi: 1997, In: Seiber, J.N., J.A. Knuteson, J.E. Woodrow, N.L. Wolfe, M.V. Yates, S.R. Yates (Eds), *Fumigants: Environmental Fate, Exposure, and Analysis*, ACS Symposium Series 652, pp. 104-115.

Kaul, P., S. Gebauer, R. Neukampf and H. Ganzelmeier, 1996. *Nachrichtenbl. Deut. Pflanzenschutzd.* **48**, 21-31.

Kubiak, R. , Maurer, T and K.W. Eichhorn: 1993, *Sci. Total Environm.* **132**, 115-123.

Kubiak, R., T. Müller, T. Maurer and K.W. Eichhorn: 1995. *Int. J. Environ. Anal. Chem.* **58**, 349-358.

Leistra, M.: 1972, *Diffusion and sorption of the nematicide 1,3-dichloropropene in soil*, Agricultural Research Reports, 769, Centre for Agricultural Publishing and Documentation, Wageningen, The Netherlands.

Leistra, M.: 1998, *Extent of photochemical transformation of pesticides on soil and plant surfaces*, Environmental Planning Bureau Series 5, DLO-Winand Staring Centre, Wageningen, The Netherlands.

Leistra, M. and S.J.H. Crum: 1990, *Water Air Soil Pollut.* **50**, 109-121.

Lentjes, P.G. and J. Denneboom: 1996, *Data and program description of ISBEST version 2.0 (In Dutch)*, Technical Report 31, DLO Winand Staring Centre, Wageningen, The Netherlands.

Liang, T.T. and E.P. Lichtenstein: 1976, *J. Agric. Food Chem.* **24**, 1205-1210

Linders, J.B.J.H., J.W. Jansma, B.J.W.G. Mensink and K. Otermann: 1994, *Pesticides, benefaction or pandora's box? A synopsis of the environmental aspects of 243 pesticides*. RIVM Report 679101014, 204 pp.

Majewski, M.S. and P. D. Capel, 1995. *Pesticides in the atmosphere: distribution, trends and governing factors*, Ann Arbour Press, Michigan, USA.

Majewski, M.S., D.E. Glotfelty and J. N. Seiber: 1989. *Atmos. Environ.* **23**, 929-938.

Majewski, M.S., D.E. Glotfelty, K.T. Paw U and J.N. Seiber: 1990. *Environ. Sci. Technol.* **24**, 1490-1497.

Majewski, M., R. Desjardins, P. Rochette, E. Pattey, J. Seiber and D. Glotfelty: 1993. *Environ. Sci. Technol.* **27**,121-128.

Majewski, M.S., McChesney, M.M., Woodrow, J.W., Prueger, J.H. and J.N. Seiber: 1995, *J. Environ. Qual.* **24**, 742-752.

Majewski, M.S.: 1999, *Water, Soil and Air Pollut.* this issue.

Maurer, T., and R. Kubiak: 1994, *Ecoinforma* **9**, 567-579.

Maybank, J., K. Yoshida, and R. Grover: 1974, *Can. J. Plant Sci.* **54**, 541-546.

McDowell, L.L., G.H. Willis, L.M. Southwick and S. Smith: 1987, *J. Pestic. Sci.* **21**, 83-92.

Miller, P. C. H. M., E. C. Hislop, C. S. Parkin, G. A. Matthews and A. J. Gilbert: 1993, *The classificiation of spray generator performance based on wind tunnel assessments of spray drift.* ANPP-BCPC 2nd International Symposium on Pesticide Application Techniques, Strasbourg, 22-24th September 1993, 109 - 116

Millington, R.J. and J.P. Quirk: 1960, In F.A. van Beren et al., *Transport in porous media*, Trans. Int. Congress Soil Sci. Soc. 7th, Volume 1, Madison, WI, Elsevier, Amsterdam, pp 97-106.

Müller, T., Staimer, N. and R. Kubiak: 1998. *Pestic. Sci.* **53**, 245-251.

Nordby, A. and R. Skuterud: 1975, *Weed Res.* **14**, 385-395.

Parkin C.S., A.J. Gilbert, E.S.E. Southcombe and C.J. Marshall: 1994, *Crop Protection* **13**, 281-285.

Parmele, L.H., E.R. Lemon and A.W. Taylor, 1972, *Water, Air Soil Pollut.* **1**, 433-451.

Scholtz M.T, and E.C. Voldner: 1992. *Estimation of Pesticide Emissions to the Air Resulting from Agricultural Applications, Atmospheric Chemistry*: Proceedings of the 9th World Clean Air Congress, Montreal, August 30 - September 4, Air and Waste Management Association, 12 pp.

Scholtz, M.T., A.C. McMillan, C. Slama, :Y.-F. Li, N. Ting, and K. Davidson: 1997. *Pesticide Emissions Modelling: Development of a North American Pesticide Emissions Inventory*, Final Report, Canadian Global Emissions Interpretation Centre, CGEIC-1997-1, Mississauga, Ontario, Canada.

Siebering, H. and M. Leistra: 1979, In: D. Mulder (Ed) *Soil disinfestation*, Elsevier, Amsterdam, pp 135-161.

Smelt J.H. and M. Leistra: 1974, *Pestic. Sci.* **5**, 401-407.

Smelt, J.H. and R.A. Smidt: 1994. *A field method to measure the rate of volatilization of fumigants from soil*, Proceedings 5th International Workshop on the Environmental Behaviour of Pesticides and Regulatory Aspects, Brussels, April 26-29, European Study Service, Belgium, p 449.

Smelt, J.H., R.A. Smidt, F. van den Berg, A.M. Matser, A. Stork and H. Ophoff: 1997. *Volatilization of fenpropimorph and clopyralid after spraying onto a sugar beet crop*, Report 136, DLO-Winand Staring Centre, Wageningen, The Netherlands.

Smidt, R.A., G. Bor and R.C.M. Merkelbach: 1999, *ISBEST 3.0*, Technical Report Winand Staring Centre Wageningen, The Netherlands (in prep.)

Smit, A.A.M.F.R., F. Van den Berg and M. Leistra: 1997, *Estimation method for the volatilization of pesticides from fallow soils*, Environmental Planning Bureau series 2, DLO Winand Staring Centre, Wageningen, The Netherlands.

Smit, A.A.M.F.R., M. Leistra and F. van den Berg: 1998, *Estimation method for the volatilization of pesticides from plants*, Environmental Planning Bureau series 4, DLO Winand Staring Centre, Wageningen, The Netherlands.

Southcombe, E. S. E., P. C. H. Miller, H. Ganzelmeier, J. C. van de Zande, A. Miralles and A. J. Hewitt: 1997, *The International (BCPC) Spray Classification System including a drift potential factor*. Proceedings Brighton Crop Protection Conference - Weeds, 371-380.

Spencer, W.F., and M.M. Cliath: 1973. *J. Environ. Qual.* **2**, 284-289.

Spencer, W.F., and M.M. Cliath: 1974, *J. Agric. Food Chem.* **22**, 987-991.

Spencer, W.F., Farmer, W.J. and W.A. Jury: 1982. *Environ. Toxicol. Chem.* **1**, 17-26.

Spencer, W.F. and M.M. Cliath: 1990. In Kurtz, D.A. (ed.), *Long Range Transport of Pesticides*, Lewis Publishing Co., Chelsea, MI, pp. 1-16.

Stork A., H. Ophoff, J.H. Smelt and F. Führ: 1998a. In: F. Führ, R.J. Hance, J.R. Plimmer and J.O. Nelson (eds.), *The Lysimeter Concept – Environmental Behaviour of Pesticides*, ACS Symposium Series 699, American Chemical Society, Washington DC, USA, pp. 21-39.

Stork, A., H. Ophoff, J.H. Smelt and F. Führ: 1998b. *Z. PflKrankh. PflSchutz. Sonderh.* **XVI**, 745-755.

Symons, P.E.K.: 1977, *Residue Rev.* **68**,1.

Taylor, A.W., D.E. Glotfelty, B.C. Turner, R.E. Silver, H.P. Freeman and A. Weiss: 1977. *J. Agric. Food Chem.* **25**, 542-548.

Taylor, A.W. and D.E. Glotfelty: 1988. In Grover, R. (ed.), *Environmental Chemistry of Pesticides*, Vol. 1, CRC Press, Boca Raton, FL, pp. 89-129.

Taylor, A.W. and W.F. Spencer: 1990, In: H.H. Cheng (ed.), *Pesticides in the Soil Environment – Processes, Impacts and Modeling*, Soil Sci. Soc. Amer. Book Series 2, Madison, Wisconsin, USA, pp. 213-269.

Thompson, N. and A.J. Ley: 1983, *J. Agric. Engineering Res.* **28**, 419-435.

Tomlin, C: 1997, *The Pesticide Manual (11th ed.)*. British Crop Protection Council, Bracknell, Berkshire, UK.

Troeh, F.R. J.D. Jabro and D. Kirkham: 1982, *Geoderma* **27**, 239-253.

Turner, B.C., D.E. Glotfelty and A.W. Taylor: 1977. *J. Agric. Food Chem.* **25**, 548-550.

United States Department of Agriculture: 1994, *State Soil Geographic (STATSGO) Data Base: Data Use Information*, Natural Resources Conservation Service, National Soil Survey Center, Miscellaneous Publication No. 1492, 109 pp.

Van den Berg, F.: 1992. *Emission of fumigants from soil and dispersion in air*. Ph. D Thesis, Agricultural University, Wageningen, The Netherlands.

Van den Berg, F., G. Bor, R.A. Smidt, A.E. Van de Peppel-Groen, J.H. Smelt, T. Müller and T. Maurer: 1995, *Volatilization of parathion and chlorothalonil after spraying onto a potato crop*, Report 102, DLO-Winand Staring Centre, Wageningen, The Netherlands.

Van den Berg, F. and J.J.T.I. Boesten: 1999. *PESTLA 3.4. Description and User's Guide*, Technical

Document 43, DLO-Winand Staring Centre, Wageningen, The Netherlands (in press).

Van den Berg, F., J.H. Smelt, J.J.T.I. Boesten and W. Teunissen: 1999, *J. Environ. Qual.* **28**, 918-928.

Van Os, E.A., H.J. Holterman and G. Klomp: 1993, Management of emission flows of pesticides from glasshouses, paper on ISHS Symposium "Engineering as a tool to reduce pesticide consumption and operator hazards in horticulture", Ulvik, Norway, 8-13 August, 1993.

Wagenet, R.J. , J.L. Hutson and J.W. Biggar: 1989, *J. Environ. Qual.* **18**, 78-84.

Walklate, P.J.: 1992, *J. Agric. Engineering Res.* **51**, 263-283.

Walter, U., M. Frost, G. Krasel and W. Pestemer. 1996. *Assessing volatilization of pesticides: a comparison of 18 laboratory methods and a field method.* Heft 16, BBA, Braunschweig, Germany.

Wang, D., S.R. Yates and J. Gan: 1997, *J. Environ. Qual.* **26**, 1072-1079.

Wang, D., S.R. Yates and W.A. Jury: 1998, *J. Environ. Qual.* **27**, 821-827.

Warren, L.E.: 1972, *Volatility and drift of herbicides*, paper presented at the Montana Herbicide Workshops, October 1972.

Whang, J.M., C.J. Schomburg, D.E. Glotfelty and A.W. Taylor: 1993, *J. Environ. Qual.* **22**, 173-180.

Willis, G.H., L.L. Mc Dowell, S. Smith and L.M. Southwick: 1992, *J. Agric. Food Chem.* **40**, 1086-1089.

Wood, G.W. and D.R.K. Stewart: 1976, *Bull. Environ. Contam. Toxicol.* 15, 623.

Woodrow, J.E., J.N. Seiber and L.W. Baker: 1997. *Environ. Sci. Technol.* **31**, 523-529.

Yates, S.R., F.F. Ernst, J. Gan, F. Gao, and M.V. Yates, 1996. *J. Environ. Qual.* **25**, 192-202.

Yates, S.R. , D. Wang, F.F. Ernst and J. Gan, 1997a. *Environ. Sci. Technol.* 31, 1136-1143.

Yates, S.R., J. Gan, F.F. Ernst, D. Wang and M.V. Yates: 1997b, In: Seiber, J.N., J.A. Knuteson, J.E. Woodrow, N.L. Wolfe, M.V. Yates, S.R. Yates (Eds), *Fumigants: Environmental Fate, Exposure, and Analysis*, ACS Symposium Series 652, pp. 116-134.

TRANSFORMATIONS OF PESTICIDES IN THE ATMOSPHERE: A STATE OF THE ART

ROGER ATKINSON[1], ROB GUICHERIT[2], RONALD A. HITES[3], WOLF-ULRICH PALM[4], JAMES N. SEIBER[5], PIM DE VOOGT[6*]

[1]Air Pollution Research Center, University of California, Riverside CA 92521, USA
[2]TNO Institute of Environmental Sciences, Energy Research and Process Innovation, P.O.Box 342, NL-7300 AH Apeldoorn, The Netherlands
[3]Department of Chemistry, Indiana University, Bloomington IN 47405, USA
[4]Fraunhofer-Institut für Toxikologie und Aerosolforschung 30625 Hannover, Germany
[5]Western Regional Research Center, Agricultural Research Service, USDA, Albany, CA 94710, USA
[6]Amsterdam Research Institute for Substances in Ecosystems (ARISE) Environmental and Toxicological Chemistry, University of Amsterdam, Nieuwe Achtergracht 166, NL-1018 WV Amsterdam, The Netherlands
(corresponding author. E-mail: pim@mtc.chem.uva.nl)*

Abstract. The current knowledge about transformation rates and products of pesticides in the atmosphere is reviewed. Reactive species and their concentrations in the atmosphere are presented. Reactions of pesticides with these species (including photolysis) in the gas and the particulate phase are evaluated from available experimental data. The potential of estimation methods is discussed. Experimental techniques for laboratory and outdoor measurements are reviewed. Finally, an estimation is made of uncertainties in atmospheric lifetimes due to chemical or physical reactions. It is concluded that the most important transformation of pesticides in the atmosphere is due to reaction with OH radicals. Very few experimental data for pesticides are available though. The levels of uncertainty in OH radical concentrations are acceptable, however, for a proper estimation of atmospheric removal rates due to reactions with OH radicals of those pesticides for which experimental transformation rates (of homologues) are available.

Keywords: atmospheric lifetimes, pesticides, reaction rates, transformation products

1. Introduction

Long-range transport of pesticides will occur when such compounds have sufficiently long atmospheric lifetimes (Jury *et al.*, 1987) and when they will be lifted into the troposphere, i.e. above the atmospheric boundary layer (de Voogt and Jansson, 1993). The atmospheric lifetime is the net result of emission and removal processes. In April 1998, the Health Council of The Netherlands organised a Workshop on 'Fate of pesticides in the atmosphere;

Water, Air, and Soil Pollution **115:** 219–243, 1999.

implications for risk assessment' in Driebergen, The Netherlands, where these processes were discussed. During the workshop a working group was formed to review the state of current knowledge on *transformation of pesticides* in the atmosphere. The working group agreed to the following arrangement of discussion topics:

- Reactions in the gas (or vapour) phase
    - reactive species
    - experimental data for pesticides
    - estimation methods
- Reactions in the particle phase
    - reactive species
    - experimental data for pesticides
    - estimation methods
- Reaction products
    - gas phase
    - particle / aerosol phase
    - reactions of first generation products
- Experimental techniques

The working group addressed these topics with the specific aim to discuss uncertainties associated with (the lack of) available data, as these have serious implications for risk assessment. The current paper summarises the discussions and conclusions drawn in this working group.

2. Reactions in the Gas Phase

2.1 REACTIVE SPECIES

As for other volatile organic compounds, the potential removal and transformation processes for pesticides present in the atmosphere in the gas phase (from volatilization during or after application) involve wet and dry deposition, and photolysis, reaction with the hydroxyl (OH) radical, reaction with the nitrate (NO_3) radical, reaction with ozone (O_3), and possibly reaction with gaseous nitric acid (HNO_3) in urban areas where gaseous nitric acid concentrations can be significant (Atkinson *et al.*, 1992; Atkinson, 1995). Reaction with chlorine (Cl) atoms may also be important in certain locations during certain times of the year (Zetzsch and Becker, 1989), as observed for anthropogenic volatile organic compounds (VOCs) in the Arctic during the early springtime (Jobson *et al.*, 1994; Ariya *et al.*, 1998). Furthermore Cl_2

has been observed at levels up to 150 ppt during nighttime in coastal air (Spicer *et al.*, 1998). Dry and wet deposition were not considered by this working group. The processes leading to the presence of O_3, OH radicals and NO_3 radicals in the troposphere are briefly discussed below, as is photolysis.

Presence of O_3 in the Troposphere. Because of the presence of high mixing ratios of O_3 in the stratosphere, there is net transport of O_3 by eddy diffusion from the stratosphere into the troposphere (Roelofs and Lelieveld, 1997). In addition, O_3 is formed photochemically in the troposphere from the interactions of VOCs and oxides of nitrogen ($NO + NO_2$; NO_x) in the presence of sunlight (Roelofs and Lelieveld, 1997). These sources of tropospheric O_3 are balanced by *in situ* photochemical destruction and by dry deposition at the Earth's surface (Roelofs and Lelieveld, 1997). The result of these processes is the presence of ozone throughout the troposphere with mixing ratios at "clean" remote sites at ground level in the range of (1-4) x 10^{-8} (Oltmans and Levy, 1994); O_3 mixing ratios in polluted urban areas often exceed 1 x 10^{-7}.

For the (few) gaseous pesticides studied to date (see Table I), their gas-phase reactions with ozone are of no importance as a tropospheric loss process. Only those pesticides containing carbon-carbon double bonds are expected to react rapidly with O_3 (Atkinson and Carter, 1984).

Formation of Hydroxyl Radicals in the Troposphere. The presence of relatively low levels of O_3 in the troposphere is important because photolysis of O_3 in the troposphere occurs in the wavelength region 290-330 nm to form the excited oxygen, $O(^1D)$, atom (Atkinson *et al.*,1997a). $O(^1D)$ atoms are either deactivated to ground-state oxygen, $O(^3P)$ atoms, or react with water vapour to generate OH radicals.

$$O_3 + h\nu \rightarrow O_2 + O(^1D) \qquad (\lambda \lesssim 330 \text{ nm})$$

$$O(^1D) + M \rightarrow O(^3P) + M \qquad (M = N_2, O_2)$$

$$O(^1D) + H_2O \rightarrow 2\,OH$$

Direct spectroscopic measurements of OH radical concentrations close to ground level show peak daytime OH radical concentrations in the range (2-10) x 10^6 molecules cm^{-3} for mid-latitude northern hemisphere sites during August and September (Brauers *et al.*, 1996; Mount *et al.*, 1997; Mather *et al.*, 1997). These measurements show a distinct diurnal profile, with a maximum OH radical concentration around solar noon (for clear-sky conditions overestimating OH by about 40 % on the boundary layer or inside clouds and underestimating OH by about 30% near/above clouds - Mauldin III *et al.*, 1998).

A diurnally, seasonally and annually averaged global tropospheric OH radical concentration of 1.0 x 10^6 molecules cm^{-3} (24-hr average) has been estimated from the emissions, atmospheric concentrations and atmospheric chemistry of methyl chloroform (Prinn *et al.*, 1995; Hein *et al.*, 1997; Krol *et al.*, 1998). While the photolysis of O_3 results in the formation of OH radicals only during daylight hours, it has been suggested that OH radical formation from the reactions of O_3 with alkenes (including biogenic alkenes such as the monoterpenes) could be significant during both daytime and nighttime (Paulson and Orlando, 1996). Model calculations confirm that, in addition to exhibiting a diurnal profile, the OH radical concentration depends on season and latitude (Hein *et al.*, 1997) with mean monthly surface OH radical concentrations (24-hr averages) at 35 °N of ~2 x 10^5 molecules cm^{-3} in January and ~2.0 x 10^6 molecules cm^{-3} during July, compared to 24-hr average concentrations at the Equator of ~(1.2-1.4) x 10^6 molecules cm^{-3} in January and ~(1.0-1.2) x 10^6 molecules cm^{-3} in July (Dentener and Crutzen, 1993). This July/January OH radical concentration ratio (~10 at 35 °N) will increase with increasing latitude.

Tropospheric OH radical concentrations derived from modelling observed ^{14}CO concentration data also show a pronounced seasonal and latitudinal distribution, with high OH radical concentrations in Equatorial regions and low concentrations in polar regions (Derwent and Volz-Thomas, 1990). Furthermore, OH radical concentrations obtained from the seasonal variations of non-methane VOCs at 43 °N show a summer/winter ratio of 9 ± 2 (Goldstein *et al.*, 1995), in general agreement with the modelling results of Hein *et al.* (1997) mentioned above.

Formation of Nitrate Radicals in the Troposphere. The presence of NO in the troposphere from natural and anthropogenic sources is followed by the reactions (Atkinson *et al.*, 1997a)

$$NO + O_3 \rightarrow NO_2 + O_2$$

$$NO_2 + O_3 \rightarrow NO_3 + O_2$$

leading to the formation of NO_3 radicals. Because the NO_3 radical photolyses rapidly, with a lifetime due to photolysis of ~5 s for overhead sun, and reacts rapidly with NO (Atkinson *et al.*, 1997a), NO_3 radical concentrations remain low during daylight hours, but can increase to measurable levels during nighttime. Measurements made over the past ~20 years show nighttime NO_3

radical concentrations at ground level over continental areas ranging up to 1 x 10^{10} molecules cm^{-3} [a mixing ratio of 430 x 10^{-12}] (Atkinson *et al.*, 1986; Mihelcic *et al.*, 1993; Platt and Heintz, 1994). A 12-hr average nighttime concentration of 5 x 10^{8} molecules cm^{-3} (a mixing ratio of 20 x 10^{-12}) has been proposed as a reasonable value for lifetime calculations (Atkinson, 1991).

Photolysis. In order for photolysis of a pesticide to occur, the pesticide must absorb light in the relevant wavelength region and, having absorbed light radiation, must undergo chemical change, either through decomposition or isomerization. Because of absorption of short-wavelength solar radiation by O_2 and O_3 in the stratosphere, photolysis in the troposphere requires the pesticide to absorb radiation at wavelengths between 290 nm and ~800 nm, the latter being the longest wavelength which will break a chemical bond. Absorption of light is defined by the Beer-Lambert Law:

$$\ln (I_0/I) = \sigma \cdot [A] \cdot l$$

where I_0 is the incident light intensity, I is the transmitted light intensity, σ is the absorption cross-section (in cm^2 $molecule^{-1}$), [A] is the concentration of the chemical (in molecule cm^{-3}) and l is the pathlength (in cm). The quantum yield, ϕ, is defined as

$$\phi = \text{number of molecules transformed/number of photons absorbed}$$

and under tropospheric conditions $\phi \leq 1$. The absorption cross-section, σ, and the quantum yield, ϕ, are in general a function of the wavelength. For the photolysis reaction,

$$A + h\nu \rightarrow \text{products}$$

the photolysis rate constant, k_{phot}, is given by

$$k_{phot} = \int_{290}^{800} J_\lambda \sigma_\lambda \phi_\lambda \, d\lambda \qquad s^{-1}$$

where J_λ is the photon flux at wavelength λ, and σ_λ and ϕ_λ are the absorption cross-section and quantum yield at wavelength λ.

While the photon flux can be accurately calculated as a function of time of day, season and latitude, the measurement of absorption cross-sections and quantum yields (both as a function of wavelength) is difficult. As a result, absorption cross-sections and photolysis quantum yields for pesticides present in the gas phase are only poorly understood. Furthermore, absorption cross-sections and photolysis quantum yields for complex chemical compounds cannot presently be estimated.

Lifetimes of Chemicals in the Atmosphere. For chemical compounds which are removed from the atmosphere by more than one process (for example, by both physical and chemical pathways), the overall lifetime, $\tau_{overall}$, is given by,

$$1/\tau_{overall} = 1/\tau_{physical} + 1/\tau_{chemical}$$

with $1/\tau_{chemical} = 1/\tau_{phot} + 1/\tau_{OH} + 1/\tau_{NO3} + 1/\tau_{O3} + etc.$, $\tau_{phot} = (k_{phot})^{-1}$, and for the OH radical, NO_3 radical, O_3 and Cl atom reactions $\tau_X = (k_X[X])^{-1}$, where τ_X = lifetime of the chemical due to its reaction with X, with X = OH, NO_3, O_3 or Cl, and k_X = reaction rate of the reaction with X .

2.2 EXPERIMENTAL DATA FOR PESTICIDES

Table I provides an overview of experimental data available from the open literature for gaseous pesticides. More data may be present in pesticide evaluation files, which are, however, not generally accessible.

The rate constants given in Table I for the gas-phase reactions of OH radicals, NO_3 radicals and O_3 with pesticides were measured using established absolute rate or relative rate techniques, and should be reliable within the error limits cited in the associated publications (noting that the rate constants measured by Brubaker and Hites (1998b) are extrapolated from measurements at elevated temperatures, as noted in comment f). Photolysis rates measured in air in environmental chambers can be subject to complications caused by the formation and presence of OH radicals in these systems. Because trifluralin and phorate are expected to react rapidly with the OH radical [with rate constants $> 1 \times 10^{-10}$ cm^3 $molecule^{-1}$ s^{-1} (Winer and Atkinson, 1990; Kwok and Atkinson, 1995)], then for an OH radical concentration of 1×10^7 molecules cm^{-3} (a possible mid-day value, see section on OH radical formation above) the first-order decay rate would be $> 1 \times 10^{-3}$ s^{-1} and in the range of those measured by Mongar and Miller (1988) and Hebert *et al.* (1998a). Further experiments are required to ascertain that the measured decay rates of trifluralin and phorate were due to photolysis and not (at least in part) to reaction with the OH radical, by, for example, conducting photolyses in the presence of sufficient alkane to scavenge OH radicals. Palm *et al.* (1999) have recently succeeded in separating OH reactivity and photolysis for the aerosol-borne pesticide pyrifenox.

Table I.

Overview of published data on atmospheric reaction rates of pesticides (reaction rate constants are in units of cm^3 $molecule^{-1}$ s^{-1}, at room temperature)

compound	k_{OH}	k_{NO3}	k_{O3}	k_{phot} (s^{-1})	Ref.
methyl bromide	2.9×10^{-14}				a
1,2-dibromo-3-chloropropane	4.3×10^{-13}		$<3 \times 10^{-20}$		b
cis-1,3-dichloropropene	8.4×10^{-12}		1.5×10^{-19}		c,d (L)
trans-1,3-dichloropropene	1.4×10^{-11}		6.7×10^{-19}		c,d (L)
EPTC	3.2×10^{-11}	9.2×10^{-15}	$<1.3 \times 10^{-19}$		e (L)
cycloate	3.5×10^{-11}	3.3×10^{-14}	$<3 \times 10^{-19}$		e (L)
α-hexachlorocyclohexane	1.4×10^{-13}				f (L)
γ -hexachlorocyclohexane	1.9×10^{-13}				f (L)
hexachlorobenzene	2.7×10^{-14}				f (L)
trifluralin				$\sim 3 \times 10^{-4}$	g (L)
				$\sim 6 \times 10^{-4}$	h (F)
phorate				$\sim 2 \times 10^{-3}$	i (L)
parathion				$\sim 6 \times 10^{-3}$	h (F)
phosphine	1.6×10^{-11}				j (L)
chloropicrin				5.7×10^{-5}	k (L)
methyl isothiocyanate				6.7×10^{-6}	l (L)
				see comment	m (L)

References and Comments to Table I.

L: laboratory measurement; F: derived from field or ambient atmospheric measurement.

a. Atkinson *et al.*, 1997b

b. Tuazon *et al.*, 1986

c. Tuazon *et al.*, 1984

d. Tuazon *et al.*, 1988

e. Kwok *et al.*,1992; EPTC is $(CH_3CH_2CH_2)_2NC(O)SCH_2CH_3$ and cycloate is CH_3CH_2(cyclo-C_6H_{11})-$NC(O)SCH_2CH_3$.

f. Brubaker and Hites 1998b. Extrapolated to 298 K from rate constants measured over the temperature range 346-386 K.

g. Mongar and Miller, 1988. Observed photolysis rate in 17000 litre outdoor chamber, during summertime mid-day.

h. Woodrow *et al.*, 1978

i. Hebert *et al.*, 1998a. Observed photolysis rate in 12800 liter outdoor chamber, during summertime mid-day. Phorate is $(CH_3CH_2O)_2P(=S)SCH_2SCH_2CH_3$.

j. Fritz *et al.*, 1981

k. Carter *et al.*, 1997 For overhead sun.

l. Alvarez and Moore, 1994. Measured value of $\phi = 0.98 \pm 0.24$ at 308 nm; photolysis rate (s^{-1}) calculated assuming $\phi = 1.0$ and for a zenith angle of 40°. Half-life calculated to be 39 hr at 40 °N (Geddes *et al.*, 1995).

m. Geddes *et al.*, 1995. Half-life measured to be ~33 hr at 40 °N during summertime.

2.3 ESTIMATION METHODS FOR REACTION RATE CONSTANTS

OH Radicals. Apart from the chlorofluorocarbons and Halons not containing (an) H atom(s), all gaseous organic compounds react with the OH radical, and reaction with the OH radical is often the dominant tropospheric chemical loss process for organic compounds. Therefore, a knowledge of the OH radical reaction rate constant is necessary to estimate the atmospheric lifetime of the chemical (or the upper limit thereof). Early efforts at conducting structure-activity correlations for gas-phase oxidation of organics are summarized in Mill (1980). Structure-reactivity relationships (SAR) as proposed by Atkinson (1986) [with the latest published update being by Kwok and Atkinson (1995)] are often used nowadays, as in the *Atmospheric Oxidation Program* from Syracuse Research Corporation (Meylan and Howard, 1993, 1995). When SAR-derived OH radical reaction rate constants are compared with the measured values, approximately 90% of the rate constants for ~485 compounds were predicted to within a factor of two of the experimental values (Kwok and Atkinson, 1995). However, there is no real knowledge of how well this estimation method performs for organic compounds for which data are not available, and it is expected that the predictions will be more uncertain the more complex the chemical is (*i.e.*, how many functional groups it contains), and especially if the chemical contains halogen atoms and/or N- and S- atoms. For example, for the PCBs estimated OH radical reaction rate constants for the more chlorinated congeners show agreement between predicted and measured rate constants within a factor of ~2 (Atkinson, 1996; Anderson and Hites, 1996a), while recently measured OH radical reaction rate constants for α- and γ-hexachlorocyclohexane, extrapolated to room temperature (Brubaker and Hites, 1998b) are lower by factors of 4 and 3, respectively, than the estimated 298 K rate constants. Furthermore, measurements of the OH radical reaction rate constants for dibenzo-*p*-dioxin and dibenzofuran were factors of 2.5 and 8, respectively, lower than estimates (Kwok *et al.*, 1994), requiring that relevant parameters in the estimation method be revised to yield better fits of predicted *versus* experimental rate constants (Kwok *et al.*, 1995; Atkinson, 1996).

Not suprisingly, therefore, it has been recommended (Kwok and Atkinson, 1995) that this estimation method only be used for chemicals for which experimental data for homologs already exist and which have been compared with predictions of the estimation method. Therefore, the estimation method should not be used for chemicals which contain functional groups for which the estimation method has not been tested (and should only be used if such tests indicate reasonable agreement between predictions and experimental values). At present, it is possible that OH radical reaction rate constants can be reasonably reliably estimated for organo-phosphorus compounds of general structure

$(RX)_3P{=}X$, where R = alkyl and X = O and/or S (Goodman *et al.*, 1988; Kwok and Atkinson, 1995), carbamates (Kwok *et al.*, 1996), and thiocarbamates (Kwok *et al.*, 1992). For example, Goodman *et al.* (1988) showed that in $(CH_3X)_3P{=}X$, where X = O and/or S, the compounds containing P=S structural units reacted much more rapidly than those with P=O structural units, suggesting that the OH radical reacts rapidly with P=S units but not with P=O units (see also Kwok and Atkinson, 1995). To what extent this can be extrapolated to more complex organo-phosphorus pesticides such as malathion is presently not known.

Koch *et al.* (1996, 1997) concluded from absolute rate studies that OH radical rate constant estimates cannot reliably be made for amines or amides, and that rate constants for these classes of compounds must be measured. Because most pesticides are complex and contain multiple functional groups, it is recommended that OH radical reaction rate constants be measured or that, at a minimum, rate constants be measured for homologs of pesticides to ascertain that the estimation method used is appropriate for this class of organic compound.

NO_3 Radicals. It appears that NO_3 radical reactions are a potentially important atmospheric loss process for compounds containing $>C{=}C<$ bond(s) or S atoms (Atkinson, 1991, 1994). This may also be the case for amines and other N-atom containing compounds, although this has not been determined to date because these compounds also react with gaseous nitric acid formed in NO_3 radical reaction systems. A correlation between reaction rate and the number and position of alkyl substituents around the $>C{=}C<$ bond has been observed for alkenes (Atkinson, 1991), but no reliable estimation method is yet available for a wide variety of organic compounds and structures. A judgement can be made, based on the structure of the pesticide, as to whether or not reaction with the NO_3 radical is likely to be important; for example, benzene and biphenyl do not react with the NO_3 radical at measureable rates and alkanes react only slowly with the NO_3 radical (Atkinson, 1991), and hence compounds containing non-fused aromatic rings and not containing $>C{=}C<$ bonds or S- or N- atoms may be considered to be unreactive or of low reactivity towards the NO_3 radical. Similarly, organo-phosphorus compounds of structure $(RX)_3P{=}X$, where R = alkyl and X = O and/or S, and thiocarbamates appear to be of low reactivity towards the NO_3 radical (Goodman *et al.*, 1988; Kwok *et al.*, 1992).

Correlations between the rate constants for reactions of organic compounds with NO_3 radicals and a number of physico-chemical parameters of the organic compound (including ionization potential and the OH radical reaction rate constant) have been presented and discussed by Wayne *et al.* (1991).

Ozone. The available data-base (Atkinson and Carter, 1984; Atkinson, 1994) indicates that O_3 reactions with organic compounds will be potentially important as a tropospheric loss process mainly for compounds containing >C=C< groups. Certain N-containing compounds (for example, amines) react slowly with O_3. An estimation method has been proposed to calculate rate constants for the reactions of O_3 with organic compounds containing >C=C< groups (Atkinson and Carter, 1984), but the reliability is low and discrepancies of an order of magnitude or greater occur, especially for strained-ring cycloalkenes. The presence of Cl atoms attached to the carbon atoms of the >C=C< bond markedly decreases the reactivity towards O_3 (Atkinson and Carter, 1984).

The simple thiocarbamates and organo-phosphorus compounds of structure $(RX)_3P=X$, where R = alkyl and X = O and/or S, do not react at measureable rates with O_3 (Goodman *et al.*, 1988; Kwok *et al.*, 1992), and this may also be the case for more complex in-use pesticides with these structural units (but needs to be confirmed).

Photolysis. In general, photolysis rates cannot be estimated reliably since both absorption cross section data and quantum yields in the vapour phase are not available, not even for homologues. Because of the experimental difficulties in attaining gas-phase absorbance cross sections, an alternative approach using condensed-phase absorbance spectrum measurements has been suggested (Mill, 1980 and Pitts, 1981). Deriving the quantum yield in this manner, however, can be misleading and under-estimate the environmental importance of direct photochemical processes and their reaction products. For example, Elend and co-workers (1994) reported that a significant bathiochromic "red" shift occurred for gaseous methylisothiocyanate (MITC) when compared to its condensed phase absorbance spectrum. These researchers further reported direct photolysis to be the dominant removal pathway in air. Geddes and co-workers (1995), in experiments conducted under xenon arc irradiation and natural sunlight also showed that MITC undergoes rapid direct photolysis in the gas-phase with an observed half-life ranging from 9-10 hours. More importantly, direct photolysis resulted in substantial yields of volatile and toxic photoproducts including the formation of methylisocyanide (MIC) that would not be anticipated via oxidative removal pathways.

The gas-phase spectra of various semi-volatile compounds have recently been determined using a tandem GC remote flow cell linked by fiber optics to a rapid-scanning UV-VIS chromatography detector. This instrument is capable of acquiring gas-phase UV spectra for moderate to low volatility substances with a 2-nm effective bandpass. This instrument has proven successful in determing absorbance spectra of gas-phase aromatic compounds (Bornhop *et al.*, 1991). The influence of band-broadening at the higher temperatures required for assessing many pesticides will need to be characterized before calculating gas-phase cross sections and quantum yields by the Beer-Lambert law.

Nitric Acid. Certain nitrogen-containing compounds, such as pyridines and amines, react with gaseous nitric acid to form the salts (for example, pyridinium nitrate from pyridine) (Atkinson *et al.*, 1987). These reactions may be important in urban areas where the concentrations of gaseous nitric acid are high (see, for example, Tuazon *et al.*, 1980).

Other Reactions. To date, there is no evidence from laboratory studies of gas-phase hydrolysis reactions of gaseous pesticides. Although no reactions of Cl atoms with pesticides have been studied to date, the reactions of Cl atoms with organic pesticides are expected to be rapid and a rate constant of 10^{-10} cm^3 $molecule^{-1}$ s^{-1} (with an uncertainty to within a factor of 5) can be assumed based on the rate constants for Cl atom reactions with the higher alkanes and alkenes (Atkinson, 1997).

3. Particle phase reactions (including the aerosol phase)

3.1 REACTIVE SPECIES

In the particle phase, reactions with OH radicals and ozone, as well as photolytic reactions, are the major chemical transformation routes. These reaction routes are more or less assumed, based on the known gas-phase reactions (Atkinson *et al.* 1992). A further distinction between adsorbed (surfacial) and absorbed (inner particle) phases is probably necessary. Photocatalytic degradation of adsorbed compounds may occur on certain semiconductive metal oxides (TiO_2, ZnO, Fe_2O_3) and has been shown to be highly variable (Güsten, 1984). In the dissolved phase, hydrolysis and reactions with OH radicals (Sedlak and Andren, 1991) are expected to be the most important. In general, particle phase reactions in the atmosphere must be considered as an area of great uncertainty.

3.2 EXPERIMENTAL DATA

The *ambient* data that are available in general only provide data on the (type of) reaction products found. For example, for several organophosphorous pesticides (malathion, parathion, methidathion and chlorpyrifos) the corresponding oxone derivatives have been observed (Seiber *et al.* 1989, Woodrow *et al.* 1977, 1978, Glotfelty *et al.* 1987; Spencer *et al.* 1980; Aston and Seiber, 1997). Very few *laboratory studies* on particle phase reactions of pesticides are presently available. Reactions of lindane with OH radicals (Behnke and Zetzsch, 1989; Zetzsch, 1991), and of terbuthylazine with OH radicals and ozone (Palm *et al.*, 1997), and pyrifenox with OH radicals as well as ozone (Palm *et al.*, 1999) were studied on silica particles. The results of the terbuthylazine experiments are possibly valid for other triazines and suggest that monitoring efforts (and analytical method development) should be aimed at similar triazine products resulting from OH radical reaction with the R-N< bond ($R=C_2H_5$), leading to either CH_3-(C=O)-N< or the N-dealkylation product, *cf.* Table III below.

For pyrifenox, which is an oxime-ether, the reaction with OH radicals results in a ketone (Palm *et al.*, 1999). Besides for pesticides, reaction rate data from laboratory studies for other semivolatile compounds such as DEHP (Behnke *et al.* 1987) on aerosols are available.

Table II provides the very few experimental rate constants available from the open literature for pesticides in the aerosol-borne state.

Table II.
Overview of published data on reaction rates of pesticides in the aerosol-borne state (reaction rate constants are in units of cm^3 $molecule^{-1}$ s^{-1}, at room temperature)

compound	k_{OH}	k_{O3}	Ref.
lindane	6.0×10^{-13}		Behnke and Zetzsch, 1989; Zetzsch, 1991
terbuthylazine	1.1×10^{-11}	$<5 \times 10^{-19}$	Palm *et al.*, 1997
pyrifenox[a]	1.8×10^{-11}	$<2 \times 10^{-19}$	Palm *et al.*, 1999

[a] Rate constant for the reaction with hydrogen peroxide was found to be negligible ($<< 1 \times 10^{-19}$)

3.3 ESTIMATION METHODS

In general there is a lack of data on pesticide reactions in the particle phase. Statements that rates of reaction in the particle phase are slower than in the gas phase (Brubaker and Hites, 1997; Scharf and Bächmann, 1993; Scheringer, 1997; van Pul *et al.*, 1998) are neither supported nor contradicted by literature data. It has been shown that no reactions occur with PAH absorbed in particles (Behymer and Hites, 1985), whereas PAH adsorbed onto particles can react (Behymer and Hites, 1985). For the latter process, however, no quantitative data are available. The distribution of PAH between adsorbed and absorbed phases is a function of temperature (Bidleman, 1988) and may possibly shift with respect to the sources of the particulate matter. Whether such observations hold for pesticides is unknown, however.

4. Reaction products

Table III shows reaction products observed to be formed from gaseous pesticides in laboratory experiments.

Ambient data concerning the transformation products of pesticides are available for some compounds. Organophosphorous pesticides generally are transformed into the more stable oxones (see above) (Seiber *et al.* 1989; Schomburg *et al.*, 1991). p-Nitrophenols, which are transformation products from e.g., dinitroanilines and dinitrophenols have been shown to be present in the atmosphere (Geissler and Schöler, 1993; Tremi *et al.* 1993). It should be noted that the nitrophenols can also result from atmospheric reactions of non-pesticide compounds (such as, *e.g.*, urban vehicular traffic, Herterich and Herrmann, 1990).

Table III.

Reaction products observed in laboratory experiments with gaseous and particle phase pesticides

Parent compound	Transformation products	Phase[a]	Reference
1,3-dichloropropene	formyl chloride, chloro-acetaldehyde	G	Tuazon *et al.*, 1984
Molinate	keto-derivatives	G	Crosby and Moilanen, 1974
Trifluralin	N-dealkylation	G	Soderquist *et al.* 1975
Methylbromide	formaldehyde	G	Orlando *et al.*, 1996
Chloropicrin	phosgene	G	Moilanen *et al.*,1978; Carter et al., 1997
CH_3NCS	CH_3NC	G	Alvarez and Moore, 1994
Terbuthylazine	N-dealkylation	P	Palm *et al.*, 1997
Pyrifenox	keto-derivative	P	Palm *et al.*, 1999
DEF	P-containing acids	G	Woodrow *et al.*, 1983
Aldrin	dieldrin	G	Crosby and Moilanen, 1974
Dieldrin	photodieldrin	G	Crosby and Moilanen, 1974
DDT	DDE	G	Crosby and Moilanen, 1977
DDE	Dichlorobenzophenone	G	Crosby and Moilanen, 1977

[a] G = gas phase reaction; P = particle phase reaction

Transformation products have been found in *e.g.*, rain (oxones, Seiber *et al.* 1993), snow (oxones, Zabik and Seiber, 1993) and fog (nitrophenol, oxones, Glotfelty *et al.*, 1987; Schomburg *et al.* 1991). It is not known, however, whether these products result from transformations of gaseous or particle-phase pesticides. In addition, neither the reactive species involved, nor the transformation rates, are known.

Dealkylation products of triazines have been reported (Scharf *et al.*, 1992; Scharf and Bächmann, 1993; Hurle and Oberwalder, 1992; Braun *et al.*, 1990; Schössner and Koch, 1991; Goolsby *et al.*, 1994). p,p'-DDE (from p,p'-DDT: Tarrant and Tatton, 1968; Agarwal *et al.*, 1987; Strachan, 1990; Fingler *et al.*, 1994; Ballschmiter and Wittlinger, 1991), o,p'-DDE (Stanley *et al.*, 1971) and dieldrin (from aldrin) (Crosby and Moilanen 1974) have been found to be present in the atmosphere on a global scale. DDE in air may also result from revolatilisation of DDE in soil, water and leaves, however (Cotham and Bidleman, 1991). Epoxides of chlorinated aromatics have been found, as well as dichlorobenzophenone, which is a transformation product from Kelthane (Walsh and Hites 1979). No data are available on reactions of first generation products.

5. Experimental techniques

5.1 LABORATORY

Experimental techniques for the measurement of reaction rates of organic compounds with the OH radical have been reviewed by Atkinson (1986, 1989). Experimental techniques for pesticides, from both laboratory and field work, were reviewed by Woodrow *et al.* (1983). A test guideline on photochemical oxidative degradation in the atmosphere is available from the OECD (1991). The majority of pesticides are of relatively low volatility, with room temperature vapour pressures often being in the range 10^{-4} Pa to 1 Pa (10^{-6} to 10^{-2} Torr). For organic compounds with vapour pressures in this range, rate constants for OH radical and NO_3 radical reactions have been measured using relative rate methods (see, for example, Kwok *et al.*, 1995; Anderson and Hites, 1996b). Several of these relative rate studies have used large volume Teflon chambers (~7000 liter volume) (see, for example, Kwok *et al.*, 1995). In these studies, OH radicals were generated by the photolysis of methyl nitrite in air at wavelengths >300 nm,

$$CH_3ONO + h\nu \rightarrow CH_3O + NO$$

$$CH_3O + O_2 \rightarrow HCHO + HO_2$$

$$HO_2 + NO \rightarrow OH + NO_2$$

resulting in OH radical concentrations of up to ~2 x 10^8 molecules cm^{-3} (a factor of 100 higher than in the ambient atmosphere) for periods of up to ~20 min. Because of the limited overall amount of OH radicals generated by this procedure, in general only rate constants ≥1 x 10^{-12} cm^3 $molecule^{-1}$ s^{-1} can be reasonably reliably determined by this technique.

NO_3 radicals have been generated by the thermal decomposition of N_2O_5.

$$N_2O_5 \sim NO_3 + NO_2$$

The concentrations of the reactants have generally been measured during the experiments by gas chromatography.

Hites and co-workers (see, for example, Anderson and Hites, 1996b) have used a different variation of the relative rate method to measure OH radical reaction rate constants, using the photolysis of O_3 at 254 nm in the presence of water vapour in N_2/O_2 or He diluent at atmospheric pressure in an ~0.2 litre quartz vessel.

$$O_3 + h\nu \rightarrow O_2 + O(^1D)$$

$$O(^1D) + H_2O \rightarrow 2\ OH$$

The concentrations of the reactants are measured by mass spectrometry, and because of detection sensitivity limitations rate constant measurements for pesticides have been carried out at elevated temperatures (to obtain higher concentrations, which are contrained by the vapour pressure of the pesticide). Rate constants have been measured over a range of temperatures (for example, 346-386 K), enabling rate constants to be calculated at room temperature (and below) by extrapolation.

The above comments are also relevant to the measurement of rate constants for the O_3 reactions with pesticides. However, in many cases it is expected that the O_3 reactions will be of no importance and that only upper limits to the rate constants will be measured. The technique often used for these compounds is to monitor the enhanced loss of the pesticide in the presence of a measured concentration of O_3 (Atkinson *et al.*, 1981); wall losses of the pesticide will again be problematic.

The measurement of photolysis rates applicable to the ambient atmosphere require that an appropriate light source be used; either radiation from a xenon arc filtered to remove short-wavelength radiation below 290 nm, or natural solar radiation. In the first case, the use of xenon arcs tends to limit the size of the reaction chamber (with one or two exceptions) which can be used (see, for example, Geddes *et al.*, 1995), hence exacerbating any potential for wall adsorption/ desorption problems. Outdoor chambers have been used by Mongar and Millar (1988) and Hebert *et al.* (1998a) to measure photolysis rates of pesticides, and this technique should be totally appropriate provided that any losses of the pesticides are shown to be due only to photolysis (and dilution if this cannot be avoided) [see the discussion above concerning the potential for formation of OH radicals during the photolysis experiments]. Small size (72 litre) chambers have been used to study the photoreaction of aldrin and dieldrin (Crosby and Moilanen 1974).

Studies of the products formed from the photolysis and reactions of pesticides with OH radicals, NO_3 radicals and O_3 can be carried out using techniques similar to those used for kinetic studies. The major problem encountered in the product studies concerns analytical methodology for the identification and quantification of the products, which are generally more oxidized (and hence of lower volatility) than the precursor pesticide. The lack of standards for product identification is a further issue. A large body of literature is available for product studies of volatile organic compounds emitted into the atmosphere from biogenic and anthropogenic sources (see, for example, Atkinson, 1994, 1997; Atkinson and Arey, 1994, 1998; and references therein).

As mentioned above, the major problem associated with the measurements of reaction rate constants for pesticides is the low vapour pressures at room temperature, and hence the potential for wall adsorption/desorption problems, especially in small reaction vessels at around room temperature [the elevated temperature experiments of Hites and co-workers (Anderson and Hites, 1996a,b; Brubaker and Hites, 1998a,b) are less prone to these wall loss problems]. While to date only few laboratories have been involved in the measurement of reaction rate constants, photolysis rates, and reaction products for low volatility compounds (with room temperature vapour pressures <1 Pa), it is likely that new and/or improved experimental techniques will be developed when (or if) the need for such data arises. An elevated temperature approach modelled on the earlier work of Crosby and Moilanen (1974) has been recently developed to establish stable gas-phase concentrations for semi to low volatility compounds (Hebert *et al.*, 1997). This approach incorporates recent sampling advancements in solid-phase microextraction (SPME) allowing repetitive gas-phase sampling over the experimental timeframe for more precise rate determinations. As in the system by Crosby and Moilanen, this vessel is designed to minimize incidental light hitting the side-walls. A collimated filtered xenon arc light source illuminates the vessel through a quartz plate. The collimated light is trapped at the anterior end of the vessel to minimize back-scattering. This vessel is water jacketed for precise temperature control and is used at atmospheric pressure. The chamber temperature can be controlled (±1°C) from ca. 300 to 375 K to increase the vapour pressure of the test substance and reference compounds thus insuring high gas-phase concentrations while greatly minimizing wall sorption within the observe chamber. The SPME fiber is inserted into the internal chamber and allowed to equilibrate with the gaseous mixture of reactants and stable reference compounds. The exposed fiber is thermally desorbed directly into a hot gas chromatograph injection port and the gas-phase constituents quantified by mass spectroscopy. Replicated experiments are conducted over elevated temperatures at 5-degree increments to ascertain the contribution of temperature dependent and thermal oxidation effects on reaction rates. Hebert *et al.* (1998a) observed that for substances that principally undergo direct photolysis, the activation energy over this range of temperatures was reported to be negligible. The kinetic results, when compared to field information (see Table IV) and other large changer studies run at environmental temperatures, suggest that this experimental approach can be used to establish environmentally relevant photolysis rates under these controlled laboratory conditions.

Table IV.

Comparison of Field to Laboratory Gas-Phase Photochemical Reaction Rate Data for Trifluralin

Investigator	Study Conditions	Reaction half-life (in minutes)
Woodrow *et al.* (1983)	Field (aerial applied tracer + trifluralin)	21-63
Mongar and Miller (1988)	Outdoor (tracer + trifluralin)	19-74
Hebert *et al.* (1997)	57-L Reaction Chamber Tracer Study	22-26

This chamber system has also been employed in generating OH reactivity data at elevated temperatures for the semi-volatile organophosphorus pesticides chlorpyrifos and diazinon (Hebert *et al.* 1998b). In this system, the two pesticides and two reference compounds with similar OH reactivities, triethyl phosphate and 2-methyl naphthalene, were simultaneously exposed to OH radicals generated by the photolysis of methyl nitrite.

Techniques for measurements using aerosols are available and involve smog chambers (Behnke *et al.* 1987, 1988; Zetzsch, 1991). For reactions in the adsorbed phase only a very limited number of investigations are known, one with the plasticiser di-2-ethylhexyl-phthalate (Behnke *et al.*, 1987), and only three for pesticides in the aerosol-borne state (Zetzsch 1991, Palm *et al.*, 1997, Palm *et al.*,1999). Therefore, different experiments and their results are impossible to compare. Silicon dioxide has been proposed as an inert carrier (Behnke *et al.*, 1987) for the investigations in the aerosol-borne state (Palm *et al.*, 1997) because of (a). its inertness; (b). the loss of pesticides on SiO_2 is proportional to the OH radical concentration in the gas phase, (c). measured rate constants do not depend on the OH radical concentrations, and (d). measured rate constants do not depend on the extent of conversion.

Outdoor

Apart from the outdoor photolysis rate chambers used by Mongar and Miller (1988) and Hebert *et al.* (1998a) [see Table I], a specific outdoor technique involves the so called release/recapture experiment (Woodrow *et al.*, 1978, 1983), where the pesticide spray is released along a broad front perpendicular to the prevailing wind. Air samplers are placed downwind at several distances to collect the pesticide drift and the samples are analyzed for both the parent compound and known transformation products. Knowing the transit time from the point of release to the downwind sampling stations (downwind distance divided by wind speed), and assuming first-order kinetics, the following rate expression can be used to compute an approximate rate constant (k):

$$R_t = R_0 \cdot e^{-kt}$$

For the reaction $A \rightarrow B$, $R_t = A/(A + B)$, the molar ratio of parent to the sum of parent plus products, where at $t = 0$, $R_t = R_0$ is unity. Instead of this approach, a stable tracer could be released along with the test pesticide, and the rate of breakdown calculated from the ratio of pesticide to tracer. However, when this approach was used, with lindane as the stable tracer, in an outdoor, open field situation (Woodrow *et al.*, 1978), the results were not as clear-cut as with the approach relating parent to product formation.

6. Uncertainties

If experimentally measured OH radical, NO_3 radical and O_3 reaction rate constants for a pesticide are available, then the major uncertainty in the estimation of the atmospheric lifetime for the pesticide due to reaction with one or more of these reactive species is in the ambient atmospheric concentration of OH radicals, NO_3 radicals or O_3. In many cases, reaction with the OH radical is the dominant atmospheric chemical loss process, and hence the ambient atmospheric concentration of the OH radical is the needed parameter. As noted in Section 2 above, the OH radical concentration varies with time of day, season, and latitude (and with cloud cover and the concentrations of a number of chemical species such as O_3 and water vapour). While modelling studies can yield the average OH radical concentration at a given latitude and time of year, variability in the OH radical concentration due to "local" effects such as cloud cover and the concentrations of other chemical species will occur. It is difficult to assess this uncertainty factor, but an uncertainty of at least a factor of 2 seems reasonable.

Uncertainties in the ambient lower tropospheric concentration of the NO_3 radical at a given time and place of an order of magnitude have been estimated, based on the variability of the ambient concentration data-base for the NO_3 radical (ranging from $<5 \times 10^7$ molecules cm^{-3} up to 1×10^{10} molecules cm^{-3}) (Atkinson, 1991). Average O_3 concentrations or mixing ratios close to ground level (on a 24-hr average basis) probably vary by no more than a factor of 2-3 from an $\sim 3 \times 10^{-8}$ mixing ratio. Photolysis rates can be calculated accurately for clear-sky conditions provided that the quantum yield has been determined or that the photolysis rate has been measured at a known spectral distribution and intensity, especially under natural solar radiation conditions. The effects of temperature changes, as reflected in changes in the reaction rate constants, are generally relatively small compared to uncertainties in the reactive species concentrations. As a fairly extreme example, ethane is a low reactivity chemical with an atmospheric lifetime of ~2 months due to reaction with the OH radical at an OH radical concentration of 1×10^6 molecules cm^{-3}, and the temperature dependence of its OH radical reaction rate constant is such that a change in ambient temperature from 298 K to 250 K decreases the rate constant (and hence increases the lifetime at a constant OH radical concentration) by a factor of 2.

7. Conclusions

Experimental data are available for a limited number of pesticides, from both laboratory chambers and field measurements.

The most important atmospheric transformation reaction of pesticides is the reaction with OH radicals. Exceptions are the reaction of isolated and activated unsaturated C-C bonds, for which the reaction with NO_3 radicals and ozone can be important.

Only for those pesticides for which reaction rates of homologues are available extrapolations or estimations may be within an uncertainty margin that is acceptable for risk assessment. The major factor in the uncertainty is the variability of the OH radical concentration in the atmosphere.

Although field data suggests that extensive atmospheric transformations may be occurring for many pesticides, direct proof is usually lacking and it isn't clear whether the transformation products observed in air were in fact formed in air or on a surface followed by revolatilisation or wind erosion. It is also not clear whether airborne transformation products were formed in the vapour or particulate phases, if formed in the air at all.

Particle phase reactions in the atmosphere must be considered as an area of great uncertainty.

No data are available on reactions of first generation products from atmospheric pesticide transformations. In particular if reactions in the atmosphere are rapid, the products should be subject to further study and/or evaluations similar to the parent pesticide.

Acknowledgements

The authors thank Douwe Jonkers, Vince Hebert and Jim Woodrow for their valuable input and advices, and the organisers of the Dutch workshop for providing the opportunity that resulted in this document.

References

Agarwal, H.C., Kauschik, C.P. and Pillai, M.K.: 1987, *Water Air Soil Pollut.* **32**, 293

Alvarez, R.A. and Moore C.B.: 1994, *Science* **263**, 205-207.

Anderson, P.N. and Hites, R.A.: 1996a, *Environ. Sci. Technol.* **30**, 1756-1763

Anderson, P.N. and Hites, R.A.: 1996b, *Environ. Sci. Technol.* **30**, 301-306.

Ariya, P. A., Jobson, B. T., Sander, R., Niki, H., Harris, G. W., Hopper, J. F. and Anlauf, K. G.: 1998, *J. Geophys. Res.* **103** , 13169-13180.

Aston, L.S. and Seiber, J.N.: 1997, *J. Environ. Qual.* **26**, 1483-1492.

Atkinson, R.: 1986, *Chem. Rev.* **86**, 69-201.

Atkinson, R.: 1989, *J. Phys. Chem. Ref. Data, Monograph* **1**, 1-246.

Atkinson, R.: 1991, *J. Phys. Chem. Ref. Data* **20**, 459-507.

Atkinson, R.: 1994, *J. Phys. Chem. Ref. Data, Monograph* **2**, 1-216.

Atkinson, R.: 1995, *Issues in Environmental Science and Technology*, Eds. R.E. Hester and R.M. Harrison, **4**, 65-89 (Royal Society of Chemistry, Cambridge, UK).

Atkinson, R.: 1996, *Issues in Environmental Science and Technology*, Eds. R.E. Hester and R.M. Harrison, **6**, 53-72 (Royal Society of Chemistry, Cambridge, UK).

Atkinson, R.: 1997, *J. Phys. Chem. Ref. Data* **26**, 215-290.

Atkinson, R. and Arey, J.: 1994, *Environ. Health Perspect.* **102** *(Suppl. 4)*, 117-126.

Atkinson, R. and Arey, J.: 1998, *Acc. Chem. Res.* **31**, 574-583.

Atkinson, R. and Carter, W. P. L.: 1984, *Chem. Rev.* **84**, 437-470

Atkinson, R., Aschmann, S.M., Winer, A.M. and Pitts, J.N. Jr.: 1981, *Int. J. Chem. Kinet.* **13**, 1133-1142.

Atkinson, R., Winer, A.M. and Pitts, J.N. Jr.: 1986, *Atmos. Environ.* **20**, 331-339.

Atkinson, R., Tuazon, E.C., Wallington, T.J., Aschmann, S.M., Arey, J., Winer, A.M., and Pitts, J.N. Jr.: 1987, *Environ. Sci. Technol.* **21**, 64-72.

Atkinson, R., Kwok, E.S.C. and Arey J.: 1992, In: *Proceedings Brighton Crop Protection Conference*, British Crop Protection Council, Farnham, p. 469-476

Atkinson, R., Baulch, D.L., Cox, R.A., Hampson, R.F. Jr., Kerr, J.A., Rossi, M.J. and Troe, J.: 1997a, *J. Phys. Chem. Ref. Data* **26**, 1329-1499.

Atkinson, R., Baulch, D.L., Cox, R.A., Hampson, R.F., Jr., Kerr, J.A., Rossi, M.J. and Troe, J.: 1997b, *J. Phys. Chem. Ref. Data* **26**, 521-1011.

Ballschmiter, K.H. and Wittlinger, R.: 1991, *Environ. Sci. Technol.* **25**, 1103-1111

Behnke, W. and Zetzsch, C.: 1989, Über die Reaktivität von Lindan gegenüber OH-Radikalen. UBA, Berlin

Behnke, W., Nolting, F. and Zetzsch, C.: 1987, *J. Aerosol Sci.* **18**, 849-852.

Behnke, W., Holländer, W., Koch, W., Nolting, F. and Zetzsch, C.: 1988, *Atmos. Environ.* **22**, 1113-1120

Behymer T.D. and Hites R.A.: 1985, *Environ. Sci. Technol.* **19**, 1004-1006

Bidleman, T.F.: 1988, *Environ. Sci. Technol.* **22**, 361-367.

Bornhop, D.J., Hlousek, L. and Hacket, M.: 1991, *Rev. Sci. Instrum.*, **63**, 191-201.

Brauers, T., Aschmutat, U., Brandenburger, U., Dorn, H.-P., Hausmann, M., Heßling, M., Hofzumahaus, A., Holland, F., Plass-Dülmer, C. and Ehhalt, D.H.: 1996, *Geophys. Res. Lett.* **23**, 2545-2548.

Braun, F., Schüssler, W., Wanzinger, M., von Brozyskowski, R.W., van de Graaff, S. and Gast, R.: 1990, *Neue Untersuchungen zur Analytik und Verbreitung von PCB und Pflanzenbehandlungsmitteln.* Bayerische Landesanstalt für Wasserforschung, Munich.

Brubaker, W.W. and Hites, R.A.: 1997, *Environ. Sci. Technol.* **31**, 1805-1810.

Brubaker, W.W. and Hites, R.A.: 1998a, *J. Phys. Chem. A* **102**, 915-921.

Brubaker, W.W. and Hites, R.A.: 1998b, *Environ. Sci. Technol.* **32**, 766-769.

Carter, W.P.L., Luo, D. and Malkina, I.L.: 1997, *Atmos. Environ.* **31**, 1425-1439.

Cotham, W.E.Jr. and Bidleman, T.F.: 1991, *Chemosphere* **22**, 165-188.

Crosby, D.G. and Moilanen, K.W.: 1974, *Arch. Environ. Contam. Toxicol.* 2, 62.

Crosby, D.G. and Moilanen, K.W.: 1977, *Chemosphere*, **6**, 167.

Dentener, F.J., Crutzen, P.J.: 1993, *J. Geophys. Res.* **98**, 7149-7163.

Derwent, R.G. and Volz-Thomas, A.: 1990, *The Tropospheric Lifetimes of Halocarbons and their Reactions with the OH Radical: An Assessment based on the Concentrations of ^{14}CO*, in *Scientific Assessment of Stratospheric Ozone: 1989* Volume II, Appendix: AFEAS Report, World Meteorological Organization Global Ozone Research and Monitoring Project - Report No. 20, Geneva.

de Voogt, P. and Jansson, B.: 1993, *Rev. Environ. Contam. Toxicol.* **132**, 1-27.

Elend, M., Zetzsch, C. and Sarafin, R.: 1994, Photolytic and photochemical degradation of MITC (methyl isothiocyanate) in air. IUPAC Congress, Washington, D.C.

Fingler, S., Tkalcevic, B., Fröbe, Z., Drevenkar, V.: 1994, *Analyst* **119**, 1135.

Fritz, B., Lorenz, K., Steinert, W. and Zellner, R.: 1981, Proceedings of the 2nd European Symposium on the Physico-Chemical Behaviour of Atmospheric Pollutants, D. Riedel Publishing Co., Dordrecht, the Netherlands, pp. 192-202.

Geddes, J.D., Miller, G.C. and Taylor, G.E., Jr.: 1995, *Environ. Sci. Technol.* **29**, 2590-2594.

Geissler, A. and Schöler, H.F.: 1993, *Vom Wasser* **80**, 357-370.

Glotfelty, D..E., Seiber, J.N. and Lilljedahl, L.A.: 1987, *Nature* **325**, 602-605.

Goldstein, A.H., Wofsy, S.C. and Spivakovsky, C.M.: 1995, *J. Geophys. Res.* **100**, 21023-21033.

Goodman, M.A., Aschmann, S.M., Atkinson, R. and Winer, A.M.: 1988, *Arch. Environ. Contam. Toxicol.* **17**, 281-288.

Goolsby, D.A., Thurman, E.M. and Battaglin, W.A.: 1994 in: Weigman D.L. (ed.) *New directions in pesticide research, development, management and policy*. Virginia Water Resources Research Center, Blacksburg.

Güsten, H., Bozicevic, Z. and Klasinc, L.: 1984, *Staub-Reinhalt. Luft* **44**, 448-451.

Hebert, V.R., Hoonhout, C. and Miller, G.C.: 1997, Use of sunlight and solar simulation to determine atmospheric pesticide photolysis rates; Paper presented at the Society of Environmental Toxicology and Chemistry (SETAC) 18th Annual Meeting, San Francisco.

Hebert, V.R., Geddes, J.D., Mendosa, J. and Miller, G.C.: 1998a, *Chemosphere*, **36**, 2057-2066.

Hebert, V.R., Hoonhout, C. and Miller, G.C.: 1998b, Use of a photoreactor to determine atmospheric reaction rates of pesticides; Paper presented at the 9th International Congress IUPAC Pesticide Chemistry Conference, London.

Hein, R., Crutzen, P.J. and Heimann, M.: 1997, *Global Biogeochem. Cycles* **11**, 43-76.

Herterich, R. and Herrmann, R.: 1990, *Environ. Technol.* **11**, 961-972.

Hurle, K. and Oberwalder, D.: 1992. UBA, report 126 05 076, Berlin.

Jobson, B.T., Niki, H., Yokouchi, Y., Bottenheim, J., Hopper, F. and Leaitch, R.: 1994, *J. Geophys. Res.*, **99**, 25355-25368.

Jury, W.A., Winer, A.M., Spencer, W.F. and Focht, D.D.: 1987. *Rev. Environ. Contam. Toxicol.* **99**, 119-164.

Koch, R., Krüger H.-U., Elend M., Palm W.-U. and Zetzsch C.: 1996, *Int. J. Chem. Kinet.* **28**, 807-815.

Koch, R., Palm, W.-U. and Zetzsch, C.: 1997, , *Int. J. Chem. Kinet.* **29**, 81-87.

Krol, M., van Leeuwen, P. J., and Lelieveld, J.: 1998, *J. Geophys. Res.* **103**, 10697-10711.

Kwok, E.S.C. and Atkinson, R.: 1995, *Atmos. Environ.* **29**, 1685-1695.

Kwok, E.S.C., Atkinson, R. and Arey, J.: 1992, *Environ. Sci. Technol.* **26**, 1798-1807.

Kwok, E.S.C., Arey, J. and Atkinson, R.: 1994, *Environ. Sci. Technol.* **28**, 528-533.

Kwok, E.S.C., Atkinson, R. and Arey, J.: 1995, *Environ. Sci. Technol.* **29**, 1591-1598.

Kwok, E.S.C., Aschmann, S.M. and Atkinson, R.: 1996, *Environ. Sci. Technol.* **30**, 329-334.

Mather, J.H., Stevens, P.S. and Brune, W.H.: 1997, *J. Geophys. Res.* **102**, 6427-6436.

Mauldin III, R. L., Frost, G. J., Chen, G., Tanner, D. J., Prevot, A. S. H., Davis, D. D. and Eisele, F. L.: 1998, *J. Geophys. Res.* **103**, 16713-16729.

Meylan, W.M. and Howard, P.H.: 1993, *Chemosphere* **26**, 2293-2299.

Meylan, W.M. and Howard, P.H.: 1995 *Atmospheric Oxidation Program*. Syracuse Res. Corp., Syracuse.

Mihelcic, D., Klemp, D., Müsgen, P., Pätz, H.W. and Volz-Thomas, A.: 1993, *J. Atmos. Chem.* **16**, 313-335.

Mill, T.: 1980, Chemical and photooxidation. In: Hutzinger, O. (ed.), *The Handbook of Environmental Chemistry*, ed. Springer-Verlag, Berlin,, Vol. 2., Part A., pp. 77-105.

Moilanen, K.W., Crosby, D.G., Humphrey, J.R. and Giles, J.W.: 1978, *Tetrahedron* **34**, 3345.

Mongar, K. and Miller, G.C.: 1988, *Chemosphere* **17**, 2183-2188.

Mount, G.H., Eisele, F.L., Tanner, D.J., Brault, J.W., Johnston, P.V., Harder, J.W., Williams, E.J., Fried, A. and Shetter, R.: 1997, *J. Geophys. Res.* **102**, 6437-6455.

OECD: 1991, *Test guideline on photochemical oxidative degradation in the atmosphere*, UBA, Berlin.

Oltmans, S.J. and Levy, H., III.: 1994, *Atmos. Environ.* **28**, 9-24.

Orlando, J.J., Tyndall, G.S. and Wallington, T.J.: 1996, *J. Phys. Chem.* **100**, 7026-7033.

Palm, W.-U., Elend, M., Krüger, H.-U. and Zetzsch, C.: 1997, *Environ. Sci. Technol.* **31**, 3389-3396.

Palm, W.-U., Elend, M., Krüger, H.-U. and Zetzsch, C.: 1999, *Chemosphere*, **38**, 1241-1252.

Paulson, S.E. and Orlando, J.J.: 1996, *Geophys. Res. Lett.* **23**, 3727-3730.

Pitts, J.N.: 1981, Experimental protocol for determining absorption, cross-sections of organic chemicals; EPA Report No. 60013-81-051, Washington.

Platt, U. and Heintz, F.: 1994, *Israel J. Chem.* **34**, 289-300.

Prinn, R.G., Weiss, R.F., Miller, B.R., Huang, J., Alyea, F.N., Cunnold, D.M., Fraser, P.J, Hartley, D.E. and Simmonds, P.G.: 1995, *Science* **269**, 187-192.

Roelofs, G.-J. and Lelieveld, J.: 1997, *Tellus* **49B**, 38-55.

Scharf, J. and Bächmann, K.: 1993, *Nachrichtenbl. Deut. Pflanzenschutzd* **45**, 82.

Scharf, J., Wiesiollek, R. and Bächmann, K.: 1992, *Fres. J. Anal. Chem.*, **342**, 813-816.

Scheringer, M.: 1997, *Environ. Sci. Technol.* **31**, 2891-2897.

Schomburg, C.J., Glotfelty, D.E. and Seiber, J.N.: 1991, *Environ. Sci. Technol.* **25**, 155-160.

Schössner, H. and Koch, A.: 1991, *Forum Städte-Hyg.* **42**, 109-112.

Sedlak, D. L. and Andren, A. W.: 1991, *Environ. Sci. Technol.* **25**, 1419-1426.

Seiber, J.N., McChesney, M.M. and Woodrow, J.E.: 1989, *Environ Toxicol. Chem.* **8**, 577-588.

Seiber, J.N., Wilson, B.W. and McChesney, M.M.: 1993, *Environ. Sci. Technol.* **27**: 2236-2243.

Soderquist, C.J., Crosby, D.G., Moilanen, K.W., Seiber, J.N. and Woodrow, J.E.: 1975, *J. Agric. Food Chem.* **23**, 304.

Spencer, W.F., Adams, J.D., Shoup, T.D. and Spear, R.C.: 1980, *J. Agric. Food Chem.* **28**, 366-371.

Spicer, C. W., Chapman, E. G., Finlayson-Pitts, B. J., Plastridge, R. A., Hubbe, J. M., Fast, J. D. and Berkowitz, C. M.: 1998, *Nature* **394**, 353-356.

Stanley, C.W., Barney, J.E., Michael, R. and Yobs, A.R.: 1971, *Environ. Sci. Technol.* **5**, 430.

Strachan, W.M.J.: 1990, in: Kurtz D.A.(ed.) *Long range transport of pesticides*. Lewis Publishers, Boca Raton p.233-240.

Tarrant, K.R. and Tatton, J.G.: 1968, *Nature* **219**, 725.

Tremi, J., Mattrel, P., Fingler, S. and Giger, W.: 1993, *Water Air Soil Pollut.* **68**, 113-123.

Tuazon, E.C., Winer, A.M. and Pitts, J.N. Jr.: 1980, *Environ. Sci. Technol.* **15**, 1232-1237.

Tuazon, E.C., Atkinson, R., Winer, A.M. and Pitts, J.N. Jr.: 1984, *Arch. Environ. Contam.Toxicol.* **13**, 691-700.

Tuazon, E.C., Atkinson, R., Aschmann, S.M., Arey, J., Winer, A.M. and Pitts, J.N. Jr.: 1986, *Environ. Sci. Technol.* **20**, 1043-1046.

Tuazon, E.C., Atkinson, R., Aschmann, S.M., Goodman, M.A. and Winer, A.M.: 1988, *Int. J. Chem. Kinet.* **20**, 241-265.

van Pul, W. A. J., de Leeuw, F. A. A. M., van Jaarsveld, J. A., van der Gaag, M. A. and Sliggers, C. J.: 1998, *Chemosphere* **37**, 113-141.

Walsh, P.R. and Hites, R.A.: 1979, *Bull. Environ. Contam. Toxicol.* **22**, 305-311.

Wayne, R.P., Barnes, I., Biggs, P., Burrows, J.P., Canosa-Mas, C.E., Hjorth, J., Le Bras, G., Moortgat, G.K., Perner, D., Poulet, G., Restelli, G. and Sidebottom, H.: 1991, *Atmos. Environ.* **25A**, 1-203.

Winer, A.M. and Atkinson, R.: 1990, in: *Long Range Transport of Pesticides*, ed. Kurtz, D.A., Lewis Publishers, Chelsea, pp. 115-126.

Woodrow, J.E., Seiber, J.N., Crosby, D.G., Moilanen, K.W., Soderquist, C.J. and Mourer, C.: 1977, *Arch. Environ. Contam. Toxicol.* **6**, 175-182.

Woodrow, J.E., Crosby, D.G., Mast, T., Moilanen, K.W. and Seiber, J.N.: 1978, *J. Agric. Food Chem.* **6**, 1312-1316.

Woodrow, J.E., Crosby, D.G. and Seiber, J.N.: 1983, *Residue Rev.* **85**, 111-125.

Zabik, J.M. and Seiber, J.N.: 1993, *J. Environ. Qual.* **22**: 80-90.

Zetzsch, C.: 1991, *UWSF-Z. Umweltchem. Ökotox.* **3**, 59-64.

Zetzsch, C. and Becker, K. H.: 1989, in: *Halogenierte organische Verbindungen in der Umwelt*, VDI Berichte Nr. 745, Düsseldorf, pp. 97-127.

ATMOSPHERIC TRANSPORT AND DEPOSITION OF PESTICIDES: AN ASSESSMENT OF CURRENT KNOWLEDGE

W. ADDO J. VAN PUL[1*], TERRY F. BIDLEMAN[2], EVA BRORSTRÖM-LUNDÉN[3], PETER J.H. BUILTJES[4], SERGEY DUTCHAK[5], JAN H. DUYZER[4], SVEN-ERIK GRYNING[6], KEVIN C. JONES[7], HARRIE F.G. VAN DIJK[8], J. (HANS) A. VAN JAARSVELD[1]

[1] National Institute of Public Health and the Environment (RIVM), P. O. Box 1, NL-3720 BA Bilthoven, The Netherlands, [2] Atmospheric Environment Service, 4905 Dufferin St. Downsview, Ontario M3H 5T4, Canada [3] Swedish Environmental Research Institute, P. O. Box 47086, S-40258 Göteborg, Sweden [4] TNO Institute of Environmental Sciences, Energy Research and Process Innovation, P.O. Box 342, NL-7300 AH Apeldoorn, The Netherlands, [5] Meteorological Sythesizing Centre-East (EMEP/MSC-E), Kedrova str. 8-1, Moscow, Russia, [6] Wind Energy and Atmospheric Physics Department, Risø National Laboratory, DK-4000 Roskilde, Denmark, [7] Institute of Environmental and Biological Sciences, Lancaster University, Lancaster, LA1 4YQ, United Kingdom, [8] Health Council of the Netherlands, P. O. Box 16052, NL-2500 BB The Hague, The Netherlands

(corresponding author, e-mail: addo.van.pul@rivm.nl)*

Abstract. The current knowledge on atmospheric transport and deposition of pesticides is reviewed and discussed by a working group of experts during the Workshop on Fate of pesticides in the atmosphere; implications for risk assessment, held in Driebergen, the Netherlands, 22-24 April, 1998. In general there is a shortage of measurement data to evaluate the deposition and re-emission processes. It was concluded that the mechanisms of transport and dispersion of pesticides can be described similarly to those for other air pollution components and these mechanisms are rather well-known. Large uncertainties are present in the exchange processes at the interface between air and soil/water/vegetation. In all process descriptions the uncertainty in the physicochemical properties play an important role. Particularly those in the vapour pressure, Henry's law constant and its temperature dependency. More accurate data on physicochemical properties and particularly the temperature dependencies is needed.

Key words: atmospheric transport, surface exchange, deposition, pesticides

1. Introduction

Pesticides are emitted into the atmosphere in a direct way during application and indirectly via (re-) emission of applied or deposited material. Once in the atmosphere the pesticide is dispersed and transported by the wind. The pesticide may attach to aerosols depending on the vapour pressure of the pesticide and the available aerosol surface. The pesticide is removed from the atmosphere by (chemical) transformation and dry and wet deposition. The transport distance of a pesticide in the atmosphere is strongly dependent on the rate of these removal processes. The concentration and deposition of a pesticide at a certain location are a function of the configuration and strength of

Water, Air, and Soil Pollution **115:** 245–256, 1999.

the emissions, the diffusion by turbulence, the wind pattern and the removal processes.

The Health Council of the Netherlands organised a workshop on 'Fate of pesticides in the atmosphere; implications for risk assessment', which was held at Driebergen, the Netherlands, from 22-24 April, 1998. This paper is the report of the session on the atmospheric transport and deposition of pesticides. Central questions in this session were: How well do we know the transport and deposition processes? What are the most relevant parameters in describing these processes? What is the main source of uncertainty in the processes and in the deposition as a whole?

For each process the current knowledge is indicated and discussed. Emission processes and (chemical) transformation were considered in two separate sessions (Atkinson et al., Van den Berg et al., this issue).

2. Partitioning of a pesticide between the gaseous and particle phase

The processes involved in the removal of gases from the atmosphere are different from those governing the removal of particles. Therefore the partitioning of the pesticide between the gaseous and particle phase is an important parameter in estimating the rate of the removal processes.

Only a few measurements of the particle/gas phase partitioning of pesticides have been reported. These have been mainly for non-polar pesticides such as DDT, HCHs and chlordane (Pankow and Bidleman, 1992; Lane *et al.*, 1992; Hoff *et al.*, 1996; Millet *et al.*, 1997) and less frequently for polar pesticides (Trevisan *et al.*, 1993; Hawthorne *et al.*, 1996; Millet *et al.*, 1997). More measurements of this partitioning are available for PAHs and dioxins in urban and rural areas (Yamasaki *et al.*, 1982; Eitzer and Hites, 1989; Pankow and Bidleman, 1992; Welsch-Paulsch *et al.*, 1995; Goss and Schwarzenbach, 1998; Lohmann *et al.*, 1998).

By analogy to PAHs and dioxins the partitioning of a pesticide between the gas and particle phase can be estimated with the well-known Junge-Pankow equation (Junge, 1977; Pankow, 1987). The partitioning in this equation is a function of the liquid-phase vapour pressure of the pesticide and the available aerosol surface. However, it is questionable if the constant c of 17.2 Pa.cm for organics in the equation is also applicable to polar compounds. According to the Junge-Pankow model, pesticides having vapour pressures $>10^{-3}$ Pa will be for 95% or more in the vapour phase in rural air. This was calculated with a surface area, q, of 3.5×10^{-6} cm^2 cm^{-3}, which is characteristic for background air with some local source influence (Whitby, 1978).

Other approaches are absorption models in which the partitioning is a function of liquid-phase vapour pressure or octanol-air partition coefficient and the organic carbon content of the aerosol (Pankow, 1994a,b, Finizio *et al.*, 1997; Harner and Bidleman, 1998; Bidleman, this issue). However, it is also uncertain if these models can be extrapolated to polar compounds.

At the present time, most particle-phase distribution estimates are based on by modelling (Junge-Pankow), or determined experimentally by sampling with a filter to collect particles, followed by sorbent trap for vapour-phase pesticides. Limitations of the model include uncertainties in the parameters c and q, the inability to account for the kinetics of the sorption process, and humidity effects. Filtration sampling is subject to artifacts arising from volatilisation losses of POPs from particles on the filter and sorption of vapour-phase POPs to the filter itself. It is important to recognize that neither the model nor the measurements necessarily give the "true" value of the particle/gas distribution, and more research is needed to improve the estimates.

Comparisons of phase distributions estimated from filtration sampling with those calculated from the Junge-Pankow model have been made for PAHs and to a lesser extent for PCBs, PCDD/Fs and chlorinated pesticides (Bidleman *et al.*, 1998; Bidleman and Harner, 1998). The agreement between modelled and measured distributions is quite good for PAHs in urban and rural air. For example, the predicted particulate fraction in urban air ($q = 1.1 \times 10^{-5}$ cm^2 cm^{-3}, (Whitby, 1978)), is 0.5 at a liquid-phase vapour pressure of 2×10^{-4} Pa, compared to 0.3-0.7 found by ambient air sampling (Bidleman *et al.*, 1998). A similar comparison shows that measured particulate fractions for the chlorinated compounds are about 2-3 times lower than predicted values, indicating that the model overestimates (or the sampler underestimates) the sorption of these compounds to aerosols (Bidleman *et al.*, 1998; Bidleman and Harner, 1998).

Very little information is available on the size distribution of the particle to which the pesticides are attached. Since the pesticides will distribute dependent on the available aerosol surface, it is expected that the pesticides will be attached for the greater part to smaller particles.

Another source of uncertainty resides in estimating or measuring the liquid-phase vapour pressure of the pesticide. This property is not directly accessible for compounds that are solids at ambient temperature and must be estimated by thermodynamic calculations, extrapolation of measured vapour pressure data from above the melting point, or capillary gas chromatography. Uncertainties can be a factor of two for non-polar pesticides and even larger for polar molecules (Bidleman, this issue).

3. Atmospheric transport of pesticides

Once in the atmosphere, the pesticide is dispersed by turbulence and transported by the mean wind flow. Pesticides are dispersed and transported similar to other quantities such as water vapour or other air pollution components. This means that atmospheric transport models, which are already used for other air pollutants, like acidifying substances and ozone, can be used to describe the transport of pesticides as well (Van Jaarsveld and Van Pul, this issue). In modelling of the transport of air pollutants the dispersion by turbulence and the transport by the mean flow is parameterised using meteorological observations and numerical calculations of the flow field.

The uncertainty in the transport and dispersion of an air pollution component as such is difficult to obtain experimentally, because of the uncertainty in the emissions and the influence of the removal processes on the concentration. Estimates can be found from a large number of tracer experiments. Hanna (1993) found by comparing a number of models often used in applied dispersion modelling with observations from 28 independent tracer experiments, that the median fractional bias of the absolute concentrations is in the range of 20-27%. These values apply to short range and short term dispersion experiments of the order of a few kilometres. For larger scales and larger averaging times the uncertainty is expected to be rather similar, because the uncertainties in the meteorological situations (flow field) have a tendency to cancel each other on a larger time scale. Stohl *et al.* (1998) tested the Lagrangian FLEXPART model with measurements from three long range tracer experiments on the 100-1000 km scale, the ANATEX and CAPTEX experiments in the US, and ETEX in Europe. The model was found generally to perform well under simple meteorological conditions, but is less good in the presence of fronts. Overall the mean fractional bias between measurements and model predictions could be estimated to be 17%. These experiments were all performed with inert tracers, therefore the values do not include the uncertainty in transformation and deposition processes.

Transport models can be tested using them for substances for which much more reliable data are available. Van Jaarsveld *et al.*, (1997) estimated the uncertainty in the modelled annual deposition of SO_2 at 30%. This included the meteorological variability and uncertainties in the deposition process.

Overall, the uncertainty in the concentration of an inert species on an annual basis caused by the modelling concept and meteorological variability was estimated tentatively at 30-50%. More details on the uncertainty in transport and deposition modelling are provided by Van Jaarsveld and Van Pul (this issue).

4. Deposition processes

4.1 DRY DEPOSITION OR EXCHANGE OF PESTICIDES AT THE EARTH'S SURFACE

Pesticides in the gaseous phase are transported to the earth's surface by atmospheric turbulence. The actual deposition is the uptake of the pesticide by vegetation and diffusion into or adsorption at soil or water surfaces. Pesticides may accumulate in the soil/water/vegetation compartments dependent on the soil/water/vegetation conditions and the physicochemical properties of the pesticides. Dependent on its atmospheric concentration the accumulated pesticide may revolatilise and be transported further via the atmosphere. Since the surface concentration of the pesticide is strongly dependent on the surface temperature the pesticide will migrate towards the colder regions of the earth. This is known as the grasshopper-effect of persistent chemicals and may lead to the global fractionation of those substances (Wania and Mackay, 1993). Fractionation and cold condensation of pesticides takes place not only in arctic regions but also in mountainous areas (Buser, 1990; Blais *et al.*, 1998; McConnell *et al.*, 1998).

Estimates of the atmospheric part of the deposition or exchange process can be made using the parameterisations as used in other air pollution studies. For instance, the resistance formulations as used for ozone and SO_2 (*e.g.* Hicks *et al*, 1987), *i.e.*, the aerodynamic and laminar boundary layer resistance parameterisations, are also valid for pesticides. This means that in describing the exchange of pesticides the experimental and modelling efforts should be directed towards the uptake by and processes in the soil, water and vegetation, for example, by formulating parameterisations for the surface resistance and surface concentration.

Validation of the exchange modules for treated fields showed a close agreement between measured and modelled volatilisation fluxes (Majewski *et al.*, 1990; Scholtz *et al.*, 1997). Field data for other components which are easier to measure and which have a somewhat similar behaviour concerning the deposition/re-emission process can also be used to validate those models. Van Jaarsveld (1997), *e.g.*, showed that with a complex atmosphere-soil exchange model the diurnal pattern of emission fluxes of ammonia from a manure treated field could be described well. However, it should be noted that under these conditions mainly the atmospheric exchange part of the flux parameterisation is validated and not (or partly) the parameterisation of the soil processes.

Few direct measurements of the dry deposition and re-emission of pesticides are available. In gas chamber experiments Duyzer and Van Oss (1997) studied the uptake of a number of pesticides by water, soil and vegetation. They found that the deposition velocity could be well parameterised using descriptions similar to Jury *et al.* (1983), for soil and to Liss and Slater (1974) for water, *i.e.* within a factor of 3 for a range of 4 orders of magnitude in the physicochemical properties. The most important parameters in describing this process are the Henry's law constant and other phase partition coefficients of the pesticide (K_{ow}, K_{oc}, K_{oa}), the leaf area index (LAI) and the organic carbon content of the soil. Field experiments have shown the relationship of pesticide volatilisation from water and soil to these properties (Hoff *et al.*, 1996; Ridal *et al.*, 1996; Woodrow *et al.*, 1997; McConnell *et al.*, 1997; Hillary *et al.*, 1998; Bidleman, this issue). It was found for other organic chemicals, such as PCBs and PAHs, that the contents in soil where the input is, and was, atmospheric, correlated well with the soil organic carbon content and the K_{oa} (Cousins and Jones, 1995).This does not apply to pesticides in agricultural soils, where pesticides were directly put onto the fields. Organochlorine pesticides in agricultural soils are far from air/soil equilibrium and are usually oversaturated in the soil relative to the atmosphere. Aigner *et al.*, 1998 showed that there is no strong correlation between organochlorine residues and soil organic carbon content.

For the assessment of the deposition and re-volatilisation on a regional scale, residue data on pesticides particularly the "old" persistent ones (DDT, toxaphene etc.), are necessary, but are still lacking. Measurements have shown that the spatial variability of the residues, particularly in application areas, is quite large (Aigner *et al.*, 1998; Harner and Bidleman, 1998). Soil residue data are necessary to show the grasshopper behaviour of some pesticides.

Vegetation can remove organic pollutants from the atmosphere (Simonich and Hites, 1994; Asten and Seiber, 1996; Brorström-Lundén and Löfgren, 1998). Some parameterisations for this uptake of organics by vegetation are available (*e.g.* McLachlan and Horstmann, 1998). In the latter study filter factors for forests are defined and are primarily a function of the K_{oa}. Measurements of PCBs showed that the concentration in plants (grass) could be well described by using K_{oa} and for these compounds equilibrium between air and vegetation appeared quite rapidly and was established within 2 weeks. (Thomas *et al.*, 1998a, 1998b). Vegetation can also act as a source of organics by re-volatilisation of accumulated material (Lee *et al.*, 1998). Vegetation also affects the air-soil exchange process by reducing the turbulent exchange between the air and the soil. Also less energy at the soil surface is available for evaporation of substances.

It was stressed that dry deposition and re-emission, *i.e.* gas exchange, of pesticides are very important and in many cases the most important terms in the total budget of pesticides for water bodies (Baart *et al.*, 1995; Hoff *et al.*, 1996; Duce, 1998; Bidleman, this issue). For a large number of pesticides the dry deposition to water surfaces is dominated by atmospheric turbulence and not by the uptake or diffusion processes at the surface. This is because many of the pesticides are fairly soluble in water. Following the formulation by Liss and Slater (1977), it was calculated that for a large number of pesticides the surface resistance of the water surface is very small compared to the atmospheric resistances to the transport from air to water surface (Van Pul *et al.*, 1998). However, for pesticides which accumulate in the water body the dry deposition or exchange is strongly dependent on the surface water concentration. In some cases the flux may reverse and show re-emission of the pesticide (Bidleman *et al.*, 1995; Jacobs and Van Pul, 1996).

The dry deposition of pesticides attached to particles is strongly dependent on the physical properties of the particles themselves, of which particle diameter is the most important. As was pointed out in Section 2, little is known about the particle size distribution, but it is expected that the pesticides will be attached for the greater part to smaller particles.

Parameterisations for the deposition velocity or surface resistance for particles are rather well-known and are given by Slinn (1977), Sehmel and Hodgson (1980) and Seinfeld (1986).

4.2 WET DEPOSITION OF PESTICIDES

Pesticides may be removed from the atmosphere by precipitation via in-cloud and below-cloud scavenging (rain-out and wash-out). The rates of these processes depend, among others, on the cloud and rain drop size distribution, the type of hydrometeor, the molecular diffusion and Henry's law constant of the pesticide. In the case of gaseous pesticides the pesticide may accumulate in the cloud or rain droplets, which reduces the scavenging rate and may even lead to revolatilisation dependent on the ambient air concentration. The characteristics of the wet deposition process of particles can be used for the wet deposition of particle bound pesticides. In the case of particles the uptake process by the droplets is not limited by the concentration of the pesticide in the droplet.

Measurements for a limited number of pesticides in rain were carried out at a few sites throughout Europe (Van Dijk and Guicherit, this issue). Some of these measurements are reported in international frameworks like EMEP/CCC and OSPARCOM/CAMP (Berg *et al.*, 1996; OSPAR, 1994). There is an absolute need to expand these measurement activities towards a monitoring network that covers Europe. In the IADN (Integrated Atmospheric Deposition Network) program in the US and Canada wet deposition of pesticides is routinely measured (Hoff *et al.*, 1996; Hillary *et al.*, 1998).

The wet deposition process can be described by using the scavenging ratio and the rain amount. There are a limited number of concurrent measurements of pesticides in air and rain from which the scavenging ratio can be deduced (Brorström-Lundén, 1995; Bidleman, 1988). The scavenging coefficient of pesticides in the gas phase can be relatively well decribed as a function of the Henry's law constant (Bidleman, 1988).

It was estimated that the uncertainty in the wet deposition for gaseous pesticides is a factor 2-3. This was found to be caused mainly by the uncertainty in the Henry's law constant and its temperature dependency.

The wet deposition of pesticides attached to particles is strongly dependent on the physical properties of the particles themselves of which particle diameter is the most important. The wet scavenging of particles is an efficient process (Slinn, 1983) Typical scavenging ratios found are 10^5-10^6 (Bidleman, 1988) and are used in modelling studies (Van Jaarsveld *et al.*, 1997).

4.3 UNCERTAINTY IN ESTIMATES OF DEPOSITION FLUXES

The deposition flux of a pesticide depends on many factors, such as its physicochemical properties and properties of the receiving surface. These properties differ over a large range of pesticides and therefore it is not easy to make a general statement on the uncertainty in estimates of the deposition, particularly in those of dry deposition or re-emission fluxes.

Since the uncertainty in each deposition process is different, the uncertainty in deposition estimates of a pesticide on a certain location depends on the main deposition mechanism. As a first step it is useful to distinguish between gaseous and particle bound pesticides.

For pesticides mainly present in the gas phase the deposition mechanism to water surfaces is relatively well-known. Hoff *et al.* (1996) estimated that the uncertainty in the net gas flux (deposition and volatilisation) of a number of organic chemicals, which were far from air/water equilibrium (among others dieldrin, HCB), to the Great Lakes was 50-100%, which was dominated by the 50% uncertainty in the air-water mass transfer coefficient. However, for chemicals which were near air/water equilibrium the uncertainty in the net-flux can be as large as 50-7400%, caused by the uncertainty in the concentration differences between water and air. Note that in those cases the net flux is only a small term in the mass balance of the substance. In the case of lindane, which is deposited at water surfaces mainly via air-water gas exchange the uncertainty in the deposition on the North Sea by modelling was estimated at 30% (Van Jaarsveld *et al.*, 1997).

For gaseous pesticides which are soluble in water, wet scavenging from the gas-phase is an important deposition route at land surfaces. For these pesticides the uncertainty in the wet deposition can be taken as an approximation of the uncertainty in the total deposition and this was estimated at a factor of 2-3 (Section 4.2).

Deposition rates of gaseous pesticides to vegetation or bare soil are largely uncertain. It was estimated that the uptake of a number of pesticides by water, soil and vegetation could be described within a factor of 3 (Section 4.1, Duyzer and Van Oss, 1997). Therefore, for gaseous pesticides which are mainly dry deposited, the uncertainty in the total deposition as a first approximation is estimated also at this factor of 3. However, under field conditions, and particularly for pesticides which accumulate in soil and vegetation, the lack of information on the (top) surface concentration will add considerably to the uncertainty in the dry deposition flux.

For a pesticide which is mainly in the particle phase (2,4-D, for example), knowledge of the particle size distibution, which determines the particle dry deposition velocity and wet scavenging rate, is the main limitation. Hoff *et al.* (1994) estimate that the uncertainty in the dry deposition of B(a)P, which is mainly particle bound, is 150%. Van Jaarsveld *et al.*, (1997) estimated this uncertainty in the deposition for B(a)P at 100%.

Concluding, the uncertainty in the deposition of a pesticide is estimated tentatively at a factor of 3. This figure should be considered as an indicative value. This means that it can not be excluded that for some individual pesticides this uncertainty may deviate largely from this value. The uncertainty estimates in the deposition modelling of pesticides are presented in Van Jaarsveld and Van Pul (this issue).

5. Conclusion and recommendations

1. The gas-particle phase partitioning of pesticides can be estimated as a first approximation with the Junge-Pankow-equation. The uncertainty is estimated at a factor of 2-3.

2. The processes of transport and dispersion are similar to those for other air pollutants (such as ozone and SO_2), which have been studied more intensively over the last decades. From these studies it can be concluded that the uncertainty in the transport mechanism in models is about 30-50%, induced by the modelling concepts and meteorological variability.

3. The atmospheric part of the deposition or exchange process of pesticides can be described in a similar way to that used for other air pollutants (such as ozone and SO_2); *e.g.* resistance parameterisation.

4. There are not many measurement data available to evaluate deposition and re-emission processes. Hardly any data on water and soil concentrations on a regional scale are available for use in deposition and re-emission estimates.

5. Gaseous dry deposition and re-emission (*i.e.*, gaseous exchange) are very important terms in the budget of pesticides, particularly for water bodies.

6. Main parameters determining the exchange process of pesticides at the soil-air and vegetation-air interfaces, apart from the atmospheric transport, are: K_{ow}, H (or clustered to K_{oa}) and soil organic carbon or plant lipids content. This conclusion is partly based on measurements and parameterisations for other organic chemicals.

7. Wet deposition of pesticides in the gas phase can be well described with a scavenging coefficient which is a function of the Henry's law constant. The uncertainty in the wet deposition parameterisation of gaseous pesticides is estimated at a factor of 2-3.

8. The uncertainty in the deposition of pesticides is estimated tentatively at a factor of 3. This uncertainty is added to the above uncertainty in modelling of the transport and dispersion of pesticides. The overall uncertainty in the modelled deposition values therefore is estimated at a factor of 3-4. That is not including the uncertainty in the emissions and transformations.

9. In all process descriptions the uncertainty in the physicochemical properties play an important role. Particularly those in the vapour pressure, Henry's law constant and its temperature dependency.

10. More accurate data on physicochemical properties and particularly the temperature dependencies are needed. More process oriented data such as the exchange of pesticide vapours with atmospheric particles, soil, vegetation and water specifically for the modern, polar compounds are needed.

11. Use all available knowledge about the "old" pesticides and apply this to make estimates for new pesticides. For the validation of these estimates, measurement data are needed.

References

Aigner, E. J., Leone, A. D. and Falconer, R. L.: 1998, *Environ. Sci. Technol.* **32**, 1162-1168.

Aston, L. S. and Seiber, J. N.: 1996, *J. Environ. Sci. Health*, **B31**, 671-698.

Atkinson, R., Guicherit, R., Hites, R., Palm, W.-U., Sieber, J. N. and De Voogt, P.: 1999, Transformation of pesticides in the atmosphere: a state of the art. *Water Air Soil Pollut.* this issue.

Baart, A. C. Berdowski, J. J. M. and van Jaarsveld, J. A.: 1995, *Calculation of atmospheric deposition of contaminants on the North Sea.* TNO report R95/138, Delft, The Netherlands.

Berg, T., Hjellbrekke, A-G. and Skjelmoen, J-E.: 1996, Heavy metals and POPs within the ECE region. *EMEP/CCC-report* 8/96

Bidleman, T. F, Jantunen, L. M., Falconer, R. L. and Barrie, L. A.: 1995, *Geophys. Res. Lett.*, **22**, 219-222.

Bidleman, T. F.:1988, *Environ. Sci. Technol.* **22,** 361-367.

Bidleman, T. F., Falconer, R. L. and Harner, T.: 1998, *In: D.A. Lane (ed.) Gas and Particle Measurements of Atmospheric Organic Compounds,* Gordon and Breach Publishers, Newark, NJ (in press).

Bidleman, T. F. and Harner, T.: 1998, Sorption to aerosols. In: D. Mackay and R. Boethling (eds.) *Estimating Chemical Properties for the Environmental and Health Sciences: A Handbook of Methods.* Ann Arbor Press, Chelsea, MI (in press).

Blais, J. M., Schindler, D. W., Muir, D. C. G., Kimpe, L. E., Donald, D. B. and Rosenberg, B.: 1998, *Nature.* **395**, 585-588.

Brorström-Lundén, E. 1995, *Measurements of Semivolatile Organic Compounds in air and deposition. PhD-thesis,* University Göteborg.

Brorström-Lundén, E. and Löfgren, C.: 1998, *Environ. Poll.* **102**, 139-149.

Buser, H.-R.: 1990, *Environ. Sci. Technol.*, **24**, 1049-1058.

Cousins, I. T. and Jones, K. C.: 1998, *Environ. Pollut.* **102**, 105-118.

Duce, R. A., 1998. Atmospheric input of pollution to the Oceans. *WMO; "Marine meteorology and related oceanographic activities"* No. 39.

Duyzer, J. H. and van Oss, R. F.: 1997, *Determination of deposition parameters of a number of organic pollutants by laboratory experiments.* TNO report R97/150, Apeldoorn, The Netherlands.

Eitzer, B. D. and Hites, R. A.: 1989, *Environ. Sci. Technol.* **23**, 1389-1395.

Finizio, A., Mackay, D., Bidleman, T. F. and Harner, T. : 1997, *Atm. Environ.* **31**, 2289-2296.

Goss, K-U. and Schwarzenbach, R. P. : 1998, *Environ. Sci. Technol.* **32**, 2025-2032.

Hanna, S. R.: 1993. *Boundary-layer Meteo.* **62**, 3-20.

Harner, T. and Bidleman, T. F.: 1998, *Environ. Sci. Technol.* **32**, 1494-1502.

Hawthorne, S. B., Miller, D. J., Louie, P. K. K., Butler, R. D. and Mayer, G. G.: 1996, *J. Environ. Qual.* **25**, 594-600.

Hicks, B.B., Baldocchi, D. D., Meyers, T. P., Hosker jr., R. P. and Matt, D. R.: 1987, *Water Air and Soil Pollut.* **36**, 311-330.

Hillary, B. R., Simcik, M. F., Basu, I., Hoff, R. M., Strachan, W. M. J., Burniston, D., Chan, C. H., Brice, K. A., Sweet, C. and Hites, R. A.: 1998, *Environ. Sci. Technol.* **32**, 2216-2221.

Hoff, R. M., Strachan, W. M. J., Sweet, C. W., Chan, C. H., Shackleton, M., Bidleman, T. F., Brice, K. A., Burniston, D. A., Cussion, S., Gatz, D. F., Harlin, K. and Schroeder, W. H.: 1996, *Atm. Environ.* **30**, 3505-3527.

Hoff, R. M.: 1994, *J. Great Lakes Res.* **20**, 229-239.

Jacobs, C. M. J., and Van Pul, W. A. J.: 1996, Long-range atmospheric transport of persistent organic pollutants, I: *Description of surface-atmosphere exchange modules and implementation in EUROS.* RIVM report 722401013, Bilthoven, The Netherlands

Junge, C. E.: 1977, *In: Fate of pollutants in the air and water environment. Part I.* Advances in Environmental Science and Technology, Vol. **8**, 7-25.

Jury W. A., Spencer, W. F. and Farmer W. J.: 1983, *J. Environ. Quality*, **12**, 558-564.

Lane, D. A., Johnson, N. D., Hanley, M.-J. J., Schroeder, W. H. and Ord, D. T.: 1992, *Environ. Sci. Technol.* **26**, 126-133.

Lee, R. G. M., Hung, H., Mackay, D., Jones, K. C., 1998: *Environ. Sci. Technol.* **32**, 2172-2179.

Liss, P. S. and Slater P. G.: 1974, *Nature* **247**, 181.

Lohmann, R., Green, N. and Jones, K. C.: 1998, *Organohalogen Compounds* **36**, 493-496.

Majewski, M. S., Glotfelty, D. E., Paw U, K. T. and Seiber, J. N.: 1990, *Environ. Sci. Technol.* **24**, 1490-1497.

McConnell, L. L., Nelson, E., Rice, C. P., Baker, J. E., Johnson, W. E., Harman, J. A. and Bialek, K.: 1997, *Environ. Sci. Technol.* **31**, 1390-1398.

McConnell, L. L., LeNoir, J. S., Datta, S. and Seiber, J. N.: 1998, *Environ. Toxicol. Chem.* **17**, 1908-1916.

McLachlan, M. S. and Horstmann, M.: 1998, *Environ. Sci. Technol.* **32**, 413-420.

Millet, M., Wortham, H., Sanusi, A. and Mirabel, P.: 1997, *Environ. Sci. and Pollut. Res.* **4**, 172-180.

OSPAR: 1994, *Calculation of Atmospheric inputs of contaminants to the North Sea, 1987-1992.* Oslo and Paris Commisions, London, UK.

Pankow, J. F., 1987. *Atm. Environ.* **21**, 2275-2283.

Pankow, J. F. and Bidleman, T. F.: 1992. *Atm. Environ.* **26A**, 1071-1080.

Pankow, J. F. 1994a., *Atm. Environ.* **28**, 185-188.

Pankow, J. F. 1994b. *Atm. Environ.* **28**, 189-193.

Ridal, J. J., Kerman, B., Durham, L. and Fox, M. E.: 1996, *Environ. Sci. Technol.* **30**, 852-858.

Scholtz, M. T., McMillan, A. C., Slama, C. C., Li, Y-F., Ting, N. and Davidson, K.: 1997, *Pesticides Emission Modelling*. Development of a North American Pesticides Emissions Inventory. Canadian Global Emissions Interpretation Centre, ORTEC Corporation, Mississauga, Ontario. 242 pages.

Sehmel, G. A. and Hodgson, W. H.: 1980, A model for predicting dry deposition of particles and gases to environmental surfaces. *AIChE Symposium Series* **86**, 218-230.

Seinfeld, J. H., 1986, *Atmospheric chemistry and physics of air pollution.* John Wiley & Sons, New York.

Simonich, S. L. and Hites, R. A.:1994, *Nature* **370**, 49-51.

Slinn, W. G. N.: 1977, *Water, AirSoil Pollut.* **7**, 513-543

Slinn, W. G. N.: 1983, *Atm. Environ.* **16**, 1785-1794.

Stohl, A., Hittenberger, M. and Wotawa, G.: 1998. *Atm. Environ.* **32**: 4245-4264.

Thomas, G., Sweetman, A. J., Ockenden, W. A., Mackay, D. and Jones, K. C.: 1998a, *Environ. Sci. Technol.* **32**, 936-942.

Thomas, G., Smith, K. E. C., Sweetman, A. J. and Jones, K. C.: 1998b, *Environ. Pollut.* **102**, 119-128.

Trevisan, M., Montepiani, C., Ragozza, L., Bartoletti, C., Ioanilli, E. and Del Re, A. A. M.: 1993, *Environ. Pollut.* **80**, 31-39.

Van den Berg, F., Kubiak, R., Benjey, W. G., Majewski, M. S., Yates, S. R., Reeves, G. L., Smelt, J. H. and Van der Linden, A. M. .A.: 1999, Emission of pesticides into the air. *Water Air Soil Pollut.* this issue.

Van Jaarsveld, J. A, Van Pul, W. A. J. and De Leeuw, F. A. A. M.: 1997, *Atm. Environ.* **31**, 1011-1024.

Van Jaarsveld, J. A., 1997, *In* (ed. Gryning and Schiermeier): *Air Pollution Modelling and its Application XI.* Plenum Press, New York, 333-334.

Van Pul, W. A. J., De Leeuw, F. A. A. M., Van Jaarsveld, J. A., Van der Gaag, M. A. and Sliggers, C. J.: 1998, *Chemosphere*, **37**, 113-141.

Wania, F. and Mackay, D.: 1993, *Ambio*, **22**, 10-18.

Welsch-Paulsch, K., McLachlan, M. and Umlauf, G.: 1995, *Environ. Sci. Technol.* **29**, 189-194.

Whitby, K. T.: 1978, *Atm. Environ.* **12**, 135-159.

Woodrow, J. E., Seiber, J. N. and Baker, L. W.: 1997, *Environ. Sci. Technol.* **31**, 523-529.

Yamasaki, H., Kuwata, K. and Miyamoto, H.: 1982, *Environ. Sci. Technol.* **16**, 189-194.

IMPLEMENTING ATMOSPHERIC FATE IN REGULATORY RISK ASSESSMENT OF PESTICIDES: (HOW) CAN IT BE DONE?

D.J. BAKKER[1], A.J. GILBERT[2], D. GOTTSCHILD[3], T. KUCHNICKI[4], R.W.P.M. LAANE[5], J.B.H.J. LINDERS[6], D. VAN DE MEENT[6], M.H.M.M. MONTFORTS[6], J. PINO[7], J.W. POL[8], N.M. VAN STRAALEN[9]

[1] TNO Institute of Environmental Sciences, Energy Research and Process Innovation, P.O. Box 57, NL-1780 AB Den Helder, The Netherlands, [2] Pesticides Group, Central Science Laboratory, MAFF, Sand Hutton, York YO41 1LZ, United Kingdom, [3] Biologische Bundesanstalt für Land- und Forstwirtschaft (BBA), Fachgruppe Chemische Mittelprüfung, Messeweg 11/12, D-38104 Braunschweig, Germany, [4] Pest Management Regulatory Agency, Health Canada, Environmental Assessment Division, Sir Charles Tupper Building, 2250 Riverside Drive, Ottawa, Canada, K1A OK9, [5] National Institute for Coastal and Marine Management (RIKZ), P.O. Box 20907, NL-2500 EX, Den Haag, The Netherlands, [6] National Institute of Public Health and the Environment (RIVM), P.O. Box 1, NL-3720 BA Bilthoven, The Netherlands, [7] Californian EPA, Department of Pesticide Regulation, 1020 "N" Street, Room 161, Sacramento, California 95814-5624, USA, [8] Board for Authorisation of Pesticides (CTB), P.O. Box 217, NL-6700 AE Wageningen, The Netherlands, [9] Vrije Universiteit, Department of Ecology and Ecotoxicology, De Boelelaan 1087, NL-1081 HV Amsterdam, The Netherlands

Abstract. Atmospheric fate of pesticides and their possible effects in ecosystems beyond the immediate surrounding of the application site are not actively considered in currently used regulatory risk assessment schemes. Concern with respect to atmospheric transport and subsequent deposition of pesticides in non-target areas is however growing. In this article the results of discussions on the possibilities of implementing atmospheric fate in regulatory risk assessment are presented. It is concluded that implementing atmospheric fate in regulatory risk assessment schemes is possible and that, from a scientific point of view, these schemes should distinguish between pesticides on the basis of both their possibility/probability to reach non-target areas and on their toxicity. This implies that application of the precautionary principle or use of intrinsic pesticide properties alone is not considered justifiable. It is recommended that the risk assessment scheme should follow a tiered approach. The first tier should be entered only if the existing regulatory risk assessment procedure, including a local PEC:PNEC calculation, has been passed and involves a test for the pesticide's total atmospheric emission potential, *i.e.* its potential for becoming airborne during and after application. The second tier, which is only entered if the total emission potential is higher than a certain trigger value, should consist of a PEC:PNEC calculation for regional off-site areas (10-50 km) (tier 2A). If the pesticide's atmospheric transport potential is expected to exceed a certain value, the PEC:PNEC ratio should also be calculated for more remote areas (>1000 km) (tier 2B).

Keywords: atmospheric fate, pesticides, risk assessment, tiered approach

Water, Air, and Soil Pollution **115:** 257–266, 1999.

1. Introduction

To date, the atmospheric fate of pesticides and their possible effects in regional off-site or more remote areas are not actively considered in existing regulatory risk assessment schemes. Exceptions are the assessment of the impact of spray-drift on adjacent non-target areas and the assessment of the effects of local air-quality on workers and nearby living humans. Concern is however growing with respect to atmospheric transport and subsequent deposition of pesticides in non-target areas beyond the immediate surroundings of the application site. This concern has already lead to the inclusion of the pesticides aldrin, chlordane, chlordecone, DDT, dieldrin, endrin, heptachlor, HCB, HCH, mirex and toxaphene in the draft POP-protocol (UN-ECE, 1998) that was drawn up under the auspices of the United Nations Economic Commission for Europe's (UN-ECE) Convention on Long-Range Transboundary Air Pollution (CLRTAP) and that was accepted on June 24th, 1998 at the Ministers Conference in Aarhus, Denmark. In April 1998, the Health Council of The Netherlands organised a workshop on 'Fate of pesticides in the atmosphere; implications for risk assessment' in Driebergen, The Netherlands. During this workshop a working group was formed to discuss the possibilities of implementing atmospheric fate in regulatory risk assessment. The members concentrated on environmental risk assessment. This article summarises the discussions and conclusions.

2. Current regulatory activities on pesticides

2.1. UN-ECE POP-PROTOCOL

The objective of the UN-ECE POP-protocol is 'to control, reduce or eliminate discharges, emissions and losses of persistent organic pollutants'. Persistent organic pollutants are defined as 'set of organic compounds that: (i) possess toxic characteristics, (ii) are persistent, (iii) are liable to bioaccumulate, (iv) are prone to long-range atmospheric transport and deposition and (v) can result in adverse environmental and human health effects at locations near and far from their sources (UN-ECE, 1998). Most of the pesticides in the protocol are listed under Annex I of the protocol, which contains the substances that are scheduled for elimination. The POP-protocol, however, aims at reducing *existing* risks because several of the organochlorine pesticides in the protocol, such as DDT and HCH, have already found their way to remote areas such as the Arctic and have imposed risks on ecosystems (Barrie *et al.*, 1992; Bidleman, this issue). Furthermore it can be expected that adding additional pesticides to the internationally agreed POP-list is due to take ample time and to encounter many difficulties. For these reasons it seems better to include the atmospheric fate of pesticides in the regulatory risk assessment at the moment when a manufacturer asks permission for a pesticide to be admitted to the market and to minimise the risk of atmospheric dispersion to non-target areas beforehand.

2.2. EU DIRECTIVE 91/414/EEC

The concern about the atmospheric fate and behaviour of pesticides is also reflected in the EU Authorisations Directive 91/414/EEC (EC, 1991) under the terms of which admittance of pesticides is considered in the European Union. In Annex III of this Directive, prediction of 'the level of residues in air, to which man, animals and non-target organisms may be exposed (acute and chronic)' is asked for (EC, 1995). In this Annex III, the EPPO/CoE decision making scheme (EPPO/CoE, 1993) is referred to, in which a sub-scheme for air is being implemented. In this sub-scheme the various temporal and spatial scales of atmospheric pesticide dispersion are distinguished. Actual guidelines for implementing atmospheric fate and behaviour in risk assessment procedures or guidelines for regional off-site or more remote risk assessment could however not yet be presented in the form of an easily usable regulatory risk assessment tool. This is largely because the underlying science concerned is still under development.

3. Possibilities for implementing atmospheric fate of pesticides in regulatory practice

3.1. APPLYING THE PRECAUTIONARY PRINCIPLE

When discussing the possibilities of reducing or minimising risks of pesticides in non-target areas, 'the precautionary principle' was brought up by several participants. The interpretation of this principle seemed however to vary considerably. The most strict interpretation that was pronounced, implied that pesticides should not be present in non-target areas at all because they 'do not belong there', which is more or less an ethical point of view. In practice this would mean that either the pesticide may not be used at all or the pesticide may be used but may not leave the application areas in such quantities that it exceeds the level of detection in non-target areas. This interpretation of the precautionary principle does not account for differences in exposure or for differences in toxicity of different pesticides.

Another interpretation of the precautionary principle was one that says that levels in non-target areas should not become higher than the present ones (the stand-still principle). The stand-still principle would only be applicable to pesticides that are already in use or have been used in the past. For a new pesticide the stand-still principle therefore coincides with the aforementioned more strict interpretation of the precautionary principle. The stand-still principle does not account for differences in exposure or differences in toxicity of pesticides either.

Use of safety factors was also mentioned as a kind of precautionary principle. By applying safety factors to the results of the exposure and/or effect assessment, one can incorporate additional safety into the risk assessment procedure to prevent 'false positives' (pesticides that are unjustifiably considered safe to use).

As the members of the working group were of the opinion that regulatory risk assessment schemes should distinguish between pesticides on the basis of both their possibility/probability to reach non-target areas and their toxicity, they concluded that it would not be appropriate to incorporate atmospheric fate in regulatory risk assessment schemes on the basis of the most strict interpretation of the precautionary principle or on the basis of the stand-still principle. Use of safety factors was, however, considered justifiable and necessary in order to account for uncertainties in toxicity and environmental behaviour.

3.2. USING INTRINSIC PESTICIDE PROPERTIES

One way to distinguish pesticides that may reach non-target areas from those that will not, would be to use certain intrinsic (physico-chemical) properties as regulatory criteria. Examples of intrinsic pesticide properties that are already being used in regulatory risk assessment schemes are the half-life with respect to (bio)degradation DT_{50}, the octanol-water partitioning coefficient K_{ow} as a measure for bioaccumulation potential and the adsorption coefficient K_{om} for sorption to organic matter. In some risk assessment schemes intrinsic properties are used as cut-off values, in others they are used to calculate for example the risk that a pesticide leaches to the groundwater (CTB, 1999; Winkler *et al.*, 1999). In the case of implementing the atmospheric fate of a pesticide in a risk assessment scheme, one must find the intrinsic properties that govern the emission of the pesticide to the atmosphere during and after application and the properties that determine its atmospheric transport potential. For the properties that govern the emission to the atmosphere, cut-off values could be set. For the properties determining the atmospheric transport potential this is more difficult. The problem here is that, if one wants to protect not only remote areas at more than 1000 km distance but also regional off-site areas at 10-50 km distance, the cut-off values based on the 10-50 km distance will probably be so restrictive that hardly any pesticide will comply with the criterion. A property determining this transport potential could, however, be used for calculating the average or maximum distance of transport and the percentage of the emitted pesticide arriving at a certain distance. Intrinsic pesticide properties (co)determining the emission to the atmosphere are the saturated vapour pressure P, Henry's law constant H and the octanol-air partitioning coefficient K_{oa}. The atmospheric transport potential is determined by a pesticide's atmospheric half-life which is governed by its (photo)chemical degradation rate k_{atm} and its dry and wet deposition removal rate (Atkinson *et al.*, this issue; Van Pul *et al.*, this issue). However, intrinsic properties that determine a pesticide's potential for atmospheric emission and/or transport do not say much about the magnitude of the exposure concentration (as the volume used is important as well) and nothing at all about its toxicity.

The working group held the opinion that regulatory risk assessment schemes should distinguish between pesticides both on the basis of their possibility/probability to reach non-target areas and on their toxicity. It was therefore concluded that intrinsic pesticide properties determining the atmospheric fate of pesticides alone should not be used to determine whether or not a pesticide can be admitted to the market.

3.3. COMPARING ENVIRONMENTAL CONCENTRATIONS WITH (NO) EFFECT CONCENTRATIONS

A common way to assess the environmental risks of a pollutant is by comparing the measured or Predicted Environmental Concentration (PEC) with an effect level for a particular organism or for the ecosystem as a whole. Often the Predicted No Effect Concentration (PNEC) is chosen as the critical limit (the limit above which unacceptable effects take place). The PEC:PNEC ratio then gives an indication of the risk of harmful effects as a result of the exposure to the pollutant. For pesticides, this PEC:PNEC approach is used in several existing regulatory schemes to assess the risks at or near the site of application (EC, 1991; EC, 1995; EC, 1997; EPPO/CoE, 1993; Greig-Smith, 1992; UK PSD, undated). For risk assessment of substances other than pesticides, the PEC:PNEC approach is also often used (EC, 1996; EC, 1998; RIVM, 1998). For this reason and the fact that the PEC:PNEC approach incorporates both exposure and toxicity, it was concluded that the risk assessment of a pesticide in regional off-site and more remote areas should also be based on a PEC:PNEC approach or a procedure equivalent to it. The question, however, that directly follows this conclusion is how to determine the PEC and PNEC for regional off-site and more remote areas?

With respect to the PNEC one could answer that the risk assessment should be directed at protecting the entire regional off-site or remote ecosystem and that this aim could be attained by basing the PNEC on the sensitivity of one or more key indicator species in the ecosystem of concern or on a species sensitivity distribution. This approach would comply with existing risk assessment schemes. It can however be questioned whether it is to be expected that the sensitivity (expressed in the value of the PNEC) of species in regional off-site or remote ecosystems is much different from that of species at the site of application. If the sensitivities are more or less comparable, one might expect that the PEC:PNEC ratio in the off-site or remote area will always be smaller than the ratio at the site of application because it is to be expected that the PEC in off-site and remote areas is lower than at the site of application itself. However, as long as it is conceivable that certain key species or ecosystem functions are more sensitive under other (*i.e.* harsher) environmental conditions, it is advisable to determine a PNEC specifically for the regional off-site and remote ecosystems. Another consideration that supports this advise is the fact that biomagnification may play different roles in different ecosystems. This does not automatically imply that the manufacturer of the pesticide should be required to test the toxicity on all kinds of 'exotic' species. The most simple way to arrive at a PNEC for off-site and remote ecosystems is by applying a safety factor to the PNEC that is used in the local risk assessment. Another method would be to use models, including food chain or food web models representing the situation in the remote ecosystem, to determine 'a remote PNEC'.

Determining the PEC in regional off-site and remote areas was thought to be more difficult than estimating the 'remote PNEC'. The different spatial scales* for regional off-site (10-50 km) and remote (> 1000 km) areas and the fact that information is needed on the volume of use and the spatial pattern of use in the region (for off-site PEC) or even the world (for remote PEC) make it difficult to estimate the PEC. This information on volume and pattern of use is hard to obtain for existing pesticides and does not yet exist for new ones. A possible approach for the regional off-site area would be to assume that all of a pesticide that is emitted into the atmosphere from one hectare of agricultural area will be deposited on one hectare of non-target area such as a nature reserve. This is called the 'unit surface area approach' (Van de Meent *et al.*, 1995). A refinement would be to correct the input to the non-target area with a factor based on the ratio between total treated agricultural area and total non-treated area in the region. Another refinement would be to introduce a time-dependency, which could take into account the fact that in some cases 90% of the emission takes place during the first 24 hours and that in other cases it may take a year before 90% of the emission has occurred. The 'unit surface area approach', and to a lesser extent the refined methods, would be worst case approaches. Another approach to calculate the regional off-site PEC would be the 'scenario approach'. In this approach a hypothetical standard region consisting of both agricultural fields and non-agricultural areas is drawn up. In the agricultural fields, the pesticide of concern is assumed to be used with typical spatial and temporal variations. For this region atmospheric emissions, transport and deposition and the PEC in the non-target areas are estimated by model calculations. Both the 'unit surface area approach' and the 'scenario approach' can be applied for existing and new pesticides.

For remote areas, the calculation of a PEC becomes more uncertain but is still possible. The 'unit surface area approach' could be applied if the losses

* The spatial scales presented here are just examples to make a clear distinction between off-site, but still relatively nearby areas and really remote areas. This does however not mean that the area between 50 and 1000 km should not be considered.

due to atmospheric degradation and atmospheric deposition along the trajectory to the remote area are accounted for. The 'scenario approach' is less appropriate for calculating PECs in remote areas as the input data can only be estimated very roughly and the results would bear a large uncertainty.

3.4. TIERED APPROACH

As not every pesticide has a high potential for becoming airborne and therefore not every pesticide poses a risk for off-site or remote areas, it was concluded that regulatory PEC:PNEC risk assessment for these areas should be restricted to only those pesticides that are expected to have a certain minimum atmospheric emission potential. Such a distinction on the basis of atmospheric emission potential assessment leads to a tiered approach in which several levels of assessment are distinguished. In this tiered approach, the first tier is entered only if the existing regulatory risk assessment, including a local PEC:PNEC calculation, has been passed. The first tier involves the comparison of the atmospheric emission potential, based on intrinsic pesticide properties, with a trigger value. If the intrinsic properties indicate that the estimated total atmospheric emission potential, *i.e.* the potential for becoming airborne during and after application, is lower than the maximum that is considered acceptable, authorisation of the pesticide is possible and the second tier is not entered. In the case that the intrinsic pesticide properties indicate a total emission potential that is higher than considered acceptable, a second tier of risk assessment has to be entered. This second tier involves a PEC:PNEC based risk assessment for both off-site (tier 2A) and remote areas (tier 2B). The distinction between off-site (10-50 km) and remote (>1000 km) can be based on a trigger value for an intrinsic pesticide property determining its atmospheric transport potential such as its atmospheric half-life $DT_{50,atm}$. A method for determining the atmospheric transport potential, based on intrinsic compound properties, is given by Van Pul *et al.* (1998) and could be used in the second tier. Within the second tier, first the off-site (10-50 km) risk assessment (tier 2A) is performed. If this off-site risk assessment results in a PEC:PNEC ratio that is higher than acceptable, authorisation is not possible. If the PEC:PNEC ratio is acceptable, the pesticide is subsequently checked for its transport potential. If the transport potential is not higher than the trigger value, a remote risk assessment is not necessary and the authorisation of the pesticide is possible. Only if the transport potential is higher than the trigger value, the remote risk assessment (tier 2B) has to be performed. Here again the PEC:PNEC ratio determines whether authorisation of the pesticide is possible. The tiered approach discussed here, is depicted schematically in figure 1. Another tiered approach based on volatility and behaviour in air is currently under development in Germany. This approach is based on the experience of authorisation according to guidelines of the German Biologische Bundesanstalt für Land- und Forstwirtschaft (BBA, 1990).

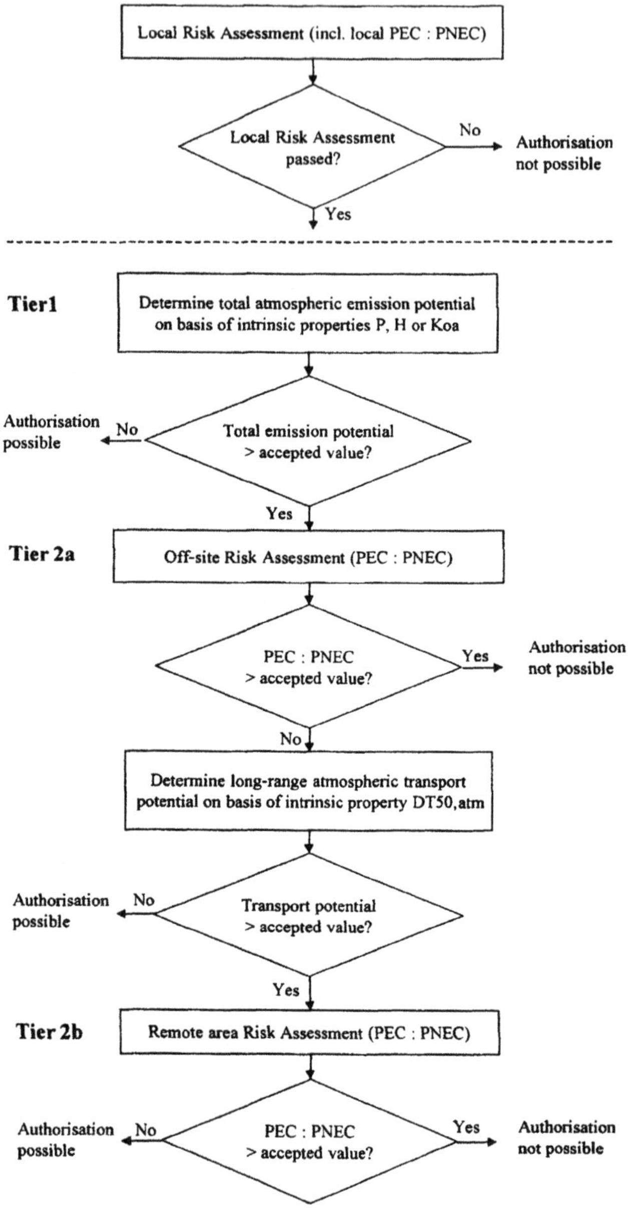

Figure 1. Tiered approach for implementing atmospheric fate in regulatory risk assessment of pesticides. The tiered scheme is entered only if the local risk assessment has been passed. (The abbreviations are explained in the main text.)

4. Conclusions and recommendations

The main conclusion of the members of the working group is that it is feasible to implement atmospheric fate in regulatory risk assessment of pesticides. They further conclude that this would, from a scientific point of view, not be acceptable on the basis of the precautionary principle or solely on the basis of intrinsic pesticide properties. They recommend to follow a tiered approach. The first tier, which is entered only if the existing regulatory risk assessment procedure, including a local PEC:PNEC calculation, has been passed, should involve a test for the pesticide's total atmospheric emission potential. This potential should be estimated with the help of intrinsic pesticide properties such as the saturated vapour pressure, Henry's law constant and the octanol/air partitioning coefficient. Trigger values for the total emission potential could be established to be able to determine whether the second risk assessment tier should be entered. This second tier should consist of a PEC:PNEC calculation for regional off-site areas (10-50 km) (tier 2A) and, if the pesticide's atmospheric transport potential is expected to exceed a certain value, also for more remote areas (>1000 km) (tier 2B). This atmospheric transport potential can be estimated on the basis of intrinsic pesticide properties including the (photo)chemical degradation rate and properties that govern the deposition velocity. We recommend to investigate which (combination of) intrinsic pesticide properties predict the atmospheric emission and transport potentials best and to work out further the PEC:PNEC based risk assessment procedures for regional off-site and remote areas.

5. References

Atkinson, R., Guicherit, R., Hites, R. A., Palm, W.-U., Seiber, J. N. and De Voogt, P.: 1999, *Water Air Soil Pollut.*, this issue.

Barrie, L.A., D. Gregor, B. Hargrave, R. Lake, D. Muir, R. Shearer, B. Tracey and T. Bidleman, 1992. *Sci. Total Environ.* **122**, 1-74.

BBA, 1990. *Testing the volatility behaviour and the fate of plant protection products in air. In: Richtlinien für die Prüfung von Pflanzenschutzmitteln im Zulassungsverfahren.* Biologische Bundesanstalt für Land- und Forstwirtschaft. Saphir Verlag Ribbesbüttel, Germany. (in German).

Bidleman, T. F.: 1999, *Water Air Soil Pollut.*, this issue.

CTB, 1999. *Guidelines for Authorisation of Pesticides, version 0.1.* College voor de Toelating van Bestrijdingsmiddelen, Wageningen, The Netherlands. (in Dutch, in press).

EC, 1991. *EC Directive 91/414/EEC (Authorisations Directive) concerning the placing of plant protection products on the market.*

EC, 1995. *EC Directive 95/36/EC establishing data requirements for environmental behaviour in annex II and annex III of Directive 91/414/EEC.*

EC, 1996. *Technical Guidance Documents in support of Commission Directive 93/67/EEC on Risk Assessment for New Notified Substances and Commission Regulation (EC) No. 1488/94 on Risk Assessment for Existing Substances (Parts I, II,III and IV)*. EC catalogue numbers CR-48-96-001, 002, 003, 004-EN-C. Office for Official Publication of the European Communities, 2 rue Mercier, L-2965 Luxembourg, ISBN 92-827-8011-2.

EC, 1997. *EC Directive 97/57/EG establishing AnnexVI to Directive 91/414/EEC concerning the placing plant protection products on the market.*

EC, 1998. *EC Directive 98/8/EC of the European Parliament and of the Council of 16 February 1998 concerning the placing of biocidal product on the market.*

EPPO/CoE, 1993. European and Mediterranean Plant Protection Organisation/Council of *Europe decision making schemes for the environmental risk assessment of plant protection products*, EPPO Paris.

Greig-Smith, P.W., 1992. *Environ. Toxicol. Chem.* **11**, 1673-1689.

RIVM, 1998. *Uniform System for the Evaluation of Substances 2.0 (USES 2.0)*. National Institute of Public Health and the Environment (RIVM), Ministry of Housing, Spatial Planning and the Environment (VROM), Ministry of Health, Welfare and Sport (VWS). RIVM report 679102044 (eds. Linders, J.B.H.J and D.T. Jager), Bilthoven, The Netherlands.

UK PSD, undated. *The Registration Handbook: Pesticides, Biocides, Plant Protection Products. A guide to the policies, procedures and data requirements relating to their control in the United Kingdom.* Available from the Pesticide Safety Directorate, Mallard House, Kings Pool, 3 Peasholme green, York YO1 2PX, United Kingdom.

UN-ECE, 1998. *Draft Protocol to the Convention on Long-Range Transboundary Air Pollution on Persistent Organic Pollutants*, EB.Air/1998/2, Executive Body for the Convention on Long-Range Transboundary Air Pollution, Economic Commission for Europe, Economic and Social Council, United Nations.

Van de Meent, D., J.H.M. de Bruijn, F.A.A.M. de Leeuw, A.C.M. de Nijs, D.T. Jager and T.G. Vermeire, 1995. In: *Risk Assessment of Chemicals. An Introduction.* (eds. C.J. van Leeuwen and J.L.M. Hermen), 103-145. Kluwer Academic Publishers, Dordrecht, The Netherlands.

Van Pul, W.A.J., F.A.A.M. de Leeuw, J.A. van Jaarsveld, M.A. van der Gaag and C.J. Sliggers, 1998. *Chemosphere* **37**, 113-141.

Van Pul, W. A. J., Bidleman, T. F., Brorström-Lundén, E., Builtjes, P. J. H., Dutchak, S., Duyzer, J. H., Gryning, S.-E., Jones, K. C., Van Dijk, H. F. G. and Van Jaarsveld, J. A.: 1999, *Water Air Soil Pollut.*, this issue.

Winkler, R., B. Stein, D. Gottschild and M. Streloke, 1999. *Nachrichtenbl. Deut. Pflanzenschutzd.*, **51**, 38-43 (in German).

LIST OF PARTICIPANTS

Roger Atkinson
Statewide Air Pollution Research Center
University of California
Riverside CA 92521
USA
Phone:+1-404-787-4191
Fax: +1-404-787-5004
E-mail: roger.atkinson@ucr.edu

Dick J Bakker
RDR van Catsweg 180
2805 BB Gouda
THE NETHERLANDS
Phone: +31-182-581979
E-mail: bakker.montizaan@wxs.nl

William G Benjey
US Environmental Protection Agency
National Exposure Research Laboratory
Atmospheric Modeling Division (MD-80)
Research Triangle Park
North Carolina 27711
USA
Phone: +1- 919-541-0821
Fax: +1- 919-541-1379
E-mail: benjey@hpcc.epa.gov

Terry F Bidleman
Atmospheric Environment Service
4905 Dufferin Street
Downsview, Ontario
M3H 5T4, CANADA
Phone: +1-416-739-5730
Fax: +1-416-739-5708
E-mail: terry.bidleman@ec.gc.ca

Eva Brorström-Lundén
Swedish Environmental Research Institute
(IVL)
P.O. Box 47086
SE-40258 Gothenburg
SWEDEN
Phone: +46-31-460-080
Fax: +46-31-482-180
E-mail: eva.bl@ivl.se

Peter JH Builtjes
TNO Institute of Environmental Sciences,
Energy Research and Process Innovation
P.O. Box 342
NL-7300 AH Apeldoorn
THE NETHERLANDS
Phone: +31-55-5493038
Fax: +31-55-5493252
E-mail: p.j.h.builtjes@mep.tno.nl

Pim de Voogt
University of Amsterdam
Department of Environmental and
Toxicological Chemistry
Nieuwe Achtergracht 166
NL-1018 WV Amsterdam
THE NETHERLANDS
Phone: +31-20-5256565
Fax: +31-20-5256504
E-mail: pim@mtc.chem.uva.nl

Water, Air, and Soil Pollution **115:** 267–272, 1999.

Sergey Dutchak
Meteorological Synthesizing Center East of EMEP
Kedrova str.,8.k.1
Moscow
RUSSIAN FEDERATION
Phone: +7-095-125-52-81
Fax: +7-095-310-70-93
E-mail: msce@glasnet.ru

Jan H Duyzer
TNO Institute of Environmental Sciences, Energy Research and Process Innovation
P.O. Box 342
NL-7300 AH Apeldoorn
THE NETHERLANDS
Phone: +31-55-5493944
Fax: +31-55-5493252
E-mail: j.h.duyzer@mep.tno.nl

Andy J Gilbert
Pesticides Group
Central Science Laboratory
Sand Hutton
York YO4 1LZ
UNITED KINGDOM
Phone: +44-1904-462000 (office 462234)
Fax: +44-1904-462111
E-mail: a.gilbert@csl.gov.uk

Dietmar Gottschild
Biologische Bundesanstalt für Land- und Forstwirtschaft
Fachgruppe Chemische Mittelprüfung
Messeweg 11/12
D-38104 Braunschweig
GERMANY
Phone: +49-531-2993512
Fax: +49-531-2993004
E-mail: d.gottschild@bba.de

Sven-Erik Gryning
Riso National Laboratory
Department of Wind Energy and Atmospheric Physics
DK-4000 Roskilde
DENMARK
Phone: +45-4677-5005
Fax: +45-4677-5970
E-mail: sven-erik.gryning@risoe.dk

Robert Guicherit
TNO Institute of Environmental Sciences, Energy Research and Process Innovation
P.O. Box 342
NL-7300 AH Apeldoorn
THE NETHERLANDS
Phone: +31-55-5493493
Fax: +31-55-5493252
E-mail: r.guicherit@wxs.nl

Ronald A Hites
Indiana University
School of Public and Environmental Affairs
and Department of Chemistry
SPEA 410
Bloomington, IN 47405
USA
Phone: +1-812-855-0193
Fax: +1-812-855-1076
E-mail: hitesr@indiana.edu

Kevin C Jones
Lancaster University
Environmental Science Department
Lancaster LA1 4YQ
UNITED KINGDOM
Phone: +44-1524-593-972
Fax: +44-1524-593-985
E-mail: k.c.jones@lancaster.ac.uk

Douwe A Jonkers
Ministry of Housing, Spatial Planning and the Environment
P.O. Box 30945
NL-2500 GX The Hague
THE NETHERLANDS
Phone: +31-70-3394276
Fax: +31-70-3391317
E-mail:
douwe.jonkers@ddwl.dgm.minvrom.nl

Roland Kubiak
Staatliche Lehr- und Forschungs Anstalt
Fachbereich Ökologie
Breitenweg 71
67435 Neustadt
GERMANY
Phone: +49-6321-671-285
Fax: +49-6321-671-222
E-mail: slfa.kubiak@t-online.de

Ted Kuchnicki
Pest Management Regulatory Agency
Environmental Assessment Division
Room E777
Sir Charles Tupper Building
2250 Riverside Drive
Address Locator: 6607E1
Ottawa, CANADA
K1A OK9
Phone: +1-613-736-3733
Fax: +1-613-736-3710
E-mail: tkuchnicki@pmra.arla.hc.sc.gc.ca

Remi WPM Laane
RIKZ
P.O. Box 20907
NL-2500 EX Den Haag
THE NETHERLANDS
Phone: +31-70-3114293
Fax: +31-70-3114330
E-mail: r.w.p.m.laane@rikz.rws.minvenw.nl

Jan BHJ Linders
RIVM
CSR
P.O. Box 1
NL-3720 BA Bilthoven
THE NETHERLANDS
Phone: +31-30-2743164
Fax: +31-30-2744401
E-mail: jan.linders@rivm.nl

Michael S Majewski
US Geological Survey
6000 J Street
Placer Hall
Sacramento
California 95819
USA
Phone: +1-916-278-3086
Fax: +1-916-278-3091 or 3071
E-mail: majewski@usgs.gov

Mark HMM Montforts
RIVM
CSR
P.O. Box 1
NL-3720 BA Bilthoven
THE NETHERLANDS
Phone: +31-30-2742529
Fax: +31-30-2744401
E-mail: mark.montforts@rivm.nl

Wolf-Ulrich Palm
Fraunhofer-Institut für Toxikologie und Aerosolforschung
Nikolai-Fuchs-Str. 1
30625 Hannover
GERMANY
Phone: +49-511-5350-528
Fax: +49-511-5350-155
E-mail: palm@ita.fhg.de

Judy Pino
California Environmental Protection Agency
Dept. of Pesticide Regulation
1020 "N" Street, Room 161
Sacramento, California 95814-5624
USA
Phone: +1-916-324-4023 (work) or 1-530-823-9941 (home)
Fax: +1-916-324-4088
E-mail: jpino@cdpr.ca.gov

J Werner Pol
CTB
P.O. Box 217
6700 AE Wageningen
THE NETHERLANDS
Phone: +31-317-471850
Fax: +31-317-471899
E-mail: j.w.pol@ctb.agro.nl

Graham L Reeves
Dow AgroSciences
Letcombe Regis,
Wantage
Oxon. OX12 9JT
UNITED KINGDOM
Phone:+44-1235-774684
Fax: +44-1235-774803
E-mail: glreeves@dow.com

James N Seiber
University of Nevada, Reno
Center for Environmental Sciences and Engineering
MS-199
Reno Nevada, 89557
USA
Phone: +1-702-784-6460
Fax: +1-702-784-1142
E-mail: jseiber@med.unr.edu

Johan H Smelt
DLO Winand Staring Centre for Integrated Land, Soil and Water Research
P.O. Box 125
NL-6700 AC Wageningen
THE NETHERLANDS
Phone: +31-317-474200
Fax: +31-317-424812
E-mail: smelt@sc.dlo.nl

Hugo E van de Baan
Ministry of Agriculture, Nature Management and Fisheries
P.O. Box 20401
NL-2500 EK Den Haag
THE NETHERLANDS
Phone: +31-70-3784881
Fax: +31-70-3786157
E-mail: h.e.van.de.baan@dl.agro.nl

Dik van de Meent
RIVM
Laboratory for Ecotoxicology
P.O. Box 1
NL-3720 BA Bilthoven
THE NETHERLANDS
Phone: +31-30-2743130
Fax: +31-30-2744413
E-mail: d.van.de.meent@rivm.nl

(Fred)erik van den Berg
DLO Winand Staring Centre for Integrated Land, Soil and Water Research
P.O. Box 125
NL-6700 AC Wageningen
THE NETHERLANDS
Phone: +31-317-474669
Fax: +31-317-424812
E-mail: f.vandenberg@sc.dlo.nl

A (Ton) MA van der Linden
RIVM
Soil and Groudwater Research Laboratory
P.O. Box 1
NL-3720 BA Bilthoven
THE NETHERLANDS
Phone: +31-30-2743342
Fax: +31-30-2744419
E-mail: Ton.van.der.Linden@rivm.nl

Harrie FG van Dijk
Health Council of the Netherlands
P.O. Box 16052
NL-2500 BB The Hague
THE NETHERLANDS
Phone:+31-70-3407520
Fax: +31-70-3407523
E-mail: hfg.van.dijk@gr.nl

J (Hans) A van Jaarsveld
RIVM
Air Research Laboratory
P.O. Box 1
NL-3720 BA Bilthoven
THE NETHERLANDS
Phone: +31-30-2742818
Fax: +31-30-2287531
E-mail: hans.van.jaarsveld@rivm.nl

W Addo J van Pul
RIVM
Air Research Laboratory
P.O. Box 1
NL-3720 BA Bilthoven
THE NETHERLANDS
Phone: +31-30-2743421
Fax: +31-30-2287531
E-mail: addo.van.pul@rivm.nl

Nico M van Straalen
Vrije Universiteit Amsterdam
Department of Ecology and Ecotoxicology
De Boelelaan 1087
NL-1081 HV Amsterdam
THE NETHERLANDS
Phone: +31-20-4447070
Fax: +31-20-4447123
E-mail: straalen@bio.vu.nl

Scott R Yates
U.S.D.A. - ARS
U.S. Salinity Laboratory
450 W. Big Springs Road
Riverside, CA 92507
USA
Phone: +1-909-369-4803
Fax: +1-909-342-4964
E-mail: yates@ucracl.ucr.edu

THE ORGANISING COMMITTEE

Robert Guicherit - *chair*
Dick J. Bakker - *member*
Pim de Voogt - *member*
Douwe A. Jonkers - *consultant*
Hugo E. van de Baan - *consultant*
Frederik van den Berg - *member*
W. Addo J. van Pul - *member*
Harrie F. G. van Dijk - *scientific secretary*

Secretarial and technical support

Joke Hoorens van den Berg - de Vlieger
Marlon Debidien
Rob van Zeventer
Marja van Kan
Jeannette van Kan
Take Alberts
Kees Spaans
Theo Maes

ABOUT THE HEALTH COUNCIL OF THE NETHERLANDS

The Health Council of the Netherlands (Gezondheidsraad) has a long history. It has existed in one form or another since 1902, with the Health Council in its present form established by the 1956 Health Act, lastly amended in 1997. The Health Council is a statutory advisory body to the government. It informs the government on the current level of knowledge in the field of public health. This includes identifying developments which are relevant to government policy. The Council's recommendations encompass subjects which affect the health of the population, such as medicine, health care, environmental protection, nutrition, occupational hygiene and living conditions.

The Health Council is headed by a president and two vice-presidents and boasts approximately 170 members, selected from the scientific and health care communities. Members are nominated by the President of the Council on the basis of their personal expertise and are appointed by Royal Decree.

The Health Council never meets in plenary session. Its reports are prepared by autonomous, multidisciplinary *ad hoc* committees appointed by the President of the Health Council. In most cases, the committee is required to answer ministerial queries. However, the Council has also the power to prepare advisory reports on its own initiative and regularly exercises this right.

Committees consist of both Council members and external experts and around forty committees are at work at any one time. Within the Health Council, there are also eight standing committees, each of which deals with a broader subject area than those covered by the *ad hoc* committees. The standing committees review draft committee reports and bring topics to the fore which merit Health Council assessment. As such they are involved in the preparation of the Council's work programme. The Minister of Health establishes the definitive version of this programme, before sending it to parliament.

The Health Council and its committees are supported by a secretariat, which consists of a 30-strong scientific staff and a technical and clerical team of similar proportions. The secretariat is managed by the Executive Director of the Council. The Health Council is financed entirely by the government. Depending upon its work programme, the Council is awarded a budget made up of contributions from the Ministry of Health, Welfare and Sport, the Ministry of Housing, Spatial Planning and the Environment, the Ministry of Agriculture, Nature Management and Fisheries and the Ministry of Social Affairs and Employment.

Each year, the Health Council presents some thirty publications to the government. Reports are public and are accompanied by an executive summary in English. In many instances, an entire report is translated. The Council's international relations are kept informed by means of an English-language

Water, Air, and Soil Pollution **115:** 275–276, 1999.

newsletter entitled 'Network', published three times a year. The Health Council has its seat in The Hague.

Health Council of the Netherlands
P.O. Box 16052
NL-2500 BB The Hague
Parnassusplein 5
NL-2511 VX The Hague
The Netherlands
telephone: +31 (0)70 340 7520
fax: +31 (0)70 340 7523
e-mail: gr@gr.nl